Coconut Biotechnology:
Towards the Sustainability of the 'Tree of Life'

Steve Adkins • Mike Foale
Roland Bourdeix • Quang Nguyen
Julianne Biddle
Editors

Coconut Biotechnology: Towards the Sustainability of the 'Tree of Life'

Springer

Editors
Steve Adkins
School of Agriculture and Food Sciences
The University of Queensland
St Lucia, QLD, Australia

Roland Bourdeix
Biological Systems Department
Centre de coopération internationale en recherche agronomique pour le développement (CIRAD)
Montpellier, QLD, Australia

CIRAD – UMR AGAP, CIRAD
(Agricultural Research for Development)
Montpellier, France AGAP, University
Montpellier, CIRAD, INRA
Montpellier SupAgro, Montpellier, France

Julianne Biddle
School of Agriculture and Food Science
University of Queensland
St. Lucia, QLD, Australia

Mike Foale
School of Agriculture and Food Sciences
The University of Queensland
St Lucia, QLD, Australia

Quang Nguyen
School of Agriculture and Food Sciences
The University of Queensland
St Lucia, QLD, Australia

School of Biotechnology
International University
Vietnam National University HCM
Ho Chi Minh City, Vietnam

ISBN 978-3-030-44987-2 ISBN 978-3-030-44988-9 (eBook)
https://doi.org/10.1007/978-3-030-44988-9

This Springer imprint is published by the registered company Springer Nature Switzerland AG
The registered company address is: Gewerbestrasse 11, 6330 Cham, Switzerland

Preface

Coconut is often referred to as the 'tree of life'. Primarily grown on 12 million hectares across more than 90 tropical and subtropical countries, coconut is one of the world's most highly valued palm crops. This species contributes directly to the amenity and income for 20 million smallholder farmers and their dependents, providing food, health benefits, structural products as well as aesthetic beauty to the landscape. Apart from coconut water and sugar, beneficial effects of various oil products have been increasingly acknowledged worldwide by users, becoming one of the most attractive functional foods in recent years.

The present volume has 12 chapters contributed by academics and researchers from numerous countries including Australia, France, Vietnam, Mexico, Indonesia, the Philippines, Iran, China, Britain, Sri Lanka, Canada and Papua New Guinea. The volume covers most of the recent developments in coconut biotechnology, starting from the latest methods used for the identification of genetic diversity, pests and disease-causing agents to several applications for conservation and safe movement of germplasm, clonal propagation, improving the value of food products and promoting sustainable livelihoods.

The editors of this book express their deep gratitude to all the chapter contributors for sharing their work in this volume. We are also thankful to Springer Nature for giving us a chance to compile this important book. We hope that the contents presented in this book will be useful to the readers involved in coconut research and for all those interested in this amazing plant.

St Lucia, QLD, Australia Steve Adkins

Contents

Contributors

Steve Adkins School of Agriculture and Food Science, University of Queensland, St Lucia, OLD, Australia

Jelfina C. Alouw Indonesian Center for Estate Crops Research and Development, Bogor, Indonesia

H. D. Dharshani Bandupriya Department of Plant Sciences, Faculty of Science, University of Colombo, Colombo, Sri Lanka

Luc Baudouin CIRAD, UMR AGAP, Montpellier, France

AGAP, Université de Montpellier, CIRAD, INRA, Montpellier Supagro, Montpellier, France

Amirhossein Bazrafshan School of Agriculture and Food Science, University of Queensland, Brisbane, OLD, Australia

Julianne Biddle School of Agriculture and Food Science, University of Queensland, St Lucia, QLD, Australia

Roland Bourdeix CIRAD – UMR AGAP, CIRAD (Agricultural Research for Development), Montpellier, France

AGAP, University of Montpellier, CIRAD, INRA, Montpellier SupAgro, Montpellier, France

Gilles Coppens d'Eeckenbrugge CIRAD – UMR AGAP, CIRAD (Agricultural Research for Development), Montpellier, France

AGAP, University of Montpellier, CIRAD, INRA, Montpellier SupAgro, Montpellier, France

Ivan Córdova Centro de investigación Científica de Yucatán (CICY), Mérida, México

Fabian M. Dayrit Department of Chemistry, Ateneo de Manila University, Quezon City, Philippines

Mike Foale School of Agriculture and Food Science, University of Queensland, St Lucia, OLD, Australia

Meldy L. A. Hosang Indonesian Center for Estate Crops Research and Development, Bogor, Indonesia

Vincent Johnson Bioversity International, Montpellier, France

Jean Louis Konan National Centre for Agricultural Research, Abidjan, Ivory Coast

Arturo López-Villalobos Department of Biology, Purdue University, West Lafayette, IN, USA

Zhihua H. Mu School of Agriculture and Food Science, University of Queensland, Brisbane, QLD, Australia

Wayne Myrie Coconut Industry Board (CIB), Kingston, Jamaica

Maria Narvaez Centro de investigación Científica de Yucatán (CICY), Mérida, México

Quang Nguyen School of Agriculture and Food Sciences, The University of Queensland, St Lucia, QLD, Australia

School of Biotechnology, International University, Vietnam National University-HCM, Ho Chi Minh City, Vietnam

German Nic-Matos Centro de investigación Científica de Yucatán (CICY), Mérida, México

Hengky Novarianto Indonesian Coconut and Palm Research Institute (IPRI), Manado, Indonesia

Carlos Oropeza-Salín Biotechnology Unit, Centro de Investigación Científica de Yucatán, Mérida, Yucatán, Mexico

Carlos F. Ortíz Colegio de Postgraduados (CP), Puebla, Mexico

Messias G. Pereira Universidade Estadual do Norte Fluminense, Rio de Janeiro, Brazil

Chandrika Perera Faculty of Agriculture, University of Peradeniya, Peradeniya, Sri Lanka

Lalith Perera Coconut Research Institute of Sri Lanka, Bandirippuwa Estate, Lunuwila, Sri Lanka

Eder Ramos Colegio de Postgraduados (CP), Puebla, Mexico

Luis Sáenz-Carbonell Biotechnology Unit, Centro de Investigación Científica de Yucatán, Mérida, Yucatán, Mexico

Luis Sáenz Centro de investigación Científica de Yucatán (CICY), Mérida, México

Uron Salum School of Agriculture and Food Science, University of Queensland, Brisbane, OLD, Australia

Sisunandar Universitas Muhammadiyah, Purwokerto, Indonesia

Regi J. Thomas The Central Plantation Crops Research Institute, Kasargod, Kerala, India

Valentin Luis Fredrik Wolf CIRAD – UMR AGAP, CIRAD (Agricultural Research for Development), Montpellier, France

AGAP, University of Montpellier, CIRAD, INRA, Montpellier SupAgro, Montpellier, France

About the Editors

Steve Adkins Professor Steve Adkins is Professor of Plant Physiology at The University of Queensland (UQ). He obtained a degree in Botany and Zoology from the University of London and a PhD in Weed Physiology from the University of Reading in England in 1981 and has served as a postdoctoral fellow at the University of Saskatchewan, Saskatoon, in Canada (1981–84) and at Murdoch University, Perth, Australia (1984–88). Professor Adkins joined UQ in 1988 and has spent the last 30 years studying various tropical and subtropical crops and pastures, their weeds and the native plant community. He has held several leadership roles at UQ since 2010, including Deputy Director and Acting Director in the UQ Centre for Plant Architectural Informatics. In these roles, he has led initiatives that have improved teaching quality and the student experience, instituted guidelines and funding schemes for supporting the career development of RHD students and ECRs, and established several new cross-cutting research networks in collaboration with key external partners. Professor Adkins has served as Treasurer and for two terms as the President of the Asian-Pacific Weed Science Society. His research focus is tropical plants, especially coconut, and conservation using ex situ seed banking and tissue culture. He has been a principle investigator and scientific advisor on more than 50 scientific projects worth more than $12 million. Professor Adkins has published more than 180 peer-reviewed papers in international journals, including *Proceedings of the National Academy of Sciences*, and supervised more than 50 higher degree research and 40 honours students to completion.

Mike Foale Mike Foale conducted 10 years of agronomic research into coconut production in the Solomon Island, commencing in 1959, as the agronomist of the Joint Coconut Research Scheme shared by Levers Plantations (Unilever) and the Solomon Island's government. The successful Maren hybrid was released in the 1970s. Mike was also a Coconut Consultant to ACIAR from 1983 until 1992, and later joined the University of Queensland as a Senior Research Fellow specialising in coconut research. His publications include The Coconut Odyssey – ACIAR 2003.

Roland Bourdeix Holding a bachelor's degree in biological sciences, Dr. Bourdeix continued his studies at the University of Paris Sud Orsay up to the master's level in the field of Genetics and Plant Breeding. He was then recruited by the French Agricultural Research Centre for International Development (CIRAD). Dr. Bourdeix was the first French student at CIRAD to conduct a doctoral dissertation (PhD) being based permanently in a southern country, Côte d'Ivoire, West Africa. Until 2000, he continued to work in this African country, in the Department of Genetics of the Coconut Research Station 'Marc Delorme' from CNRA (National Agricultural Research Centre of Côte d'Ivoire). Since the 1990s, he conducted training activities at the University of Cocody. Dr. Bourdeix also led a number of projects and expert missions in more than 30 countries in the tropics, especially on behalf of Bioversity International and the COGENT (the International Network for Coconut Genetic Resources), which brings together 41 coconut-producing countries. The themes of his research have gradually evolved from genetics to a multidisciplinary approach, integrating ethnology and multifunctional landscape management; to understand the diversity of crops, one needs to look not only for plants but also the people who cultivate them and their cultural specificities. From 2000 to 2014, Dr. Bourdeix worked on behalf the Coconut Research Programme of CIRAD and the research Unit 'Bio cultural interactions' of CEFE (Centre for Evolutionary and Functional Ecology). In 2014, he joined the Research Unit AGAP (Genetic Improvement of Mediterranean and Tropical Plants) and its Scientific Team DDSE (Dynamics of diversity, societies and environments).

Quang Nguyen Dr. Quang Nguyen is currently an academic in the School of Biotechnology, International University of Vietnam National University, Ho Chi Minh City. Dr. Nguyen has been part of international research projects regarding coconut mass propagation and improvement. He obtained a PhD from the University of Queensland in clonal propagation of elite coconut varieties. Dr. Nguyen has been fully committed to the sustainable development of the 'Tree of Life'. His research has been published in recognised plant-related research journals, including *Planta*, *Plant Physiology and Biochemistry* and others. Dr. Nguyen has delivered talks at multiple international conferences and provided technical training for needful countries over the past few years. His present endeavour is focussed on improving plant tissue culture techniques that enable large-scale and affordable production of coconut planting materials, as well as strategic conservation of elite germplasms.

Julianne Biddle Dr. Julianne Biddle grew up on a cattle farm in Central Queensland, Australia, and has experience working in science, agriculture, science communication and project management. Dr. Biddle is a Doctor of Philosophy in Ecology, Evolution and Genetics from the Australian National University (ANU), Honours in Biochemistry and Molecular Biology and a Bachelor of Science in Advanced Studies, Biochemistry and Molecular Biology, Cell Biology and Biological Sciences. Dr. Biddle's most recent research at the University of Queensland has focussed on coconut biotechnology and demand-led plant breeding in Africa. Dr. Biddle is newly appointed to the role of Director Multilateral Engagement, Research Strategy, at the Australian Centre for International Agricultural Research (ACIAR) and is also the Alternate Member from Australia on the CGIAR System Council.

Chapter 1
Towards the Sustainability of the "Tree of Life": An Introduction

Uron Salum, Mike Foale, Julianne Biddle, Amirhossein Bazrafshan, and Steve Adkins

1.1 Introduction

This book on the potential role of biotechnology in preventing a disastrous decline in the production of edible products of the coconut has been produced at a critical moment to show how this decline might be arrested. This first chapter provides detailed information on the very large human population immersed in coconut production, their productivity and income, and the place of coconut in national economies and trade, serving to build the case that the coconut industry needs urgent support. The anticipated deeply negative impact on almost 100 million producers and workers in the industry adds great urgency to the task of developing the affordable means to regenerate the coconut resource and stabilize the economic and social future for all involved.

1.1.1 Background

Cocos nucifera L., the coconut, is justifiably referred to as the *tree of life*, especially in the Asian and Pacific Regions. It is a unique palm that inhabits coastal and nearby land areas across most of the tropics. Nearly all parts of the palm, from the crown to the roots, serve the human economy. Its contribution to both the food and non-food chains has a significant impact on the socio-economic welfare of large rural populations in the tropical world which depend on it. Coconut is grown in over 90 countries, but only about 50 countries

M. Foale (✉) · J. Biddle · A. Bazrafshan · S. Adkins
School of Agriculture and Food Science, University of Queensland, Brisbane, OLD, Australia
e-mail: m.foale@uq.edu.au

U. Salum
Chief Executive, International Coconut Community,
Jakarta, Indonesia

S. Adkins et al. (eds.), *Coconut Biotechnology: Towards the Sustainability of the 'Tree of Life'*, https://doi.org/10.1007/978-3-030-44988-9_1

to date utilize it commercially for income generation apart from known traditional and household uses. While a significant volume is consumed domestically, the exported products and by-products reach consumers in over 110 importing countries (International Coconut Community 2019). Coconut farms are mostly village-based smallholder in nature and operation, occupying a land area measuring less than 1 ha in most instances. Country statistics indicate that over 95% of the global coconut palm population is owned by smallholder farmers (International Coconut Community 2010–2018).

1.2 Current Status of Coconut Production

The global annual coconut production in 2017 was 11.82 million t, and 2018 volumes were expected to be 12.13 million t, in copra equivalent terms. The 18 member countries of the International Coconut Community (ICC) contributed over 87% of this global production, with 10.27 million t in 2017 (International Coconut Community 2010–2018). Production has not fluctuated much in the last decade (Fig. 1.1). Besides the devastating effects of extreme climatic events, and damage by pests and diseases, the production levels shown are largely considered to be the outcome of very little or no replanting in most countries. Current records indicate an alarming increase in the proportion of senile palms, reaching between 50% and 70% in most countries.

The slight increase recorded from 2017 is attributed largely to favourable weather conditions enabling crop recovery from prior years of prolonged drought. The successful recovery effort in the Philippines after the 2013 Typhoon Haiyan devastation is notable, as production has been revitalized through the National Coconut Productivity Program (United Coconut Association of the Philippines 2019).

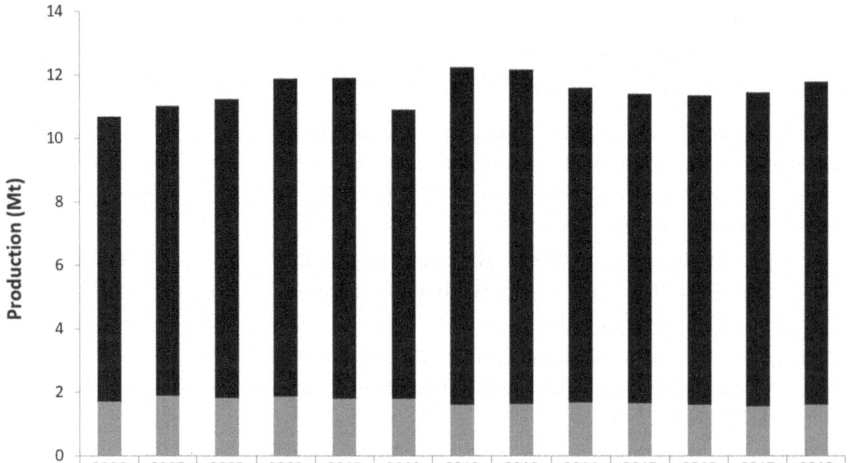

	2006	2007	2008	2009	2010	2011	2012	2013	2014	2015	2016	2017	2018
ICC Countries	8.97	9.14	9.44	10.01	10.11	9.12	10.65	10.54	9.94	9.76	9.76	9.90	10.2
Other Countries	1.7	1.89	1.81	1.87	1.8	1.8	1.6	1.63	1.67	1.65	1.59	1.55	1.6

Fig. 1.1 The global production of coconut from 2006 to 2017 (million t in copra equivalent terms) (International Coconut Community 2010–2018)

Country statistics show that Indonesia has the largest area under coconut at 3.6 million ha, followed by the Philippines and India with 3.0 and 2.0 million ha, respectively (Table 1.1). Productivity ranking shows India highest at an average of 10,000 nuts ha^{-1} $year^{-1}$ against the rest of the Community at an average of 4500 nuts ha^{-1} $year^{-1}$.

Higher productivity in India, Malaysia, Vietnam, and Sri Lanka is a direct result of increased replanting with high-yielding varieties, better crop management practices, and adoption of viable coconut-based farming systems (Ramanandam et al. 2018). This is an indication of the scope for an increase in productivity through improvements in varietal selection, crop husbandry, and farm management.

Copra and coconut oil are the major products traded in the global market, providing an important source of vegetable oil with both edible uses and industrial applications (International Coconut Community 2019).

Table 1.1 The production status of the International Coconut Community member countries listed alphabetically by country name

Country	Estimated households (growers)	Planted area (ha '000)	Estimate nut production ('000,000)	Nuts ha^{-1} (average)	Estimated number of palms ('000)	Estimated number of senile palms (ca. 50%)
Federated States of Micronesia	18,000	18	60	2197	2160	1,080,000
Fiji	120,000	64	159	2387	7440	3,720,000
India	12,000,000	2141	23,904	10,119	256,920	128,460,000
Indonesia	5,900,000	3610	14,356	4530	433,200	216,600,000
Jamaica	10,000	16	100	6156	1920	960,000
Kenya	90,000	177	254	1462	21,240	10,620,000
Kiribati	20,000	20	198	2730	2400	1,200,000
Malaysia	200,000	88	505	7464	10,560	5,280,000
Marshall Islands	15,000	8	38	4375	960	480,000
Papua New Guinea	300,000	221	1483	6710	26,520	13,260,000
Philippines	3,500,000	3502	13,825	4196	420,240	210,120,000
Samoa	40,000	99	267	2697	11,880	5,940,000
Solomon Islands	50,000	38	100	2631	4560	2,280,000
Sri Lanka	50,000	440	3011	6623	52,800	26,400,000
Thailand	290,000	206	686	4859	24,720	12,360,000
Tonga	25,000	31	72	2423	3720	1,860,000
Vanuatu	50,000	92	699	4512	11,040	5,520,000
Vietnam	60,000	159	1471	7834	19,080	9,540,000
Total	22,738,000	10,928	61,90	4661	1,311,360	655,680,000

A palm density of 120 ha^{-1} is used to estimate the number of palms per country (International Coconut Community 2010/18)

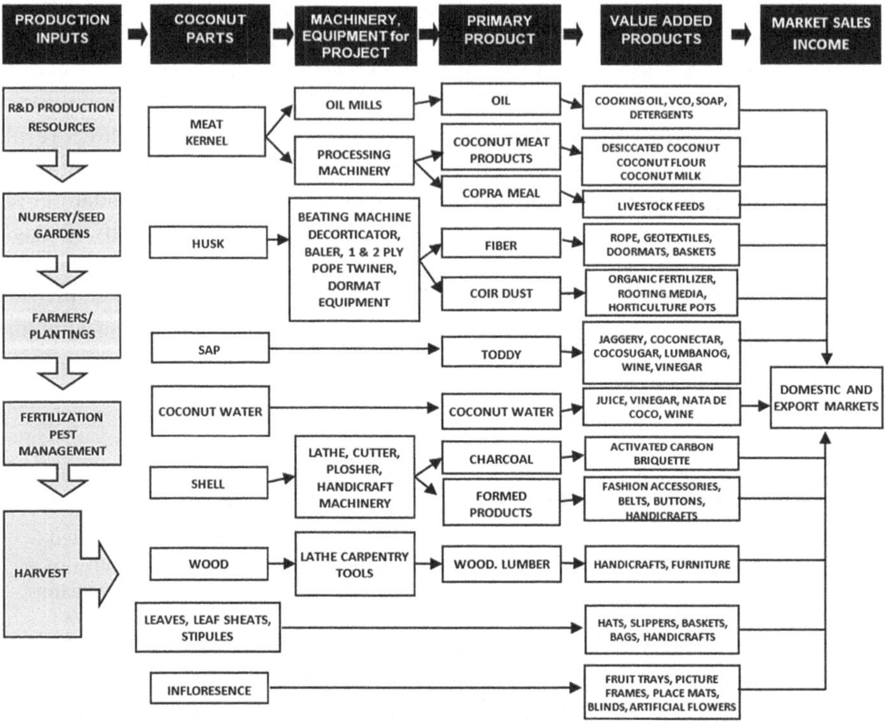

Fig. 1.2 A flow chart depicting an overview of the coconut production value chain present in most coconut-growing countries

The prices of the fresh nut and its products are largely dependent on the price of coconut oil, which is in turn affected by the price of other competing vegetable oils. Coconut is not only a source of oil but also a source of diverse food items, a beverage, a food supplement, a dairy alternative, an alternate sugar, and much more. It is the buoyant market for these diverse products that makes an ideal crop to ensure the sustainable income of farmers in this era of climate uncertainty and an increased focus on food security and poverty reduction. The range of diversified products from coconut is demonstrated in the supply value chain (Fig. 1.2).

A high volume of copra production in most countries, purchased by copra mills, is processed into crude coconut oil (CNO) and shipped to be refined, bleached, and deodorised, for distribution to consumer markets. Over 50% of CNO is utilized by the oleochemical industry to produce cleaning agents (soaps, detergents, surfactants) and more recently for the production of biofuel in some countries (International Coconut Community 2010–2018).

Statistics indicate an increasing demand for desiccated coconut (DC) and virgin coconut oil (VCO) over the last decade (Fig. 1.3). The production volume of DC, presented alongside copra and coconut oil, is an indication of the shift to higher value products that would, in the long term, reduce the production of coconut oil.

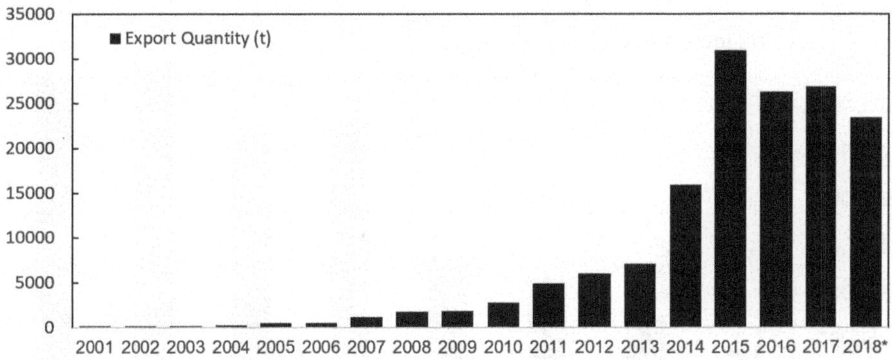

Fig. 1.3 The total Philippine export volumes of virgin coconut oil (million t) from 2001 to 2017

Income increases as producers switch to production of VCO, which has the fastest growing market among all coconut products, because of increased awareness and experience of its health benefits related to its high levels of antioxidants and medium-chain triglycerides (MCT). Such high value niche markets offer good prospects to improve the income level for the farmer. The growth in the Philippines' export of VCO has occurred in the past 15 years with an exponential growth factor of 5 between 2011 and 2015 (Fig. 1.3).

The coconut water market has expanded rapidly, because of growing health consciousness globally. It has achieved the status of a healthy natural drink through its own merit and strong marketing. Exports from the Philippines increased from a mere 647,000 L in 2008 to 1.8 million L in 2010 and then soared up to 85 million L in 2017. The total export value of coconut water in 2017 was US$ 91.3 M. The export price of coconut water also showed an increasing trend from US$ 0.80 L^{-1} in 2008 to 1.03 L^{-1} in 2012, then reaching a plateau slightly higher at 1.07 L^{-1} in 2017. Brazil is the major exporter of coconut water in the world market, while the Philippines, Thailand, Indonesia, Sri Lanka, India, Vietnam, and Malaysia also export significant volumes (International Coconut Community 2010–2018).

The Philippines' export destinations for coconut water in 2017 included the USA (Fig. 1.4), the UK, Canada, Australia, the Netherlands, Singapore, China, Hong Kong, and the United Arab Emirates. The USA was the Philippines' biggest market, which accounted for more than 45.7 million L or 53.5% of the total exported, followed by the UK and the Netherlands which absorbed 16.8% and 5.3%, respectively, of total export volume.

Supply to the global market of coconut sap sugar is dominated by Indonesia (Fig. 1.5), Thailand, and the Philippines. There is an increasing demand for coconut sugar as an alternative sweetener, both in domestic and international markets, as it has a Glycaemic Index (GI) of less than 35, compared to pure sucrose at 100. This is beneficial for people who suffer from diabetes, indicating that there is a robust domestic and export market potential for coconut sugar. A great advantage of coconut sugar production is that it can be undertaken by village or small- to medium-scale enterprises, including cooperatives involving women. Indonesia currently has

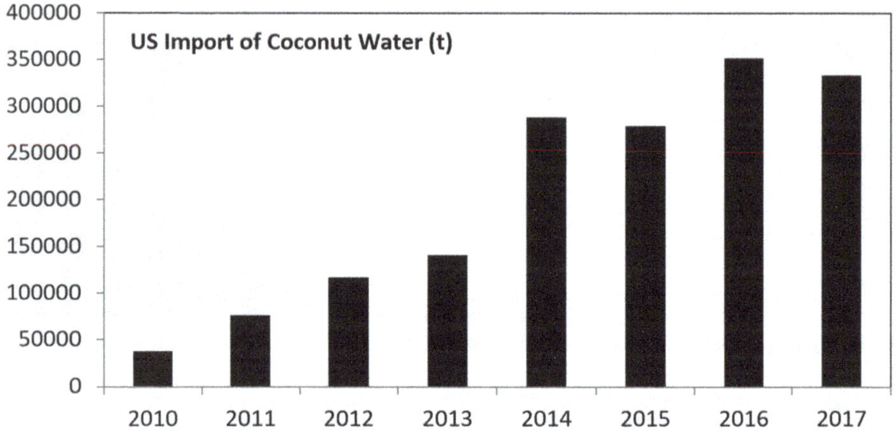

Fig. 1.4 The total volume of coconut water imported by the USA (million L) per year from 2010 to 2017 (International Coconut Community 2010–2018)

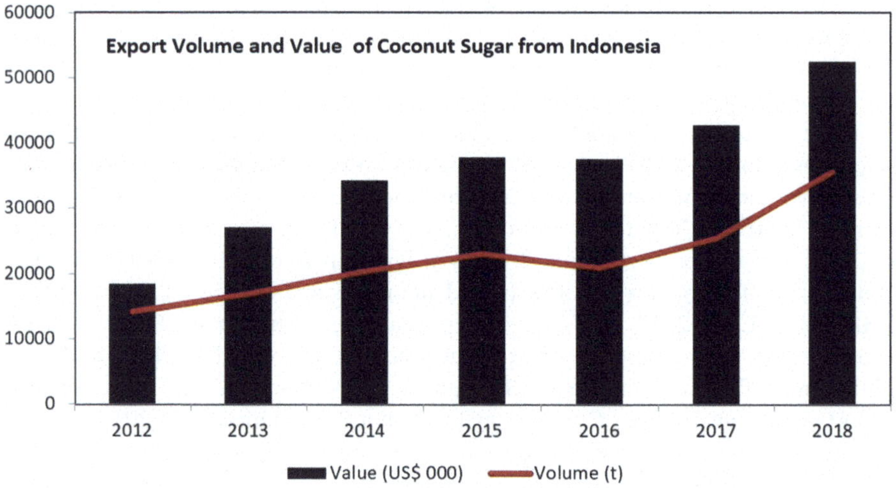

Fig. 1.5 The total export volume in t (line graph) and value in US$ (bar graph) of coconut sugar from Indonesia per year from 2012 to 2018 (International Coconut Community 2010–2018; BPS Statistics 2019)

well over 100,000 farmer households that rely on coconut sugar for their primary income (BPS Statistics 2019). Over 120,000 t of coconut sugar is produced each year with over 80% of production consumed domestically. Naturally any increase in the harvest of coconut sap for sugar further reduces the potential for nut production by the palm while tapping continues.

Competition is increasing for raw material to meet the demand for the fast-emerging, high-value products of coconut. Over 10 billion fresh mature coconuts are exported to external factories from Indonesia each year (BPS Statistics 2019).

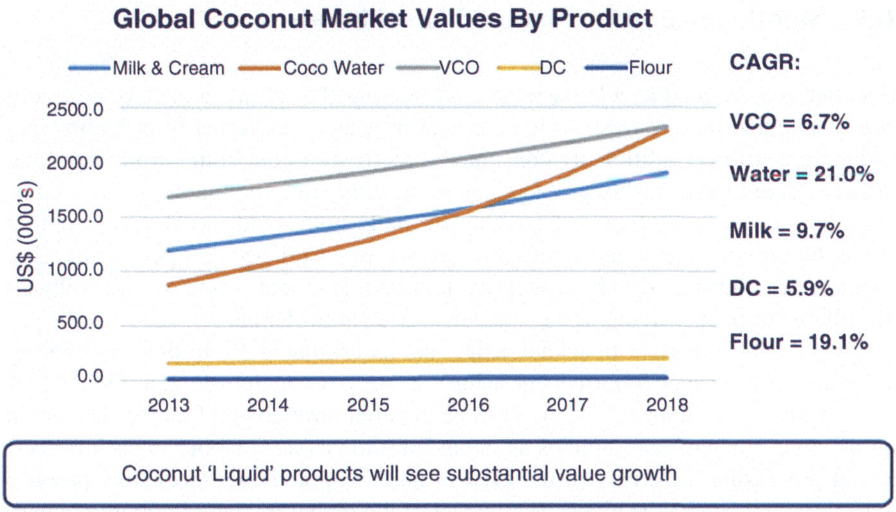

Fig. 1.6 A line graph of the total individual market values in US$ for high-value coconut products in the Philippines over time between 2013 and 2018 (Franklin Baker 2019)

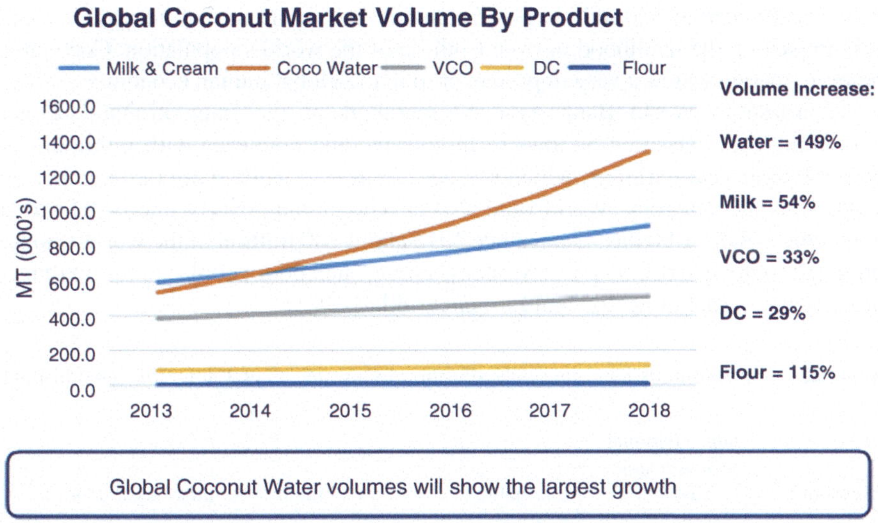

Fig. 1.7 A line graph of the total individual market volume in million t for high-value coconut products in the Philippines over time between 2013 and 2018 (Franklin Baker 2019)

This is to the detriment of local processors who, reportedly, operate at as little as 50% capacity as a result. The trend in market value in the Philippines, across a 5-year period for water, milk, VCO, DC, and flour, shows the three "liquid" products to be undergoing a substantial growth in value (Fig. 1.6) and production (Fig. 1.7) (Franklin Baker 2019).

1.3 Significance to Income and Livelihoods

Coconut is consumed as a household food in areas that produce or have access to both young and mature fresh fruit. Safe healthy water and kernel from both young and mature fruit is traditionally consumed directly as a nourishing drink and food. Milk extracted from freshly grated coconut, usually from the mature fruit, is commonly used in various forms of cooking for daily household meals. Consumption varies by country and ranges between two to four fruit per day per family. The coconut has contributed to food security, nourishment, health, and overall wellness of millions of people living in the coconut world for millennia.

It is equally important to acknowledge that an estimated 30 million rural households are fully engaged as farmers growing coconuts. An additional 60 million households would be economically dependent on coconut through gainful employment on farms (an average of two families assisting full time as farm labour to each farmer) and at processing factories and involved in marketing activities (including produce transportation) across the globe. The impact of the global market in importing countries could involve consumers numbering up to nearly 500 million households utilizing fresh coconut, coconut oil, and their by-products as food, food supplements, soap and cosmetics, and medical treatment for various physical ailments. The overall impacted population, fully engaged, involved with and active in the coconut value chain is estimated at 700 million people, and at the present rate of increase it could soon impact on the livelihood of over 1 billion of the world's population. Coconut is therefore recognized as a very important crop in the global human economy.

According to World Bank data, 12 members of the International Coconut Community are classified as low- to middle-income countries with minimum of US$ 1005 per year earnings, which is US$ 2.73 day^{-1}, while 6 are in the upper- to middle-income category (World Bank 2019). The global poverty line is defined at an income of US$ 1.90 day^{-1}, and since 2015 about 700 million of the world's population are living in extreme poverty. Many among such poor people are rural village-based farmers, including coconut farmers (Table 1.2).

Table 1.2 New thresholds for country classification by income as of 1 July 2017 (World Bank 2017)

Threshold	Gross national income/capita (US$)	International Coconut Community country classifications
Low-income	< 1,005	Federated States of Micronesia, India, Indonesia, Kenya, Kiribati, Papua New Guinea, the Philippines, Solomon Islands, Sri Lanka, Timor Leste, Vanuatu, and Vietnam
Lower-middle-income	1,006–3,955	
Upper-middle-income	3,956–12,235	Fiji, Malaysia, Marshall Islands, Samoa, Thailand, Tonga,
High-income	> 12,235	None

The income for an Indonesian family, farming 1 ha of coconut and expecting to produce 4500 nuts ha^{-1}, would reach 3,600,000 Indonesian rupiah based on a value of 800 rupiah per nut (BPS Statistics 2019). This converts to US$ 248 $year^{-1}$ or only US$ 0.68 day^{-1}. Therefore, such a family would be already living below the global poverty line of US$ 1.90 day^{-1} potential earnings.

Despite the experience of low market prices, however, millions of families continue in coconut activities as the only source of income to enable the provision of nourishment, shelter, clothing, sanitation, basic education, and primary healthcare. These are the basic needs required to be met in achieving and maintaining an acceptable quality of life.

1.4 Critical Challenges

1.4.1 Loss of Genetic Diversity

The natural spread of the coconut throughout the islands and coastlines of the tropical world of the Indian and Pacific Oceans has, through natural selection, generated populations adapted to the biohazards presented by the local ecosystem (Harries 1978). More details are presented later, describing the tolerances often possessed by coconut populations to co-located microorganisms and insects (Chap. 2).

The objective of preserving pristine genetic diversity is becoming less achievable since the global commercialization of coconut and establishment of plantations based on seeds gathered from diverse sources. There is still a case, however, for protecting populations known to have remained free of genetic mixing while seeking to identify traits that may be unique and valuable. The method of preservation of these genotypes could reside in sustaining the isolation of the source population, while the technique of cryopreservation becomes available as a backup.

1.4.2 The Threats from Pests and Pathogens

Major pests threatening coconut palms at present are coconut rhinoceros beetle (CRB), coconut scale insect (CSI), coconut hispine beetle (CHB), black palm weevil, black-headed caterpillar, and a few others with lower incidences of damage. Among the diseases of most serious concern are the phytoplasma-related ones, such as the lethal yellowing disease (LYD), Weligama disease, Bogia coconut disease, as well as coconut bud rot. Cadang-Cadang in the Philippines appears to be confined geographically by successful quarantine regulation (Hanold and Randles 1991).

The CRB, especially the new Guam biotype, has been devastating some coconut populations in the Pacific especially in the Solomon Islands, Fiji, and Samoa, though it is present in most other countries. The CSI threatened 70% of the Philippines' coconut production at one stage until brought under control to an infestation rate of 25% of the palm population. The outbreak of CHB damage in Southeast Asia also appears to have been constrained by the introduction of a parasite.

The LYD has the potential to destroy many palms if not brought under control. It has been spreading so quickly that it has reached the Caribbean and Latin Americas, the west coast of Africa, and some Asian countries, making it a global threat to the coconut industry. When the strain of coconut bud rot in the country of destination for some germplasm has been found to differ from that of the source country, then in some cases of transferred seed, there have been serious losses (Blaha et al. 1994).

1.4.3 Senility of Existing Palms

Based on annual reports by ICC member countries, 50% of existing coconut palms have reached the senile age of 60 years and are showing a declining level of productivity to as low as 40% compared to younger palms (International Coconut Community 2010–2018). The estimated population of senile palms is over 700 million, which includes the member countries of the ICC (Table 1.1).

According to reports, all countries face the challenge presented by aging palms (International Coconut Community 2018). Without a structured effort in replanting, there will be an inevitable slump in production, especially where there has been no significant planting in the last 20–30 years. Even if serious replanting were to begin immediately, the expected slump in production could last for 10–15 years following such plantings.

1.4.4 Lack of Planting Material

Demand for planting material surpasses the potential supply of good-quality plant material if robust, locally adapted hybrids are sought. Between the existing hybrid coconut seed gardens in the Philippines, India, Indonesia, Sri Lanka, and other countries, the seednut production capacity is up to a mere 7 million seeds annually, equivalent to an annual field planting of just 34,000 ha. Farmers are presently forced to select from their own best palms to begin underplanting, i.e., a young seedling is planted between existing senile palms.

The MATAG hybrid (Malayan Yellow and Red Dwarf × Tagnanan Tall) is one of the most favoured varieties in the Philippines and Malaysia and very often desired by other countries. Regrettably MATAG is in such short supply that in Malaysia seednuts are offered at a selling price between US$ 13 and 15. The Philippine Coconut Authority seed gardens in Zamboanga, Albay, and Davao can only produce up to 1 million seednuts annually, which meets local demand to plant a mere 5,000 ha. Kenya and Malaysia recently took shipments of hybrid seednuts from Deejay Farms in India at a very high cost. Ambekele Seed Garden operated by the Coconut Research Institute in Sri Lanka delivers 500,000 seednuts a year, which is far below the needs of the country. A similar situation exists in the large growing countries of Indonesia, India, Thailand, and all Pacific countries.

For more than 20 years, Côte d'Ivoire has been mass producing two hybrids (Bourdeix et al. 2005). The improved PB121 (Malayan Yellow Dwarf × West African Tall) is one of the most productive hybrids worldwide, although they produce smaller fruits than the MATAG hybrid. The improved PB113 (Cameroon Red Dwarf × Rennell Island Tall) produces large fruits with a very thin husk. Côte d'Ivoire also has the potential to produce MATAG and other hybrids such as Sri Lanka Green Dwarf × Vanuatu Tall, tolerant to lethal yellowing disease in Ghana (Dare et al. 2010).

1.5 Role of Biotechnology Towards Sustainability of Coconut: Past and Present

Biotechnology would have a critical role to enable, scientifically, the reproduction and multiplication of the desired varieties of coconut as well as ensuring that the characteristics of early bearing, high-yield potential and disease resistance are mass produced with the objective of replenishing coconut populations. These techniques will complement the field breeding experiments that are still needed to create and select the best clones and the conventional methods of seednut production by assisted pollination used in field seed gardens (De Nucé De Lamothe and Wuidart 1992).

There are numerous ways in which biotechnology is currently having an impact for the industry, and its uses are expanding rapidly in laboratories around the globe. A timeline (Fig. 1.8) compares the key advances in plant biotechnology with those in coconut biotechnology and shows that coconut research has lagged behind that for other species, although developments are accelerating. This book discusses the work conducted so far, its applications, challenges, and the further work that is required to enable the wide-scale use of the different biotechnologies for industry improvement.

Chapter 3 discusses how biotechnology can be used to improve the value, processing, and utilization of coconut for food products, nutraceuticals, pharmaceuticals, fuels, and other novel materials. Chapter 4 examines how biotechnology is currently utilized for germplasm conservation both in situ and ex situ and how cryopreservation and a multifunctional approach should be applied to improve the conservation of coconut genetic resources. Germplasm collection methods, including molecular identification approaches, are considered in Chap. 5, with a focus on the conservation of pest-/disease-tolerant varieties or varieties with other elite traits. For

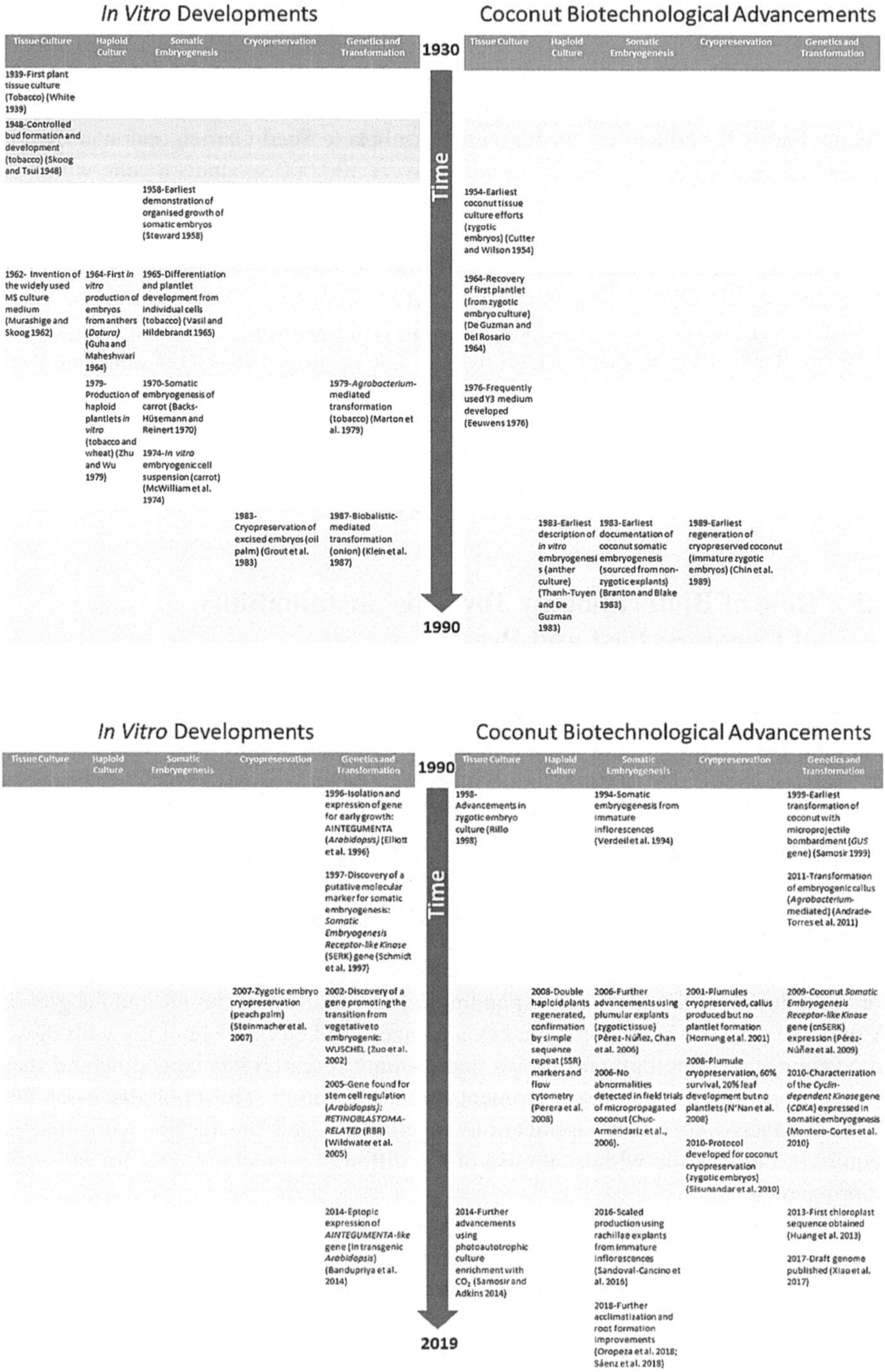

Fig. 1.8 A timeline of key plant in vitro biotechnological advances compared with important coconut biotechnological advances between 1930 and 2019

these efforts, global coordination by the International Coconut Genetic Resources Network (COGENT) is essential to reduce conservation duplication. Chapter 6 examines how since the introduction of molecular markers at the end of the twentieth century there has been a substantial advancement in plant genetics, how the development of specific molecular markers for coconut has been undertaken, and how these can be used for breeding, genetic improvement, and ultimately the conservation of coconut. Although coconut genetics has trailed other important crop species, the growing compendium of research is discussed in Chap. 7: Genome Studies for Effective Management and Utilization of Coconut Genetic Resources.

In many countries, one of the greatest challenges faced by the industry is coconut pests and/or diseases. Chapter 8 considers how current technologies are being utilized for the prevention, diagnosis, control, and treatment of these biotic factors and how future biotechnology research could be applied to improve outcomes for the industry. Chapter 9 further elaborates on this by covering LYD and LY-type diseases. These diseases have had devastating impacts in numerous locations including countries in Africa, Latin America, and the Caribbean. As there is an urgent need to control these diseases, this chapter expands on the successful use of molecular detection techniques for disease diagnosis and molecular markers that are providing promising results for identifying resistant germplasm.

Tissue culture has many applications in coconut, including conservation (germplasm movement, storage, and regeneration), propagation of elite varieties such as Makapuno and Kopyor types with an endosperm that prevents conventional seedling growth (embryo rescue/embryo culture), and cloning for the mass production of plants with desired traits (micropropagation). Chapter 10 covers the foundation coconut tissue culture method: embryo culture including coconut embryo morphology and physiology, culture types, conditions, and applications. Building on this, Chap. 11 reports on coconut micropropagation for worldwide replanting needs. This method holds much hope for the coconut industry as it has the potential to provide true-to-type clones of desired germplasm. Based on the present state of availability and a lack of planting material in most countries, biotechnology applications are needed to enable coconut micropropagation through proven tissue culture techniques. Thousands of containers of tiny plantlets can be transported in a cost-effective manner to planting locations where they undergo acclimatization and transfer into well-managed nurseries. Selected genotypes with desired traits such as high productivity, disease/pest resistance, or resilience to climatic conditions can be propagated on mass to provide valuable planting material for the renewal of coconut plantations.

Finally, Chap. 12 considers using coconut biotechnology for innovative coconut breeding programs. Present strategies (conventional breeding) are contrasted with biotechnological tools which have the potential to overcome biological constraints and poor returns on investment. The chapter discusses how molecular marker technology has been applied to characterize coconut genetic diversity and for marker-assisted selection and the recent sequencing of the coconut genome.

Recent advances in coconut biotechnology can now have a critical role in enabling the multiplication of desired varieties that possess ideal characteristics (early bearing, high-yield potential and disease resistance) with the objective of replenishing large coconut populations. An obvious challenge for the application of biotechnology faced by the industry is a lack of infrastructure in many coconut-growing regions. Due to the nature of the palm, it is often grown in regions where the main barrier is geography. Geography can pose challenges for transport and infrastructure, for example, countries in the Pacific Region where coconut is produced on hundreds of islands. The consensus from most of the chapter authors is that international research collaboration is essential to gain outcomes in coconut biotechnology, to overcome limitations and avoid duplication.

1.6 The Future: How Biotechnology Might Aid the Sustainability of the "Tree of Life"

Whereas the major proportion of coconut production shifted from large-scale plantations to smallholder farmers during the twentieth century, a sustained level of global production is shown by the data presented here in this chapter. Although there has been limited replanting until now, it is the remarkable longevity of the coconut which, despite yield decline, has continued to be productive. However, serious yield decline is now anticipated. The smallholder farmer continues to cling to the almost senile resource due in part to uncertainty about the likely resistance of the next generation of palms to possible threats. The farmer also has a wish that it might achieve greater yield.

Biotechnology, as presented in this volume, is on the verge of being able to alleviate many of the farmers' concerns by offering critical control over the genetic suitability of that new generation. The combination of clonal propagation and the identification of critical genetic markers for resistance to threats specific to a region will enable "prescription" plants to be supplied.

An important question will be the affordability of the plantlets obtained by cloning. Presently the international cost of seednuts from the Brazilian Green Dwarf variety, for instance, is about US$ 1.0 plus transportation. Clones remain more expensive to produce and are expected to be sold between US$ 3.0 and 7.0 per plantlet. Hybrid seednuts are generally sold at international level between US$ 4.0 and 12.0 per nut (J.L. Konan, personal communication).

The ICC will assume a major role in persuading governments to support coconut industry authorities locally to enable this anticipated major renewal of national coconut industries. Besides the anticipated need for a subsidy to assist the farmer meet the cost of the superior plantlets, generated by significant investment locally in the biotechnology capability to produce them, attention will also be needed to support the nutritional and other management needs of the new plantations, enabling their potential high yield to be achieved.

References

Blaha G, Hall G, Warokka JS (1994) Phytophthora isolates from coconut plantations in Indonesia and Ivory Coast: characterization and identification by morphology and isozyme analysis. Mycol Res 98(12):1379–1389

Bourdeix R, Konan JL, N'Cho YP (2005) Coconut. A guide to traditional and improved varieties. Editions Diversiflora, Montpellier

BPS Statistics (2019) The official channel YouTube of statistics Indonesia. YouTube

Dare D, Andoh-Mensah E, Owusu-Nipah J (2010) Evaluation of some basic traits of a promising coconut hybrid: Sri lankan green dwarf crossed to vanuatu tall (SGD × VTT). J Sci Technol 30(3):6

De Nucé De Lamothe M, Wuidart W (1992) La production de semences hybrides de cocotier: cas des semences hybrides Nain × Grand. Oléagineux 47(2):93–102

Franklin Baker (2019) Products. Franklin Baker. 2019. https://www.franklinbaker.com/products

Hanold D, Randles JW (1991) Coconut Cadang-Cadang disease and its viroid agent. Plant Dis 75(4):330–335

Harries HC (1978) The evolution, dissemination and classification of *Cocos nucifera* L. Bot Rev 44(3):265–319

ICC (2010–18) Statistics. International Coconut Community

ICC (2019) Statistics. International Coconut Community

Ramanandam G, Padma E, Kalpana M et al (2018) Evaluation of promising hybrids and varieties of coconut in East Coast Region of Andhra Pradesh. Int J Pure Appl Biosci 6(6):207–221

United Coconut Association of the Philippines (2019) Philippine export of coconut products up in March. News

World Bank (2017) World Bank country income ranking. https://www.ranzcp2019.com.au/wp-content/uploads/2019-World-Bank-Country-Income-Rating.pdf

World Bank (2019) World Bank country income ranking. https://www.ranzcp2019.com.au/wp-content/uploads/2019-World-Bank-Country-Income-Rating.pdf

Chapter 2
Biology, Ecology, and Evolution of Coconut

Mike Foale, Julianne Biddle, Amirhossein Bazrafshan, and Steve Adkins

2.1 Botanical Description

Coconut (*Cocos nucifera* L.) is the sole member of the *Cocos* genus within the family Palmae (Child 1964; Nayar 2016). It conforms to the common palm anatomy of a single non-branching stem (trunk) supporting a loose hemispherical crown of fronds, the bases of which form a very compact array at their points of attachment to the trunk. Both male and female flowers are borne in the same inflorescence that arises in the axil of each frond, there being many male flowers on the distil half of the many rachillae and a relatively small number of female flowers located near the inflorescence base.

Fronds emerge at regular intervals throughout the year in an environment where there is no seasonal variation in mean temperature but more slowly during any cooler season or during a marked dry period. The number of fronds, and therefore the number of fruit bunches, varies from around 12, where there are some climatic constraints, to 17 in a particularly favourable environment with a mean monthly temperature of 28 °C and the absence of any significant soil water deficit. The position of each frond is located at 140 degrees, on a horizontal circular plane, from the fronds immediately above and below it, resulting in the fifth frond above or below a reference frond being displaced by just 20 degrees either to the right or to the left of the circle (Foale 2003). The angular sequence of frond location around the trunk in any palm population is divided into half being clockwise and half anticlockwise.

There are two distinct palm types with respect to the timing of the activity of the male and female flowers, general robustness and height. In the case of the most widespread type of palm, known as the Tall, the male flowers open and shed pollen well before the female flowers are receptive to pollen, so that cross-pollination is normal, rendering the palm genetically heterozygous (De Taffin 1998). In rare

---

M. Foale (✉) · J. Biddle · A. Bazrafshan · S. Adkins
School of Agriculture and Food Science, University of Queensland, Brisbane, OLD, Australia
e-mail: m.foale@uq.edu.au

S. Adkins et al. (eds.), *Coconut Biotechnology: Towards the Sustainability of the 'Tree of Life'*, https://doi.org/10.1007/978-3-030-44988-9_2

cases, where the rate of frond emergence is high, pollen released from a younger inflorescence could reach some receptive female flowers of the preceding inflorescence, thereby achieving self-pollination. The Tall usually has a distinct "bole" or enlarged trunk at the base narrowing at ca. 1 m above the ground to a robust structure of around 300–350 mm in diameter which retains a constant width for 15 m or more before narrowing very gradually with advancing age (Fremond et al. 1966).

The other palm type is known as the Dwarf, as its rate of trunk extension is much less than the Tall; it also lacks a well-developed bole due apparently to first flowering taking place about 2 years earlier than in the Tall. The trunk diameter is generally 200–250 mm, and the "gap" between frond attachment scars on the trunk is less than half that of the Tall. The release of pollen and the onset of receptivity of the female flowers generally coincide, resulting in a high proportion of self-pollination and a genetically homozygous population. Many Dwarf populations have orange or yellow fruit, due to recessive genes for these colours. There are also brown and green fruit Dwarf populations and a further variant that is found in the South Pacific, which has a robust trunk, but a very low rate of height attainment and a very compact crown. This variant is not usually self-pollinating and remains true-to-type only when surrounded by similar Dwarfs.

Although there are two main coconut types, hybrids are generated by intervarietal crossing, generally to take advantage of the favourable morphological traits of the two different physical forms (Fig. 2.1). Hybrids are generally named depending on the original varieties (Wuidart and Rognon 1981). If the female flower used is a Tall type and is crossed with pollen from a Dwarf type, the resultant offspring is called a T × D hybrid: A Dwarf female parent and Tall male parent produce a D × T hybrid. For example, a hybrid of Malayan Red Dwarf (MRD) female parent crossed with a male Rennell Island Tall is named MAREN.

Once pollinated, the female flower, already around 20–25 mm in diameter, grows rapidly to form the full-sized fruit within 6–7 months. At that stage, the fruit comprises a husk (thin shiny exocarp and thick fibrous mesocarp) enclosing the endocarp or shell of the nut. This nut is spherical or slightly elongated in shape and contains only liquid endosperm, known as water. From that age until maturity, the solid endosperm (kernel) forms, being jellylike at first but gradually hardening until the time when abscission takes place. Once the fruit reaches maturity, the husk begins to dry out, and eventually the fruit falls naturally.

There is a single embryo embedded in the solid endosperm below the "soft eye", at the proximal end of the nut. The embryo will emerge in response to the stimulus of water entering the adjacent dry husk and eventually will form a robust seedling (Fig. 2.2), except that in some coconut populations, particularly from the Pacific region, the embryo germinates even before the husk has fully dried and before the fruit has fallen. In some cases, the seedling has emerged through the husk at the falling stage.

There is a rare genetic variant known as Makapuno or Kopyor fruits. The kernel develops a gelatinous form which still has an imbedded embryo; it is incapable of natural germination (Fig. 2.3).

Fig. 2.1 Comparison of Dwarf and Tall palms

Fig. 2.2 Healthy coconut seedlings

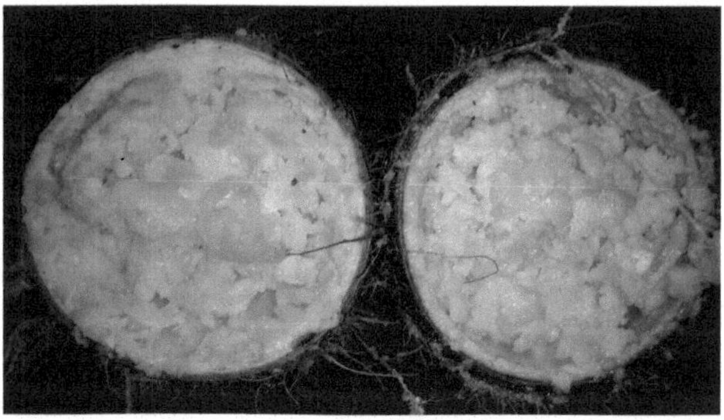

Fig. 2.3 Endosperm of the Kopyor elite type

The presence of the husk prevents observation of actual germination (first extension of the embryo through the germ pore) which can only be observed if the nut has been dehusked. Germination is stimulated to occur more rapidly following moisture stimulus if the nut is subjected to a warm unsaturated atmosphere for up to 4 weeks and then placed in an enclosed, moisture-saturated environment (Foale 1993). Following germination, the seedling is totally dependent for many weeks upon energy reserves extracted from the kernel by its spongy haustorium which expands rapidly into the vacuole of the nut. The haustorium absorbs first the liquid endosperm and then solubilizes the kernel by enzyme secretion, transferring nutrients and energy to provide support for the growth of the seedling. Within 4 months, the haustorium has usually expanded to completely fill the vacuole, thereby achieving its maximum rate of support for the development of the shoot. At about that time, the first photosynthetically active leaf unfolds, followed at intervals of about 6 weeks by subsequent new leaves, which have an increasing surface area. Gradually the relative contribution of energy from the kernel diminishes due to the photosynthetic activity of the expanding leaves (Foale 1968). By 6 months of age, although perhaps half of the endosperm remains, the seedling becomes functionally autonomous, except when experiencing severe water deficit, such as when transplanted with bare roots. In that case, the energy from the residual endosperm is drawn upon to support the seedling until photosynthetic activity resumes when the water deficit is relieved.

Under favourable conditions, new fronds will emerge from the seedling at an interval that shortens from 6 weeks in the first year to a little less than 4 weeks by year 6. In the Dwarf, flowering begins at around 3–4 years and in the Tall around 5–6 years. The earlier flowering of the Dwarf results in the lower trunk having a narrower diameter than the trunk of the Tall, as a smaller proportion of new biomass is allocated to the trunk once flowering begins. This is maintained permanently as the trunk extends, while the Tall produces enough biomass before flowering begins to support a much thicker base or bole on the trunk, and it continues permanently to allocate more biomass to the trunk than is allocated by the Dwarf (Fig. 2.1).

The life of the frond on a healthy palm lasts about 2 years, so a frequency of one new frond per month generates a crown comprising 24 fronds. In a very favourable environment, frond emergence is a little more frequent, the crown becomes larger, and the yield of fruit is greater. There is a tendency for the fruit yield to oscillate significantly between years due to variation in the success of pollination, which is affected by weather extremes and insect activity. A year of high production often results in lower production in the following year, because the energy reserve within the palm has been lowered by the demand of the previous high yield (Abeywardena 1962). The peak yield of a Tall is achieved within 20 years, when leaf area and the corresponding light interception are at a maximum. This tends to hold steady for the next 20 years and then gradually declines. However, the palm is still productive at 60 years of age and beyond, even though light interception has by then fallen to half the maximum. As age advances, the trunk diameter and the trunk extension interval between fronds fall, allowing an increasing proportion of new biomass to support fruit production, so that the decline in yield with age is not as great as the decline in light interception. The Dwarf flowers earlier, peaks at round 15 years, and declines beyond 35 years. A negative legacy of its more delicate trunk is that it is more likely to be destroyed or damaged by violent wind than the more robust Tall, which can sway markedly without breaking.

There is great genetic diversity affecting the composition of the fruit, in terms of the proportions of biomass put in to the husk, shell, and kernel (Whitehead 1966). The notional wild-type fruit has around 60% of its biomass in the husk, 15% in the shell, and 25% in the kernel, while the extreme domesticated type of fruit has only 30% of biomass in the husk, 20% in the shell, and 50% in the kernel (Harries 1978). Owing to the widespread introgression between natural (wild-type) and introduced (domesticated) coconut populations, intermediate values for the fruit components are commonly observed. Fruit form and colour also vary (Fig. 2.4), being most commonly angular and green or brown (dominant colours genetically) in Tall popu-

Fig. 2.4 Diversity of size, colour, and composition of coconut fruit resulting from selection for either high production or attractive cosmetic traits

lations and yellow or orange in Dwarfs. These latter two colours are due to recessive genes, while there are intermediate colours between non-recessive green and brown in the heterozygous Tall. There are also distinct populations of the Dwarf with both green- and brown-coloured fruit.

2.2 Origin, Evolution, and Distribution

Fossil evidence shows that the palm family first appeared on the Gondwana landmass ca.100 million years ago and then diversified as the land fragments now known as South America, Africa, southern India, and Australia, along with many smaller island fragments, drifted apart once Gondwana broke up around ca. 85 million years ago (Harries 1978). It has been postulated that while the land fragments east of the African continent drifted through the warm ocean to the north of Gondwana, known as the Tethys Sea, the modern coconut species evolved. It developed a remarkable, buoyant, sea-going fruit that enabled colonization of strand environments on large land masses but also on atolls (Harries and Clement 2014), many of which have become submerged while others have emerged in geologically recent times in response to change in the depth of the ocean.

It has been demonstrated that a quiescent state can be induced for many months in the coconut seed, enabling a long journey on an ocean current that can "deliver" the fruit to a remote strand, be it located on an atoll or a more substantial land mass. Success in colonizing the challenging niche of the tropical strand was achieved through the evolution of a relatively massive endosperm compared to all other palms and plants, except for the double coconut [*Lodoicea maldivica* (J.F.Gmelin) Persoon], that could support the growth of the seedling for many months. This enabled enough root growth to extend to the freshwater table that was to be found beneath every well-watered strand, beyond the usual reach of saltwater borne on the tide. An example of this spontaneous colonization can be found today on North Keeling coral island, an outlier of the Cocos Keeling atoll in the Indian Ocean, where there is no evidence of human settlement ever having taken place (Foale 2003; Leach et al. 2003).

The coconut can be considered to have a two-phase evolutionary history, the first in geological time and the second in response to human selection (Gunn et al. 2011; Teulat et al. 2000). Some present-day coconut populations have been observed to have a very high proportion of husk in the fruit. This imparts the high degree of buoyancy essential to preserve viability during long-distance floating, enabling the fruit to successfully establish itself on a distant strand. A humid tropical environment renders such locations suitable on the east coast of Africa and neighbouring islands, on subcontinental Asia and Southeast Asian islands, and all the tropical islands to the east, including remote Polynesia and the west coast of Mexico in the eastern Pacific, where the coconut is known to have established itself spontaneously (Sauer 1971). The duration of this period of wild coconut flourishing must surely amount to some millions of years, including eras of great fluctuation of sea level and the accompanying formation and inundation of hospitable coral atolls. The coconut was introduced by European colonizers to West Africa, the Americas, and the Caribbean only after Portuguese navigators first took fruit from India to the Cape Verde Islands in the tropical Atlantic Ocean in 1501 (Harries 1978).

The beginning of the modern phase of the evolution of coconut populations in response to human selection is somewhat uncertain but perhaps dates back much more than 100,000 years, to when hominin and then human colonization of tropical coasts is believed to have begun (Ingicco et al. 2018). As the kernel of the coconut seed provides not only a significant amount of high-energy food, it is reasonable to propose that it would almost certainly have provided nourishment to early colonizers many tens of thousands of years ago. But perhaps more importantly for voyagers, in more recent millennia, the coconut provided a significant secure reserve of portable drinkable water. It is clear from its presence on some quite remote yet inhabited islands, such as Rotuma (Fiji) and Rennell (Solomon Islands), that selection of ever larger fruit to supply water and food on anticipated long sea journeys was carried out over a long time (Baudouin and Lebrun 2009; Harries 1978).

Upon arriving at what appeared to be a suitable coast to colonize, perhaps already host to a wild coconut population, voyagers would be expected to have propagated any precious seeds remaining from the provisions on their vessel. The new coconut population that grew from their own fruit would, due to cross-pollination with the wild palms, become parents of subsequent introgressed populations, of which there are countless examples throughout the tropical world (Harries 1978).

As the human population increased in the tropical world, the coconut, a staple in the diet of all regional cultures, was propagated inland, well beyond the original strand zone. Human settlements, which extended into the lowland interior of large islands and the Asian tropical subcontinent, cultivated the coconut successfully to altitudes up to 1,000 m in some cases (Child 1964). Productivity is more closely governed by rainfall away from the supportive freshwater table found under the strand and adjacent dunes. The inland coconut therefore has a less dominant role in the diet than on the coast or on atolls where it has traditionally provided up to 60% of dietary energy (Prior et al. 1981).

Apart from the robust link between human settlements and supportive coconut plantings, the palm achieved a far greater role within the last two centuries in response to demand from outside the tropics for coconut oil, initially for soap-making in the developing world dating from the 1830s. Within a few decades, demand had expanded to edible coconut oil to supplement the supply of animal lipids that fell short of the needs of the growing populations of industrializing countries. Such was the high market value then placed on coconut oil that investment by industrialists in plantations proliferated throughout the tropical world, especially in Southeast Asia, adjacent island groups there, and throughout the South Pacific, as well as Central America, the Caribbean, and East Africa. Plantation establishment, which began around 1880, peaked between 1900 and 1925 and then declined in response to the Great Depression of the 1930s, declining further because of World War 2. The final collapse of large plantations producing coconut oil came about because of the expansion through superior marketing strategy of soybean oil (*Glycine max* (L.) Merr.) and palm oil (*Elaeis guineensis* Jacq.) production.

However, the greatly expanded distribution of the coconut during the plantation era has been sustained despite the effect of this declining demand for coconut oil in the late twentieth century that was due in part to a competitive market moving in favour of unsaturated oils. Many former plantations were taken over by traditional land owners or made available to tenants who incorporated them into a semi-subsistence lifestyle. There appears to be a comprehensive occupation by the coconut palm of suitable strand environments throughout the tropics, as well as significant plantings on inland locations where the original rain forest has been cleared, often for the initial purpose of logging and then planting annual food crops. The most significant recent expansion of coconut on a plantation scale is in Brazil in response to strong demand from the USA for coconut water (FAO 2018).

2.3 Ecological Adaptation

The prolonged period of evolution of the coconut, imparting to it the ability to occupy the strand of enormously diverse coasts from East Africa to eastern Polynesia, has resulted in its exposure to diverse biological communities which occupy the adjacent inland zone. The remarkable ability of specific coconut populations, each surviving in a unique habitat for many thousands of generations, to resist or adapt to foraging insects, lethal phytoplasmas, and non-lethal viruses, has been described (Randles et al. 1986). Other examples abound, but a good case study is provided by the resistance of Papua New Guinea and Solomon Islands populations to the leaf beetle (*Brontispa longissima* Gestro) which has decimated populations elsewhere (Chap. 8, this volume).

Ecological adaptation to climatic extremes has also been identified. The West Coast Tall of India has the notable ability to withstand a seasonal extended period of severe water deficit more successfully than the average palm of South Pacific origin (Rajagopal et al. 1993). The Tall population of Hainan Island, adjacent to southern China, withstands annual brief periods when the daily mean temperature falls below 10 °C, whereas introduced genotypes perish or suffer fruit fall. It is clear that natural selection has acted for millions of years, given the long interval between successive coconut generations occupying a specific site. In a natural setting, it is likely that a very small number of individual plants make up the population of each generation upon which natural selection favours adaptive mutations. Once a major adaptive trait has emerged, be it to a biohazard or a climatic extreme, its successful proliferation is assured.

The introduction of new germ plasm by humans poses an adaptive threat and may limit the contribution of the new population to the succeeding introgressed population until "hybrid" forms are endowed with resistance to the local hazards (Harries 1978). Over many generations, heterozygous progeny, which carry the appropriate genetic code to cope with the local biotic or physical threat, will survive, ensuring a more diverse population with respect to other traits such as fruit size and composition (Fig. 2.1).

2.4 Defining Current Threats

Given the geological time scale of coconut evolution in habitats which possess the essential physical traits of the hospitable strand, including a sandy adjacent beach and a freshwater table within reach of the root system, populations throughout the coconut zone that preceded human colonization would be expected to have much in common. Subsequent evolution in specific locations would have imparted adaptations in response to selection pressure applied by hazards present in the neighbouring biosphere, including fungal (CIRAD 1992), viral, phytoplasmic, and viroid-like pathogens (Hanold and Randles 1991) and insect species. Further dispersal of such adapted populations would be expected to deliver these adaptations to other coconut populations due to natural introgression, further enhancing their genetic diversity. However, while many adaptive traits could be of great value in establishing the palm in a new location, where there is a common threat, a population may in some cases be of limited value away from its adaptive home, due to lack of historical exposure to different threats present in that new habitat.

Evidence that human-applied selection pressure has succeeded while colonizing a coconut habitat and seeking particularly to increase the yield of kernel and water in the fruit has been found throughout the coconut world (Harries 1978). Examples of coconut populations producing the extreme wild-type fruit are now rare. The population of North Keeling coral island, in the Cocos Keeling Island group, is an example of this (Leach et al. 2003). It appears elsewhere that, almost universally, human colonizers have delivered, on arrival, a genotype possessing non-wild-type fruit which has introgressed with the pre-existing wild population. Extreme examples of modified types with very large fruit can be found in the Philippines (San Ramon), Solomon Islands (Rennell), Fiji (Rotuma), and Samoa (Niu Vai), all being islands where there were evidently few strand niches to support a large wild population preceding human colonization (Whitehead 1966). In that case, introgression was dominated by the introduced genotype. Examples of the opposite extreme, where the wild type has maintained at least partial dominance, include many of the atolls of the South Pacific, for example, in Kiribati and Tuvalu (Foale 1987).

The general objective of preserving genetic diversity requires clear definition, in the context of seeking to ensure that populations known to possess robust adaptation to recognized biotic threats are to be identified and conserved. There is little to be gained simply by conserving germ plasm from multiple geographic locations that do not differ in terms of their exposure to the principal biotic threats of the region. Priority needs to be limited to genotypes representative of regions where particular adaptations can be anticipated, which could well be several neighbouring islands expected to have shared germ plasm, exchanged naturally over geological time.

The other category of desirable germ plasm is that which can contribute a specific trait to a hybrid, combining it with a parent that has regional value. Such germ plasm includes several Dwarf types, which impart early flowering to offspring, and any variety which has a high ratio of kernel to fruit weight. No genotypes have yet been shown to have superior light use efficiency, a trait that would impart potentially higher yield regardless of fruit characteristics. On the other hand, the case for identifying superior water use efficiency is open, as adaptation to a prolonged dry season does appear to exist in locations where this is normal (Rajagopal et al. 1993).

The need to include one local parent in a hybrid pair has been demonstrated in Indonesia, for example, where a hybrid between Malayan Yellow Dwarf and West African Tall succumbed in many locations to bud rot due to the local strain of *Phytophthora* being different from that in West Africa, to which it was adapted (Blaha et al. 1994). Similarly, there are unpublished reports that the hybrid between Malayan Orange Dwarf and Rennell Tall suffered severe attack from a local species of rhinoceros beetle (*Scapanes australis* Bioisduval 1835) on New Britain island of Papua New Guinea in the 1970s. The local Tall was rarely attacked by this rhinoceros beetle, due evidently to a form of resistance. These experiences provide some guidance in deciding which germ plasm has priority for conservation in a regional collection destined to provide future material for replanting and expansion of coconut plantations in a specific region.

In summary, there is clearly much to learn at the local level about the threats to specific coconut populations. A newly established palm has a potential productive life of 50–60 years provided that it is not susceptible to any local biohazard, be it insect of microorganism. Care is therefore needed when introducing a new variety or hybrid to carry out preliminary testing of germ plasm from a distant source before large-scale establishment of what appears to be a very promising variety. It is helpful to recall that plantation management, including weeding as well as fertilizing with nutrients known to be limiting, would enable even "unimproved" genotypes to produce well (Green and Foale 1961). It would appear wise, where possible, to plant a genotype which is derived, at least in part, from the long-present coconut population of the region and therefore is very likely to be endowed with resistance to the principal biological threats. Biotechnology which enables cloning of selected outstanding individual palms in a region has the potential to meet this requirement.

References

Abeywardena V (1962) Studies in biennial bearing tendency in coconut. Ceylon Coconut Q (Sri Lanka) 13:10

Baudouin L, Lebrun P (2009) Coconut (*Cocos nucifera* L.) DNA studies support the hypothesis of an ancient Austronesian migration from Southeast Asia to America. Genet Resour Crop Evol 56(2):257–262

Blaha G, Hall G, Warokka JS et al (1994) *Phytophthora* isolates from coconut plantations in Indonesia and Ivory Coast: characterization and identification by morphology and isozyme analysis. Mycol Res 98(12):1379–1389

Child R (1964) Coconuts. Longmans, London, p 216

CIRAD (1992) Coconut Phytophthora: workshop proceedings, 26–30 October 1992, Manado, Indonesia

De Taffin G (1998) Coconut. Macmillan Education, Ithaca/New York

FAO (2018) Statistics. FAO, Rome

Foale MA (1968) The growth of the young coconut palm (*Cocos nucifera* L.). 1. The role of the seed and of photosynthesis in seedling growth up to 17 months of age. Aust J Agric Res 19(5):781–789

Foale MA (1987) Coconut germplasm in the South Pacific Islands. ACIAR technical reports series no. 4. Canberra, p 23

Foale MA (1993) The effect of exposing the germ-pore on germination of coconut. In: Nair MK, Khan HH, Gopalasundaram P, Bhaskara Rao EVV (eds) Advances in coconut research and development. Oxford and IBH Publishing Co. Ptv Ltd, New Delhi, pp 247–252

Foale MA (2003) The coconut odyssey: the bounteous possibilities of the tree of life. Australian Centre for International Agricultural Research, Canberra: ACIAR monograph 101

Fremond Y, Ziller R, de Nuce de Lamothe M (1966) Le Cocotier. Paris GP Maisonneuve et Larose

Green AH, Foale MA (1961) Improvement of coconut production on the high islands of the Pacific. In: Proceedings of the 10th Pacific science congress Honolulu, Hawaii

Gunn BF, Baudouin L, Olsen KM (2011) Independent origins of cultivated coconut (*Cocos nucifera* L.) in the Old World tropics. PLoS One 6:e21143

Hanold D, Randles JW (1991) Detection of coconut Cadang Cadang viroid-like sequences in oil and coconut palm and other monocotyledons in the south west Pacific. Ann Appl Biol 118(1):139–151

Harries HC (1978) The evolution, dissemination and classification of *Cocos nucifera* L. Bot Rev 44(3):266–319

Harries HC, Clement CR (2014) Long-distance dispersal of the coconut palm by migration within the coral atoll ecosystem. Ann Bot 113:565–570

Ingicco T, Van den Berg G, Jago-on C (2018) Earliest known hominin activity in the Philippines by 709 thousand years ago. Nature 557:233–237

Leach BJ, Foale MA, Ashburner GA (2003) Some characteristics of wild and managed coconut palm populations and their environment in the Cocos (Keeling) Islands, Indian Ocean. Genet Resour Crop Evol 50:627–638

Madhavan Nayar N (2016) The coconut: phylogeny, origins, and spread. Academic, London, pp 51–66

Prior IA, Davidson F, Salmond CE et al (1981) Cholesterol, coconuts, and diet on Polynesian atolls: a natural experiment: the Pukapuka and Tokelau island studies. Am J Clin Nutr 34(8):1552–1561

Rajagopal V, Sjhivashankar S, Kasturi Bai KV (1993) Characterisation of drought tolerance in coconut. In: Nair MK, Khan HH, Gopalasundaram P, Bhaskara Rao EVV (eds) Advances in coconut research and development. Oxford and IBH Publishing Co. Ptv Ltd, New Delhi, pp 247–252

Randles JW, Julia JF, Calvez C et al (1986) Association of a single-stranded DNA with the foliar decay disease of coconut palm in Vanuatu. Phytopathology 76:889–894

Sauer JD (1971) A re-evaluation of the coconut as an indicator of human dispersal. In: Riley CL et al (eds) Man across the sea. Texas University Press, Austin

Teulat B, Aldam C, Trehan R, Lebrun JHA, Arnold GM, Karp A, Baudouin L, Rognon F (2000) An analysis of genetic diversity in coconut (*Cocos nucifera*) populations from across the geographic range using sequence-tagged microsatellites (SSRs) and AFLPs. Theor Appl Genet 100:764–771

Whitehead RA (1966) Sample survey and collection of coconut germ plasm in the Pacific Islands. Ministry of Overseas Development/HMSO, London, p 78

Wuidart W, Rognon F (1981) La production de semences hybrides de cocotier. Oléagineux 36(3):131–137

Chapter 3
Improving the Value of the Coconut with Biotechnology

Fabian M. Dayrit and Quang Nguyen

3.1 Coconut: A Fruit with Many Uses

The fruit of the coconut tree is perhaps the most useful plant resource in the tropics. All parts of the coconut fruit have traditional uses that have been developed commercially in recent times (Foale 2003, Dayrit and Dayrit 2013). Due to its widespread household use, trade and industry statistics on coconut products reflect only part of the actual importance of the coconut. Today, coconut-based products have gone beyond the tropics and are consumed in many temperate countries and global regions such as Australia, China, Europe, North America, and the Middle East (Costello 2018). Coconut milk is the basic ingredient of traditional cuisines and desserts worldwide in the Asian tropics, while coconut flour is used in confectionery and bakery products. Coconut oil is widely used as cooking oil, hair and cosmetic oil, and domestic remedies for burns and skin ailments and in soap-making and preparation of traditional medicine. Coconut water can be either consumed fresh or converted into vinegar and nata de coco. The residues of these processes are used for animal feed and soil enhancer. The young inflorescences can be tapped directly to obtain coconut sap. This natural honey-like product can then be evaporated to prepare coco sugar or fermented to produce coconut sap wine and vinegar. These products are markedly distinct from those produced from coconut water. However, if the sap is collected, the harvest of nuts is lost. Nondairy products from the

F. M. Dayrit (✉)
Department of Chemistry, Ateneo de Manila University, Quezon City, Philippines
e-mail: fdayrit@ateneo.edu

Q. Nguyen
School of Agriculture and Food Sciences, The University of Queensland, Gatton, OLD, Australia

School of Biotechnology, International University, Vietnam National University-HCM, Ho Chi Minh City, Vietnam

S. Adkins et al. (eds.), *Coconut Biotechnology: Towards the Sustainability of the 'Tree of Life'*, https://doi.org/10.1007/978-3-030-44988-9_3

coconut, such as margarines, yoghurts, and cheese, have become more and more popular in the global market. This chapter will deal mainly with the products that can be obtained from the fruit.

3.1.1 Tradition and History of the Coconut

Various regions in the tropics utilize the coconut in food items which are culturally characteristic. In addition to its role as a basic ingredient of food, the coconut has been mentioned in historical accounts as having health and cosmetic uses (Ahuja et al. 2014). Also, evidences for the ancient use of the coconut in religious ceremonies have been found in archeological artifacts (Davis and Coningham 2018). Indeed, the coconut is inseparable from the culture and lifestyle of many peoples of tropical Asia where it was widely cultivated and shared.

The coconut grows in a wide range of tropical environments. The coconut tree can grow on sandy soil along coasts and on small islands; the highest yields have been observed on sandy, loamy, and volcanic soil and where rainfall is abundant and regular. The coconut tree can grow well up to 500 m altitude; slower growth is observed at higher altitudes and minimal growth is seen at over 1,000 m. It can be inexpensively fertilized using salt ($NaCl$) or seawater (Magat 1988). The fruit is dispersed naturally and can still germinate after floating in the sea for up to 2 months (Foale 2003).

From ancient times, the coconut tree has been grown for domestic use, and coconut cultivation has been carried out mainly by smallholder farmers. During the sixteenth century in the Philippines, the planting of coconuts was mandated by the Spanish Colonial Government to ensure availability of food for the population in times of rice (*Oryza sativa* L.) crop failure (Blair and Robertson 1906), and by the end of the nineteenth century, large-scale planting programs were promoted to convert it into a cash crop.

The coconut remained largely unknown in Europe until the nineteenth century when it entered global commerce with the development of the copra process, which made coconut oil available in Europe for the manufacture of soap and food products. Copra is the coconut kernel that is dried to about 6% moisture content in the farm or in a kiln for the main purpose of extracting the oil. The British established the first copra operation in Ceylon in the 1840s for export to Europe for the manufacture of luxury soaps. The USA expanded the production of copra in the Philippine Commonwealth in the early 1900s and soon became the largest user of copra in the world for cooking oils, margarines, and soaps, while the Philippines became the largest producer (Rice 1935). During the 1930s, the Philippines supplied ca. one-third of the copra in world trade, followed by the Dutch East Indies (Indonesia), British Malaya, and Ceylon. The interest in coconut oil encouraged more studies into the biology of the coconut tree from the 1900s to the 1930s. However, the interest in coconut gradually decreased with the rise in the USA of its domestic vegetable oil industry, such as soybean (*Glycine max* (L.) Merr.) oil (Dayrit 2008).

Although copra continues to be a major coconut commodity today, it is a low-value product which does not utilize the other coconut products maximally. Copra production requires very simple technology and the income to the farmer is low. To fully exploit the opportunities of the commercial production of these coconut food products, investments are needed for factories and technical training. However, the new markets that will be opened with new products portend a bright future for the coconut (Prades et al. 2016).

3.1.2 Overview of Coconut Food Products

Various food products can be made from the coconut depending on whether the method of processing is wet or dry. This discussion will follow the products using the dry process (Fig. 3.1) and review the typical composition of key coconut products (Table 3.1). The products – coconut milk, coconut flour, virgin coconut oil, coconut water, coco vinegar, and nata de coco – will be discussed in detail in the following sections.

The various food products that can be obtained from the coconut fruit are considered as functional food because they provide health benefits beyond basic nutrition (Mendoza 2007). The coconut meat is the edible white endosperm or kernel of the coconut. It can be consumed directly as food and is added to some coconut-based dishes. Fresh coconut meat has been noted to be satisfactory as a food, producing normal growth in humans and farm animals (Cooke 1951). Coconut cream is the emulsion that is obtained by grating the meat, and coconut milk is obtained by the addition of water. Strict cleanliness and refrigeration are needed since these rapidly become rancid. Coconut flour is obtained from the meat after pressing the oil from desiccated coconut. Coconut flour can be used to partially replace wheat flour in the

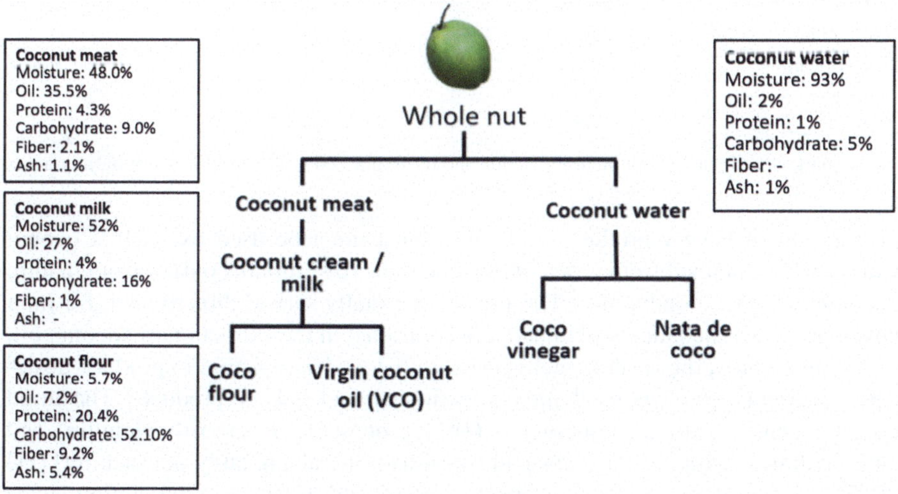

Fig. 3.1 An overview of the food products produced from the coconut fruit using the dry process. The compositions of various coconut products are indicated (see Table 3.1)

Table 3.1 The dry process produces various products which vary in terms of their composition (Cooke 1951)

	Approximate composition of coconut product, % weight			
Constituent	Fresh kernel	Coconut milk	Coconut flour	Coconut water
Water	48.0	52	5.7	93
Oil	35.5	27	7.2	2
Protein	4.3	4	20.4	1
Carbohydrate	9.0	16	52.1	5
Fiber	2.1	1	9.2	–
Ash	1.1	–	5.4	1

The process flow is shown in Fig. 3.1

Table 3.2 Amino acid and fatty acid composition of desiccated coconut (DTU 2009)

Amino acid, mg 100 g^{-1}		Fatty acid, g 100 g^{-1}	
Alanine	340	C6:0	0.4
Arginine	990	C8:0	4.7
Aspartic acid	660	C10:0	4.4
Cysteine	120	C12:0	29.8
Glutamic acid	1,400	C14:0	9.9
Glycine	340	C16:0	5.6
Histidine[a]	160	C16:1	0.3
Isoleucin[a]	290	C18:0	1.5
Leucin[a]	510	C18:1, n-9[b]	4.1
Lysine[a]	270	C18:2, n-6[b]	1.1
Methionine[a]	130	C20:0	0.6
Phenylalanine[a]	340	Total fatty acids	62.4
Proline	280		
Serine	360		
Threonine[a]	250		
Tryptophan[a]	85		
Tyrosine	210		
Valine[a]	410		

Those marked with [a] are essential amino acids and those marked with [b] are the essential fatty acids

preparation of bakery products and cakes but cannot be used by itself. Coconut water can be obtained from green unripe coconuts (6–7 months old) or from mature coconuts (11–12 months old). The former is usually served directly as refreshing beverage, while the latter is obtained as a coproduct in the extraction of coconut oil.

Commercially, the fresh kernel is dried to below 4% moisture to produce desiccated coconut, which is used for confectionary and bakery products. The dried coconut meat contains coconut oil (ca. 60%), protein (ca. 7%, mostly globulin), and carbohydrates (with ca. 7% sugar) and various minerals, notably potassium (DTU 2009). The protein in coconut meat contains all nine essential amino acids (Table 3.2).

3.2 Coconut Milk

Coconut milk is the liquid emulsion that is obtained from the grated meat of mature coconuts. It is the base ingredient of numerous traditional dishes all over the tropics, from India and Sri Lanka to Southeast Asia, and is gaining popularity as a new food ingredient in the Caribbean, Latin America, Brazil, West Africa, and North America (Reuters 2019).

3.2.1 Nutritional Value of Coconut Milk

Coconut milk is classified according to the amount of fat (minimum %) it contains: coconut cream concentrate, 29%; coconut cream, 20%; coconut milk, 10%; and light coconut milk (skimmed or defatted), 5% (Table 3.3; Codex 2015). The latter two products are used as substitutes for cow milk (*Bos taurus* L.), especially for lactose-intolerant individuals and vegans. Typical coconut cream contains 24% fat, 5.5% carbohydrates, and 2.3% protein (USDA 2018b).

Coconut milk contains about 27% coconut oil and zeatin and zeatin riboside (Van Staden and Drewes 1975). Zeatin is an adenyl cytokinin, which shows antiaging activity in human adult skin fibroblasts in vitro without inducing cell proliferation, thus avoiding potential carcinogenic effects (Rattan and Sodagam 2005). Coconut milk contains a 19 kDa protein which showed antimicrobial properties against yeasts that cause food spoilage, in particular *Debaryomyces hansenii* (Zopf) and *Candida albicans* (Algar and Mabesa 2015).

3.2.2 Recent Advancements for Improved Utilization of Coconut Milk

Coconut milk is mainly used as an affordable lactose-free substitute to animal milk. It has also been suggested that coconut milk should be mixed with cow milk to produce more attractive probiotic products (Sanful 2009). Recently, coconut milk has been developed into fermentation-based food products, such as coconut yogurt and coconut kefir.

Apart from ordinary coconut milk, skim coconut milk has been gaining popularity due to its low-fat content and attractive nutty coconut flavor (Khuenpet et al. 2016). The market for this coconut-derived product has been growing rapidly in recent times (Naik et al. 2014). Being ultra-heat treated and cream separated, skim coconut milk has higher emulsion stability and greater viscosity than the ordinary coconut milk (Khuenpet et al. 2016). It was also noted that the fat content of the skimmed milk is less than 1%, making it a favored choice for people on special diet.

Table 3.3 Selected nutrients found in coconut milk (USDA 2018b)

Component	Unit	Amount 100 g^{-1} of coconut milk
Proximates		
Energy	kcal	230.0
Water	g	68.0
Protein		2.3
Total fat		23.8
Carbohydrate		5.5
Total sugar		3.3
Dietary fiber		2.2
Ash		0.7
Minerals		
Calcium	mg	16
Magnesium		37
Phosphorous		100
Sodium		15
Potassium		263
Selenium	ug	6.2
Vitamins		
Vitamin C	mg	2.8
Choline		8.5
Folate	ug	16
Amino acids		
Alanine	mg	117
Arginine		376
Aspartic acid		224
Cystine		45
Glutamic acid		524
Glycine		108
Histidine[a]		53
Isoleucine[a]		90
Leucine[a]		170
Lysine[a]		101
Methionine[a]		43
Phenylalanine[a]		116
Proline		95
Serine		118
Threonine[a]		83
Tryptophan[a]		27
Tyrosine		71
Valine[a]		139

Those marked with [a] are essential amino acids

Recent research showed that sequential separation using ultrafiltration and nanofiltration can be applied to create high-value products from skimmed coconut milk (Ng et al. 2015). Membrane-based separation of valuable compounds, such as proteins and hormones (kinetin and zeatin), opens new opportunities for research and application. The protein-rich coconut milk has potential as a valuable source of natural growth promoters for plants and animals.

3.3 Coconut Flour

Coconut flour is obtained from the production of coconut milk. It is finely grated or ground coconut meat, which is gluten-free with a low-fat content. There are various grades of coconut flour which depend on the degree of whiteness and particle size.

3.3.1 Nutritional Value of Coconut Flour

Coconut flour has a high dietary fiber content and a low Glycemic Index (GI), making it an attractive choice for those who wish to restrict their carbohydrate intake (Trinidad et al. 2007). Coconut flour has been shown in several studies to decrease body fat, body mass index, waist circumference, and visceral adiposity and to lower blood sugar and cholesterol, even in subjects with moderately high serum cholesterol. Based on these favorable health effects, several authors have proposed that coconut flour be considered a functional food (Table 3.4) (Arumugam et al. 2014; Franco et al. 2015; Trinidad et al. 2004, 2006).

Table 3.4 List of selected components in commercial coconut flour

Component	Unit	Amount 100 g^{-1}
Proximates		
Energy	kcal	438
Protein	g	10–19
Total fat		10–22
Carbohydrate		50–70
Dietary fiber		30–60
Total sugar		19
Minerals		
Sodium	mg	94.0
Iron		6.8

Coconut flour has a high dietary fiber content and is a good source of protein (USDA 2018a; PNS/BAFPS 2010)

3.3.2 *Biotechnology for Improved Utilization of Coconut Flour and Protein Powder*

Insoluble protein, as a by-product of virgin coconut oil (VCO) processing, can be converted to value-added items, such as coconut flour and protein powder. To obtain protein powder, insoluble protein is homogenized with coconut skimmed milk prior to drying. The dehydration can be undertaken with different methods, such as drum-drying, freeze-drying, and spray-drying. Research shows that the spray-drying method scored best in multiple sensory qualities, including whiteness, powderness, crispness, milky aroma, and nuttiness (Naik et al. 2014). These products have been considered as health foods since they are low in carbohydrates and fat content, as well as being friendly to the digestive system. The nutritive powder was shown to be rich in protein (33%; w/w) and low in fat (3%; w/w). In addition, this powder possesses greater emulsifying properties when compared to skimmed milk powder and skimmed soybean powder. These attributes make coconut protein powder an attractive alternative to diary milk and artificial food additives (Naik et al. 2012).

The utilization of coconut flour and protein powder can also be enhanced with flavorsome coconut (Fan et al. 2011). Special coconut varieties, including the Aromatics, may help in improving sensory qualities of the protein powder in fitness drinks. However, due to the scarcity of such varieties, their availability is still limited. Biotechnological interventions, such as embryo culture and rapid multiplication of Aromatic coconut types, have been applied to provide sufficient elite planting materials, coping with the growing demand (Nguyen et al. 2016).

3.4 Virgin Coconut Oil

Coconut oil can be differentiated according to how it is processed. It can be consumed in its most natural form as coconut meat and high-fat coconut milk. Coconut cooking oil is obtained from copra, which is dried coconut meat and is further refined, bleached, and deodorized (RBD coconut oil). On the other hand, VCO is obtained from fresh mature coconut meat by mechanical procedures, e.g. expelling or pressing, and the application of heat only, without altering the nature of the oil. It may be purified by washing with water, settling, filtering, and centrifuging only. As an additional category, cold-pressed VCO is obtained similarly but without the application of heat (APCC 2009; Codex 2015). VCO is a relatively new commercial product having gained popularity only during the early 2000s. VCO and RBD coconut oil have the same fatty acid composition but can be differentiated using a number of parameters, such as % moisture, % free fatty acids, color, volatile organic compounds (Dayrit et al. 2008a), monoglyceride and diglyceride content (Dayrit et al. 2008b), and phenol content (Marina et al. 2009).

3.4.1 Nutritional and Medicinal Value of VCO

The nutritional and medicinal properties of VCO are attributed to its high composition of medium-chain fatty acids (C8:0, C10:0, and C12:0), especially lauric acid, phenol, and phytosterol content (Table 3.5).

VCO exhibits numerous biological properties, which have been validated in both in vitro and in vivo preclinical and clinical studies. It is important to point out that the activity reported may vary if the study is in vitro or in vivo. In the case of in vitro studies, VCO itself may not exhibit activity due to the poor solubility of the oil and/or the need to hydrolyze the triglycerides to release the fatty acids and monoglycerides which are the active compounds. These observations are consistent with the results of the pioneering studies of Kabara et al. (1972), who carried out the first systematic studies on the antimicrobial properties of fatty acids and their monoglycerides. Among his most notable conclusions were that C12:0 and its monoglyceride, monolaurin, have the highest antimicrobial activity (Dayrit 2015). The antibacterial, anti-human immunodeficiency virus (HIV), antioxidant, anti-inflammatory, hepatoprotective, dermatologic, and nutritional properties of VCO will be briefly described below.

Table 3.5 Composition of refined, bleached, and deodorized (RBD) coconut oil and virgin coconut oil

Component	Amount present	
Fatty acid[a]	*Median composition (%)*	
Caproic, C6:0	0.4	
Caprylic, C8:0	7.3	
Capric, C10:0	6.5	
Lauric, C12:0	49.2	
Myristic, C14:0	18.9	
Palmitic, C16:0	8.9	
Stearic, C18:0	3.0	
Oleic, C18:1	7.5	
Linoleic, C18:2	1.8	
Linolenic, C18:3	0.1	
Phenols[b]	*mg kg^{-1} oil*	
	RBD	*VCO*
Ferulic acid	1.4	5.0
p-Coumaric acid	1.7	0.7
Protocatechuic acid	0.2	–
Vanillic acid	–	2.0
Caffeic acid	–	0.1
Syringic acid	–	0.5
Total phytosterols[c]	8.7	8.7

[a]Codex (2015)
[b]Marina et al. (2008)
[c]Rajan et al. (2010)

3.4.1.1 Antibacterial Activity

VCO itself does not have antibacterial activity in vitro without prior hydrolysis. For example, VCO did not inhibit the growth of *Clostridium difficile*, but the lipolyzed mixture was active and C12:0 was found to be more active than C10:0 and C8:0 (Schilling et al. 2013). In another study, it was shown that free fatty acids (FFAs) which were obtained from lipase hydrolysis of VCO inhibited both gram-negative and gram-positive bacteria which cause food poisoning, especially *Bacillus subtilis*, *Escherichia coli*, *Salmonella enteritidis*, and *Staphylococcus aureus*. However, VCO itself did not show antibacterial activity (Nguyen et al. 2017).

3.4.1.2 Antiviral and Anti-HIV Activity

C12:0 was shown to possess antiviral activity by suppressing the maturation of enveloped viruses, such as vesicular stomatitis virus. The in vivo antiviral activity of C12:0 was demonstrated in the first-ever clinical trial of coconut oil and monolaurin against HIV in 14 infected patients at the San Lazaro Hospital, Manila, Philippines. By monitoring viral load, CD4 level, liver and kidney functions, and survival, it was concluded that both monolaurin and C12:0, which are produced in vivo from coconut oil, possessed anti-HIV activity (Dayrit 2000). Another study on 40 HIV-positive subjects in Indonesia showed that VCO supplementation of 15 mL taken 3x daily for 6 weeks increased CD4+ T lymphocyte levels versus the control (Widhiarta 2016).

3.4.1.3 Effect on Plasma Lipids

Although early studies reported that VCO increased plasma cholesterol, this effect depends on the organism used and the duration of the study. Using rabbits (family Leporidae), VCO showed reductions in plasma cholesterol and low-density lipoprotein-cholesterol (LDL-C) compared to the control group. The triglyceride level increased during the first 4 weeks but fell to insignificant levels during the eighth week (Zakaria et al. 2011). In a 4-week study involving 96 healthy adults, the effects of VCO, extra-virgin olive oil (EVOO), and butter on serum lipids were compared at an intake level of 50 g daily as a supplement. The results showed that LDL-C concentrations and the ratio of total cholesterol to high-density lipoprotein-cholesterol (TC/HDL-C) were significantly increased with butter compared with VCO and EVOO, with no differences between VCO and EVOO. However, VCO significantly increased HDL-C compared with EVOO and butter (Khaw et al. 2018). These studies support the conclusion that VCO has favorable effects on plasma lipid profile and possesses potential anti-hypercholesterolemic activity.

3.4.1.4 Coconut Oil and Heart Disease

Although warnings against coconut oil have been made because of its high saturated fat composition, there has been no study that unambiguously shows that coconut oil causes heart disease. In fact, in a review of 8 clinical trials and 13 observational studies, it was concluded that consumption of coconut in the context of traditional dietary patterns does not lead to heart disease (Eyres et al. 2016).

3.4.1.5 Antioxidant and Anti-inflammatory Activities

Antioxidant and anti-inflammatory activities are often linked to each other. An in vitro study compared saturated fatty acids for their antioxidant activity and inhibition of cyclooxygenase enzymes, COX-I and COX-II. For the antioxidant activity, there was an increase observed with increasing chain length from C8:0 to C14:0 and a decrease with longer chain lengths. For cyclooxygenase enzyme inhibition, the highest inhibitory activities were observed for C10:0 and C12:0 (Henry et al. 2002). In a comparison of cold and hot extraction, the antioxidant activity of VCO was attributed to the phenol content which may raise the activities of antioxidant enzymes, such as superoxide dismutase (SOD), catalase, glutathione peroxidase (GSH-Px), glutathione (GSH) concentration, and decreased lipid peroxidation in the liver (Siddalingaswamy et al. 2011). Consistent with these results, VCO was shown to increase antioxidant enzymes and to prevent lipid peroxidation (Abujazia et al. 2012).

3.4.1.6 Hepatoprotective Activity

The observed hepatoprotective activity of VCO may be due to its antioxidant action. In a study using male Sprague-Dawley rats (*Rattus* sp.), liver injury was induced by 3 g kg^{-1} paracetamol. However, pretreatment of the rats with 10 mL kg^{-1} of VCO significantly reduced liver damage (Zakaria et al. 2011).

3.4.1.7 Hair and Skin Care and Wound-Healing

Coconut oil has been traditionally used for the general care of the hair, scalp, and skin, as well as for wounds and burns, and these have all been validated by scientific studies. In a comparison with mineral oil, coconut oil was shown to penetrate the hair shaft while mineral oil did not. This study concluded that coconut oil may provide better protection from hair damage than mineral oil (Ruetsch et al. 2001). Another study on hair care which compared mineral oil and sunflower (*Helianthus annuus* L.) oil with coconut oil showed that only coconut oil was able to reduce the protein loss from both undamaged and damaged hair (Rele and Mohile 2003). Dermatological studies on the topical antimicrobial efficacy of VCO have been reported. Atopic dermatitis is a condition where dry skin is readily colonized by *Staphylococcus aureus*. A double-blind randomized clinical study which compared VCO with virgin olive oil showed that VCO was significantly better for the treatment of atopic dermatitis colonization (Verallo-Rowell et al. 2008). The burn wound-healing property of coconut oil was compared with silver sulfadiazine and a combination of both agents. The best improvement in burn wound contraction is the group treated with coconut oil which supports the use of coconut oil for burn wound-healing (Srivastava and Durgaprasad 2008).

3.4.2 Biotechnology for the Improved Processing of VCO

As beneficial properties of coconut oil have been reinforced, its associated biotechnological applications are now under way. For instance, beneficial microorganisms have been engineered to improve fermentation-based production of VCO. Probiotic bacteria such as *Lactobacillus plantarum* were shown to reduce contamination of pathogenic bacteria during the fermentation process (Satheesh and Prasad 2014). More recently, Asmoro et al. (2018) demonstrated that *Rhizopus* spp. in Ragi Tempe (also known as Tempe starter) can be applied to aid the production of VCO.

Beneficial microorganisms found in VCO can be helpful in both medicinal and nutritional applications. Intriguingly, research indicates that coconut oil can help encapsulate probiotics in yoghurt such as *Lactobacillus bulgaricus*, thereby enhancing their survival in mimicked gastrointestinal conditions (Othman et al. 2007). Furthermore, bio-yoghurt with improved flavor has been developed using coconut milk as a fermentation ingredient apart from cow milk. Results showed that the combination of coconut milk and cow milk resulted in an increased count of probiotics in yoghurt production.

3.5 Coconut Water

Coconut water is the liquid endosperm inside the coconut, which supports the development of the embryo. It is a nutritious and healthful fluid and is an excellent rehydration liquid. The composition of the coconut water varies with the age of the coconut. Traditional use of the water is usually from the green coconut (7–8 months old). Unless otherwise stated, the following discussion shall refer to water from the green coconut (tender coconut water, TCW).

3.5.1 Nutritional Value of Coconut Water

Notable among the components of coconut water is its high potassium content (Table 3.6; USDA 2018c). Potassium is a vasodilator, which increases blood flow that improves the metabolism of tissues, such as skeletal muscle and brain, and

Table 3.6 Coconut water is an excellent natural rehydration liquid that is high in potassium (USDA 2018c)

Component	Unit	Amount per 100 g^{-1}
Proximates		
Energy	kcal	19
Water	g	95.0
Protein		0.7
Total fat		0.2
Carbohydrate		3.7
Dietary fiber		1.1
Total sugar		2.6
Minerals		
Calcium	mg	24
Magnesium		25
Phosphorous		20
Sodium		105
Potassium		250
Vitamins		
Vitamin C	mg	2
Folate	μg	3

lowers blood pressure in normal and hypertensive individuals. Although the needed serum potassium concentration is low, the effects of potassium supplementation takes about 4 weeks before the effect is felt (Haddy et al. 2006).

Like coconut milk from which it is obtained, TCW also contains zeatin which has been shown to have antiaging properties (Lazim and Badruzaman 2015). Shikimic acid, a key phytochemical intermediate, was detected in TCW and was shown to have antioxidant properties, which prevented oxidative damage to liver cells (Manna et al. 2014).

In a rehydration study involving human volunteers undergoing exercise-induced dehydration, TCW was compared with a carbohydrate-electrolyte beverage and plain water. TCW was evaluated to be easier to consume in large amounts, causing less nausea and stomach upset (Saat et al. 2002). Consistent with this, a study involving eight physically active men showed that consumption of TCW improved exercise capacity in the heat without gastrointestinal stress and gave reduced urine output in comparison with plain water and flavored drink (Laitano et al. 2014). In a related application, coconut oil was shown to be an effective rehydration drink for children with mild diarrhea (Adams and Bratt 1992).

TCW has been shown to have several beneficial health effects. In a study involving 28 hypertensive human subjects, TCW gave significant decreases in mean blood pressure compared with bottled drinking water (Alleyne et al. 2005). TCW was shown to have cardioprotective potential in a study using Sprague-Dawley rats which induced myocardial infarction by isoproterenol. Cholesterol and lipoprotein concentrations were decreased, and histopathological analysis showed minimal myocardial damage. It was suggested that the beneficial effects of TCW may be due to the potassium, calcium, magnesium, and L-arginine content (Anurag and Rajamohan 2003). TCW was also shown to possess hepatoprotective and antioxidant effects in a study which used carbon tetrachloride intoxication of Sprague-Dawley rats. The rat liver did not show any fat infiltration or necrosis which indicated the hepatoprotective effect of TCW (Loki and Rajamohan 2003).

Coconut water is naturally aseptic and has been directly used in medical emergency procedures and as alternative treatment for various ailments. TCW was used as intravenous fluid during World War 2 in the Southeast Asian countries; this same procedure has since been replicated in clinical settings in Thailand, the USA (Eiseman 1954), Indonesia (Ranti et al. 1965), India (Rao et al. 1972), and the Solomon Islands (Campbell-Falck et al. 2000). Tender coconut water has also been recommended as a daily drink to prevent stone formation. In addition, TCW has been used in an endoscopic procedure called "bukolysis" to directly dissolve urinary stones in a clinical procedure (Macalalag 2011).

3.5.2 Nata de Coco

Nata de coco is a type of bacterial cellulose that is produced by the fermentation of coconut water by *Acetobacter aceti* subspecies *xylinum*. Originally produced in the Philippines, it is a gel that is often added to drinks, puddings, and other desserts. The unique structure of bacterial cellulose nata has led to the development of various nonfood applications such as biocompatible wound dressings, drug delivery systems, nanomaterials, biodegradable composites, and electronic devices (Esa et al. 2014). Nata de coco has found popularity as a health food, having a similar nutritional profile as coconut water. It is a low-calorie high-fiber aid to digestion which helps reduce serum triglycerides and total cholesterol in hyperlipidemic patients (Mesomya et al. 2006).

Traditionally, nata de coco has been inoculated using various wild-type strains of *Acetobacter xylinum* (Bernardo et al. 1998). The yield and properties of bacterial cellulose depend on both the bacterial strain and the type of carbon source, since different strains have different metabolic abilities (Tabaii and Emtiazi 2016). Based on the carbohydrate composition of coconut water, appropriate strains can be selected, mutated, or genetically engineered to produce high levels of cellulose to the required specifications, alongside optimization of media composition and fermentation parameters (Vandamme et al. 1998). By using plasmids of relative *Acetobacter* strains or broad host-range plasmids, cellulose synthase genes can be introduced into the inoculum to boost cellulose production, as has been done with the BPR2001 strain using a shuttle vector (Yoshinaga et al. 1997).

Nata de coco on its own has already proven its usefulness as facial masks and immobilization material, such as for enzymes in a urea detection device or for bacteria in bioethanol product (Amnuaikit et al. 2011; Montealegre et al. 2012; Mulyasuryani and Srihardyastutie 2011). Moreover, this previously overlooked resource, through both in situ and ex situ reinforcement, can fulfill the demand for high-performance biomaterials (Esa et al. 2014). Carboxymethyl-nata for fruit coating and hydrogel composite for drug delivery systems are just two examples showing the potential of this inexpensive resource (Treesuppharat et al. 2017).

3.5.3 Coconut Vinegar

There are two types of coconut vinegar: from coconut water and from coconut nectar or sap. This discussion shall cover vinegar from coconut water. Although vinegar from coconut water is traditionally obtained by simple uncontrolled fermentation through the agency of naturally available microorganisms, a more controlled process using selected microorganisms, such as *Lactobacillus acidophilus*, *L. casei*, or *L. plantarum*, can produce vinegars with different probiotic qualities and inhibitory activities against pathogens (Prado et al. 2015). The vinegars which are produced using different bacteria give products with different taste and odor characteristics due to their characteristic volatile metabolites (Lee et al. 2013).

Like nata de coco, coconut vinegar is an underestimated resource. Studies on its antibacterial and antitumor properties have been lacking compared to similar investigations done on rice and sugarcane (*Saccharum officinarum* L.) fermented products (Mohamad et al. 2019). Given the naturally low Brix content of coconut water, developers can easily adjust the sweetness level of coconut vinegar products, such as sports drinks (Aziz et al. 2016). Coupled with bacteria that are tolerant to temperature and to concentrations of ethanol and acetic acid in the bioreactors, coconut vinegar holds great potential for the production of health supplements and bulk solvents (Li et al. 2015; Rogers et al. 2013). Interestingly, these special bacteria can be sourced from the great diversity present in the coconut fruit itself (Lisdiyanti et al. 2003; Perumpuli et al. 2014).

3.6 Elite Coconut Varieties (Makapuno and Aromatics)

3.6.1 Nutritional Value, Uses, and Potential of Elite Coconut Varieties

It has been known that the Makapuno endosperm contains higher protein and water content compared to ordinary coconut (Adriano and Manahan 1931). Santoso et al. (1996) indicated that glutamic acid, arginine, and aspartic acid are the major amino acids, whereas lysine is more plentiful in normal coconut. Research shows that the lipid content of the endosperm of Makapuno is lower than that of the normal coconut. Also, the attractive taste has been shown to be related to the high galactomannan content, which results from an over-proliferation of microcells during the development of Makapuno endosperm (Mujer et al. 1983). Furthermore, Makapuno has high levels of important minerals, such as potassium (Santoso et al. 1996). Altogether, these special qualities make Makapuno a potential choice for food industry. On the other hand, Aromatic coconut is well appreciated due to its refreshing and fragrance-rich drinking water. The pleasant taste and aroma have been known to be related to the presence of δ-lactones (Jangchud et al. 2007). A recent study indicated that 2-methyl-1-butanol acetate and nonane are the key volatile compounds that contribute to the special aroma of both liquid and solid endosperm (Jirapong et al. 2012).

Due to their special dietary value and scarcity, elite coconuts, such as Makapuno and Aromatics, have been gaining popularity in recent times (Nguyen et al. 2016). Having a jellylike texture and nutty taste, Makapuno endosperm can be consumed fresh or converted into high-value products such as ice cream, syrups, and delicacies. However, due to its limited fruit supply, short storage life, and immature processing technology, Makapuno-derived products are still scarce in distant markets. Likewise, proper packaging of Aromatic coconut is critical for its transport. It has been shown that the storage life of Aromatic coconuts can be extended using vacuum-packing (Jangchud et al. 2007). Also, packing helps prevent microbial growth and water dehydration. A more recent study by Luengwilai et al. (2014) indicated that oxygen transmission packing and low temperature (ca. 5 °C) should be applied to Aromatic coconut to prolonging its quality during transport.

3.6.2 Biotechnology for Improved Elite Coconut Varieties

In recent years, biotechnological efforts have been made to (i) mass produce elite coconut seedlings for planting, (ii) elucidate genetic markers for useful agricultural traits, and (iii) improve planting materials through breeding. Tissue culture techniques, i.e., embryo culture and somatic embryogenesis, have been undertaken to rescue and multiply the elite but non-germinable varieties (Nguyen et al. 2016; Rillo et al. 2002; Samosir et al. 2006). A basic embryo culture protocol for Makapuno coconut usually has three stages: (i) initial germination, (ii) development of young leaves and roots, and (iii) advanced root development (Rillo et al. 2002). The embryo culture technique has been successfully undertaken in several countries including Australia, the Philippines, Vietnam, Indonesia, Fiji, PNG, etc. Different pilot plantations have also been established in many places (Adkins 2007). It often takes about 9 months for plantlets to reach the three-leaf stage and be ready for hardening off. The acclimatization of embryo-cultured plantlets can be improved by using a CO_2 enrichment protocol (Samosir and Adkins 2014).

Genetic markers have been developed to elucidate the biosynthesis pathways related to coconuts with special characteristics. Vongvanrungruang et al. (2016) suggested that a single-nucleotide substitution at position 442 of the binding site of the fragrance-related gene, betaine aldehyde dehydrogenase 2 (Badh2), resulted in the special fragrance in Aromatic coconut. This substitution has been linked to the conversion of alanine (found in the non-Aromatic) to proline (found in the Aromatic). Recently, it was suggested that DNA methylation and its associated phenotypic alterations should be considered during the tissue culture of the Makapuno (Angeles et al. 2018). However, the epigenetic changes created undesired outcomes (dismantled fruit) in oil palm (*Elaeis guineensis* Jacq.), a closely related species to coconut. Therefore, further research should be undertaken.

3.7 Improving the Food Value of the Coconut with Biotechnology: Summary

The coconut is valued for the numerous uses that can be derived from each part of the fruit. Coconut milk is a nutritious product, having all the essential amino acids. It can be prepared into products with different fat levels to suit the needs of consumers for nondairy milk. It can also be fermented into nutritious and delicious yoghurts and kefirs. From the milk can be obtained coco flour, a gluten-free low Glycemic Index flour, and VCO. VCO is a functional food with many medicinal benefits. Coconut water is a one-of-a-kind natural drink which has been used medically because of its aseptic nature and mineral composition. It has been shown to be an excellent rehydration drink. Coconut water can be converted to coco vinegar and nata de coco, two high-value food products. In addition, the coconut has two variants, Makapuno and Aromatic, that are also valued for their unique taste and aroma. Modern biotechnology can be utilized to obtain even higher value from each of these products. Truly, there is a very bright future for the coconut, where tradition can be enhanced with modern science.

References

Abujazia MA, Muhammad N, Shuid AN et al (2012) The effects of virgin coconut oil on bone oxidative status in ovariectomised rat. Evid Based Complement Alternat Med 2012:1. https://doi.org/10.1155/2012/525079

Adams W, Bratt ED (1992) Young coconut water for home rehydration in children with mild gastroenteritis. Trop Geogr Med 44(1):149–153

Adkins S (2007) Coconut tissue culture for clonal propagation and safe germplasm exchange in Indonesia, Vietnam, Papua New Guinea and the Philippines. Adoption of ACIAR project outputs: studies of projects completed in 2006–07

Adriano FT, Manahan M (1931) The nutritive value of green, ripe and sport coconuts. Philipp Agric Sci 20(3):195–199

Ahuja S, Ahuja S, Ahuja U (2014) Coconut – history, uses, and folklore. Asian Agrihist 18(3):221–248

Algar AF, Mabesa LB (2015) Isolation and partial characterization of a low molecular weight antimicrobial protein from coconut (*Cocos nucifera* L.) milk. Int Food Res J 22(5):1813

Alleyne T, Roache S, Thomas C et al (2005) The control of hypertension by use of coconut water and mauby: two tropical food drinks. W Indian Med J 54(1):3–8

Amnuaikit T, Chusuit T, Raknam P (2011) Effects of a cellulose mask synthesized by a bacterium on facial skin characteristics and user satisfaction. Med Devices (Auckl) 4:77–81
Angeles GJ, Lado PJ, Pascual DE (2018) Towards the understanding of important coconut endosperm phenotypes: is there an epigenetic control? Agronomy 8(10):225
Anurag P, Rajamohan T (2003) Cardioprotective effect of tender coconut water in experimental myocardial infarction. Plant Food Hum Nutr 58(3):1–12
Arumugam M, Raman M, Johnson B et al (2014) Dietary fiber isolate from coconut flakes – a functional food. Int J Pharm Sci Rev Res 25(2):262–267
Asmoro N, Widyastuni R, Ndrudu JJ (2018) Production of virgin coconut oil using fermentation method extraction with Ragi Tempe. In: International conference on applied science and engineering. Atlantis Press
Aziz N, Sharifudin SA, Kahar AA (2016) Azest: natural sport drink from coconut water vinegar. Paper presented at the 6th international conference on biotechnology for the wellness industry, Malacca
APCC (2009) Quality Standard for Virgin Coconut Oil. https://coconutcommunity.org/viewpdf/apcc_quality_standards_for_coconut_products/3
Bernardo EB, Neilan BA, Couperwhite I (1998) Characterization, differentiation and identification of wild-type cellulose-synthesizing acetobacter strains involved in nata de coco production. Syst Appl Microbiol 21(4):599–608
Blair EH, Robertson JA (1906) The Philippine Islands 1493–1803. AH Clark Company, Cleveland
Campbell-Falck D, Thomas T, Falck TM et al (2000) The intravenous use of coconut water. Am J Emerg Med 18(1):108–111
Cooke FC (1951) The coconut palm as a source of food. Ceylon Coconut Q 2(4):153–156
Costello H (2018) Global coconut water market and coconut oil market 2018 by demand, consumption, production, top regions, key manufacturers, growth & forecast till 2023
Codex Alimentarius (2015) Codex standard for named vegetable oils. Codex Stan 210:1–13
Davis C, Coningham R (2018) Pilgrimage and procession: temporary gatherings and journeys between the tangible and intangible through the archaeology of South Asia. World Archaeol 50(2):347–363
Dayrit CS (2000) Coconut oil in health and disease: its and monolaurin's potential as cure for HIV/AIDS. Paper presented at the XXXVII Cocotech meeting. Chennai, India
Dayrit FM (2008) A brief history of the Philippine coconut industry as reflected in the PJS, 1906 to 2005. Philipp J Sci (Centennial edition): 10
Dayrit FM (2015) The properties of lauric acid and their significance in coconut oil. J Am Oil Chem Soc 92(1):1–15
Dayrit CS, Dayrit FM (2013) Coconut oil: from diet to therapy. Anvil Publishing Inc, Manila
Dayrit F, Buenafe O, Chainani E et al (2008a) Standards for essential composition and quality factors of commercial virgin coconut oil and its differentiation from RBD coconut oil and copra oil. Philipp J Sci 136(2):119–129
Dayrit FM, Buenafe OEM, Chainani ET et al (2008b) Analysis of monoglycerides, diglycerides, sterols, and free fatty acids in coconut (*Cocos nucifera* L.) oil by 31P NMR spectroscopy. J Agric Food Chem 56(14):5765–5769
DTU (2009) Technical University of Denmark. Coconut meat, desiccated. http://www.foodcomp.dk/v7/fcdb_details.asp?FoodId=0126
Eiseman BEN (1954) Intravenous infusion of coconut water. JAMA Surg 68(2):167–178
Esa F, Tasirin SM, Rahman NA (2014) Overview of bacterial cellulose production and application. Agric Agric Sci Procedia 2:113–119
Eyres L, Eyres MF, Chisholm A et al (2016) Coconut oil consumption and cardiovascular risk factors in humans. Nutr Rev 74(4):267–280
Fan HK, Feng ML, Huang LY et al (2011) A new coconut cultivar 'Wenye 4'. Acta Hortic Sin 38(4):803–804
Foale M (2003) The coconut odyssey: the bounteous possibilities of the tree of life. ACIAR monography

Food and Agriculture Organization. Downloaded on March 25, 2019 from: http://www.fao.org/fao-who-codexalimentarius/shproxy/it/?lnk=1&url=https%3A%2F%2Fworkspace.fao.org%2Fsites%2Fcodex%2FStandards%2FCODEX%2BSTAN%2B210-1999%2FCXS_210e.pdf

Franco EP, Oliveira G, Luiz R (2015) Effect of hypoenergetic diet combined with consumption of coconut flour in overweight women. Nutr Hosp 32(5):2012–2018

Haddy FJ, Vanhoutte PM, Feletou M (2006) Role of potassium in regulating blood flow and blood pressure. Am J Phys Regul Integr Comp Phys 290(3):R546–R552

Henry GE, Momin RA, Nair MG et al (2002) Antioxidant and cyclooxygenase activities of fatty acids found in food. J Agric Food Chem 50(8):2231–2234

Jangchud K, Puchakawimol P, Jangchud A (2007) Quality changes of burnt aromatic coconut during 28-day storage in different packages. LWT-Food Sci Technol 40(7):1232–1239

Jirapong C, Uthairatanakij A, Noichinda S et al (2012) Comparison of volatile compounds between fresh and heat-processed aromatic coconut. In: Kanlayanarat S, Boonyaritthongchai P, Acedo AL (eds) Asia Pacific symposium on postharvest research, education and extension, Acta horticulturae, 943, vol 1. Int Soc Horticultural Science, Leuven, pp 111–115

Kabara JJ, Swieczkowski DM, Conlet AJ et al (1972) Fatty acids and derivatives as antimicrobial agents. Antimicrob Agents Chemother 2(1):23–28

Khaw K-T, John Sharp S, Finikarides L et al (2018) Randomised trial of coconut oil, olive oil or butter on blood lipids and other cardiovascular risk factors in healthy men and women. BMJ Open 8(3):e020167

Khuenpet K, Jittanit W, Hongha N et al (2016) UHT skim coconut milk production and its quality. SHS Web Conf 23:03002

Laitano O, Trangmar S, Soares Menezes E et al (2014) Improved exercise capacity in the heat followed by coconut water consumption. Motriz, Rio Claro 20(1):107–111

Lazim MIM, Badruzaman N (2015) Quantification of cytokinins in coconut water from different maturation stages of Malaysian coconut (*Cocos nucifera* L.) varieties. J Food Process Technol 6(11):1

Lee P-R, Boo C, Liu S-Q (2013) Fermentation of coconut water by probiotic strains Lactobacillus acidophilus L10 and Lactobacillus casei L26. Ann Microbiol 63(4):1441–1450

Li S, Li P, Feng F, Luo L-X (2015) Microbial diversity and their roles in the vinegar fermentation process. Appl Microbiol Biotechnol 99(12):4997–5024

Lisdiyanti P, Katsura K, Potacharoen W et al (2003) Diversity of acetic acid bacteria in Indonesia, Thailand, and the Philippines. Microbiol Cult Collect 19(2):91–99

Loki AL, Rajamohan T (2003) Hepatoprotective and antioxidant effect of tender coconut water on carbon tetrachloride induced liver injury in rats. Indian J Biochem Biophys 40(5):345–357

Luengwilai K, Beckles DM, Pluemjit O et al (2014) Postharvest quality and storage life of 'Makapuno' coconut (*Cocos nucifera* L.). Sci Hortic 175:105–110

Macalalag EV (2011) Buko water of immature coconut is a universal urinary stone solvent. Cocoinfo Int 18(2):12–19

Magat SS (1988) Use of salt (sodium chloride) as fertilizer for coconut. Philippine Coconut Authority, abstracted in AGRIS 1991 10 (4):1–15

Manna K, Khan A, Kr Das D et al (2014) Protective effect of coconut water concentrate and its active component shikimic acid against hydroperoxide mediated oxidative stress through suppression of NF-κB and activation of Nrf2 pathway. J Ethnopharmacol 155(1):132–146

Marina AM, Man YBC, Nazimah SAH (2009) Antioxidant capacity and phenolic acids of virgin coconut oil. Int J Food Sci Nutr 60:114–123

Mendoza EMT (2007) Development of functional foods in the Philippines. Food Sci Technol Res 13(3):179–186

Mesomya W, Varapat P, Komindr S et al (2006) Effects of health food from cereal and nata de coco on serum lipids in human. Songklanakarin J Sci Technol 28:23–28

Mohamad NE, Yeap SK, Abu N et al (2019) In vitro and in vivo antitumour effects of coconut water vinegar on 4T1 breast cancer cells. Food Nutr Res 63:1–11

Montealegre C, Dionisio RE, Sumera VL et al (2012) Continuous bioethanol production using *Saccharomyces cerevisiae* cells immobilized in nata de coco (biocellulose). 2nd international conference on environment and industrial innovation IPCBEE, p 35

Mujer CV, Arambulo AS, Mendoza EMT et al (1983) The viscous component of the mutant (Makapuno) coconut endosperm.1. Isolation and characterization. Kalikasan Philipp J Biol 12(1–2):42–50

Mulyasuryani A, Srihardyastutie A (2011) Conductimetric biosensor for the detection of uric acid by immobilization uricase on nata de coco membrane – Pt electrode. Anal Chem Insights 6:ACI-S7346

Mariana AM, Che man YB, Nazimah SAH et al (2008) Antioxidant capacity and phenolic acids of virgin coconut oil. Int J Food Sci Nutr 60(1):114–123

Naik A, Raghavendra SN, Raghavarao KSMS (2012) Production of coconut protein powder from coconut wet processing waste and its characterization. Appl Biochem Biotechnol 167(5):1290–1302

Naik A, Venu GV, Prakash M et al (2014) Dehydration of coconut skim milk and evaluation of functional properties. Cyta-J Food 12(3):227–234

Ng CY, Mohammad AW, Ng LY et al (2015) Sequential fractionation of value-added coconut products using membrane processes. J Ind Eng Chem 25:162–167

Nguyen QT, Bandupriya HDD, Foale M et al (2016) Biology, propagation and utilization of elite coconut varieties (makapuno and aromatics). Plant Physiol Biochem 109:579–589

Nguyen V, Le T, Phan H et al (2017) Antibacterial activity of free fatty acids from hydrolyzed virgin coconut oil using lipase from *Candida rugosa*. J Lipids 2017, Article ID 7170162, 7 pages

Othman NI, Shahril M, Ramli S et al (2007) Effect of coconut oil emulsion on encapsulation of Lactobacillus bulgaricus and survival in simulated gastrointest conditions. In: 10th ASEAN food conference 2007, August 21–23, Kuala Lumpur, Malaysia

Perumpuli PABN, Watanabe T, Toyama H (2014) Identification and characterization of thermotolerant acetic acid bacteria strains isolated from coconut water vinegar in Sri Lanka. Biosci Biotechnol Biochem 78(3):533–541

PNS/BAFPS (2010) Coconut flour – specification. Department of Trade and Industry, Bureau of Product Standards, Philippines (75)

Prades A, Salum UN, Pioch D (2016) New era for the coconut sector. What prospects for research? OCL 23(6):D607

Prado FC, De Dea Lindner J, Inaba J et al (2015) Development and evaluation of a fermented coconut water beverage with potential health benefits. J Funct Foods 12:489–497

Rajan RGR, Kumar PKP, Krishna AGGK (2010) Tocopherols and Phytosterols content of coconut oil blends prepared for coconut oil consumers and non coconut oil consumers. Indian Coconut J 53(4):16–20

Ranti I, Kwee T, Thio I et al (1965) Coconut water for intravenous fluid therapy. Paediatr Indones 5(3):782–792

Rao P, Rao S, Kumar S et al (1972) Intravenous administration of coconut water. J Assoc Physicians India 20(3):235–239

Rattan SIS, Sodagam L (2005) Gerontomodulatory and youth-preserving effects of zeatin on human skin fibroblasts undergoing aging in vitro. Rejuvenation Res 8(1):46–57

Rele A, Mohile R (2003) Effect of mineral oil, sunflower oil, and coconut oil on prevention of hair damage. J Cosmet Sci 54(2):175–192

Reuters (2019) Global coconut market and coconut milk market 2019 by demand, consumption, types, regions, supply, growth, key-players, market-impact and business forecast 2024. https://www.reuters.com/brandfeatures/venturecapital/article?id=88219

Rice LP (1935) Philippine copra and coconut oil in the American market. Far East Surv 4(20):156–161

Rillo EP, Cueto CA, Medes WR et al (2002) Development of an improved embryo culture protocol for coconut in the Philippines. In: Engelmann F, Batugal P, Oliver J (eds) Coconut embryo in vitro culture: part II. Proceedings of second international on embryo culture workshop,

Mérida, Yucatán, Mexico, 14–17 March 2000. International Plant Genetic Resources Institute (IPGRI), Rome, pp 41–65

Rogers P, Chen J-S, Zidwick MJ (2013) Organic acid and solvent production: acetic, lactic, gluconic, succinic, and polyhydroxyalkanoic acids. In: Rosenberg E, DeLong EF, Lory S, Stackebrandt E, Thompson F (eds) The prokaryotes: applied bacteriology and biotechnology. Springer Berlin Heidelberg, Berlin/Heidelberg, pp 3–75

Ruetsch SB, Kamath Y, Rele AS et al (2001) Secondary ion mass spectrometric investigation of penetration of coconut and mineral oils into human hair fibers: relevance to hair damage. J Cosmet Sci 52:169–184

Saat M, Singh R, Sirisinghe RG et al (2002) Rehydration after exercise with fresh young coconut water, carbohydrate-electrolyte beverage and plain water. J Physiol Anthropol Appl Hum Sci 21(2):93–104

Samosir YMS, Adkins SW (2014) Improving acclimatization through the photoautotrophic culture of coconut (*Cocos nucifera*) seedlings: an in vitro system for the efficient exchange of germplasm. In Vitro Cell Dev Biol Plant 50:493–501

Samosir YMS, Rillo EP, Mashud N et al (2006) Revealing the potential of elite coconut types through tissue culture. Paper presented at the coconut revival – new possibilities for the 'tree of life'. Proceedings of the international coconut forum held in Cairns, Australia, 22–24 November 2005. ACIAR proceedings

Sanful RE (2009) Promotion of coconut in the production of yoghurt. Afr J Food Sci 3(5):147–149

Santoso U, Kubo K, Ota T et al (1996) Nutrient composition of kopyor coconuts (*Cocos nucifera* L.). Food Chem 57(2):299–304

Satheesh N, Prasad NBL (2014) Production of virgin coconut oil by induced fermentation with *Lactobacillus plantarum* NDRI strain 184. Hrvatski časopis za prehrambenu tehnologiju, biotehnologiju i nutricionizam 9:37–42

Schilling M, Matt L, Rubin E et al (2013) Antimicrobial effects of virgin coconut oil and its medium-chain fatty acids on Clostridium difficile. J Med Food 16(12) 1079–1085

Siddalingaswamy M, Rayaorth A, Khanum F (2011) Anti-diabetic effects of cold and hot extracted virgin coconut oil. J Diabetes Mellitus 1:118–123

Srivastava P, Durgaprasad S (2008) Burn wound healing property of *Cocos nucifera*: an appraisal. Indian J Pharm 40(4):144–146

Tabaii MJ, Emtiazi G (2016) Comparison of bacterial cellulose production among different strains and fermented media. AFB 3(1):35–41

Treesuppharat W, Rojanapanthu P, Siangsanoh C (2017) Synthesis and characterization of bacterial cellulose and gelatin-based hydrogel composites for drug-delivery systems. Biotechnol Rep (Amsterdam, Netherlands) 15:84–91

Trinidad TP, Loyola AS, Mallillin AC et al (2004) The cholesterol-lowering effect of coconut flakes in humans with moderately raised serum cholesterol. J Med Food 7(2):136–140

Trinidad T, Mallillin A, Valdez HDS et al (2006) Dietary fiber from coconut flour: a functional food. Innov Food Sci Emerg Technol 74(4):309–317

Trinidad TP, Valdez DH, Loyola AS et al (2007) Glycaemic index of different coconut (*Cocos nucifera*)-flour products in normal and diabetic subjects. Br J Nutr 90(3):551–556

USDA (2018a) Branded food products database: full report (all nutrients) 45022923. Coconut flour, UPC: 897922002256. Release July 2018

USDA (2018b) National nutrient database for standard reference: basic report 12117. Nuts, coconut milk, raw. Release 1 April 2018

USDA (2018c) National nutrient database for standard reference: basic report 12119. Nuts, coconut water (liquid from coconuts). Release 1 April 2018

Van Staden J, Drewes SE (1975) Identification of zeatin and zeatinriboside in coconut milk. Physiol Plant 34(2):106–109

Vandamme EJ, De Baets S, Vanbaelen A et al (1998) Improved production of bacterial cellulose and its application potential. Polym Degrad Stab 59(1):93–99

Verallo-Rowell VM, Dillague KM, Syah-Tjundawan BS (2008) Novel antibacterial and emollient effects of coconut and virgin olive oils in adult atopic dermatitis. Dermatitis 19(6):308–315

Vongvanrungruang A, Mongkolsiriwatana C, Boonkaew T et al (2016) Single base substitution causing the fragrant phenotype and development of a type-specific marker in aromatic coconut (*Cocos nucifera*). Genet Mol Res 15(3):gmr.15038748. https://doi.org/10.4238/gmr.15038748

Widhiarta K (2016) Virgin coconut oil for HIV-positive people. Cord 32(1):50–57

Yoshinaga F, Tonouchi N, Watanabe K (1997) Research progress in production of bacterial cellulose by aeration and agitation culture and its application as a new industrial material. Biosci Biotechnol Biochem 61(2):219–224

Zakaria ZA, Rofiee MS, Somchit MN et al (2011) Hepatoprotective activity of dried- and fermented-processed virgin coconut oil. eCAM 2011, Article ID 142739, 8 pages

Chapter 4
In Situ and Ex Situ Conservation of Coconut Genetic Resources

Roland Bourdeix, Steve Adkins, Vincent Johnson, Lalith Perera, and Sisunandar

4.1 Introduction: Towards Multifunctional Approaches to Coconut Conservation

Coconut is an important crop, not only because of its economic value but also in terms of its social, cultural, symbolic, and religious importance. Conservation of coconut genetic resources is an essential element in the development of improved varieties that will achieve more sustainable and cost-effective coconut uses.

In situ conservation of coconut germplasm refers to the maintenance of coconut genetic diversity in its natural habitat or through the continued cultivation of landraces or traditional varieties in the agroecosystems where they have evolved. The

The original version of this chapter was revised. The incorrect citation (page 65) and its corresponding reference (page 75) has been updated. The correction to this chapter can be found at https://doi.org/10.1007/978-3-030-44988-9_13

R. Bourdeix (✉)
CIRAD – UMR AGAP, CIRAD (Agricultural Research for Development), Montpellier, France

AGAP, University Montpellier, CIRAD, INRA, Montpellier SupAgro, Montpellier, France
e-mail: Roland.bourdeix@cirad.fr

S. Adkins
School of Agriculture and Food Sciences, The University of Queensland, Brisbane, OLD, Australia

V. Johnson
Bioversity International, Montpellier, France

L. Perera
Coconut Research Institute of Sri Lanka, Bandirippuwa Estate, Lunuwila, Sri Lanka

Sisunandar
Universitas Muhammadiyah, Purwokerto, Indonesia

S. Adkins et al. (eds.), *Coconut Biotechnology: Towards the Sustainability of the 'Tree of Life'*, https://doi.org/10.1007/978-3-030-44988-9_4

concept of "conservation beyond genebanks" broadens this approach by gathering in situ conservation, on farm conservation and some additional multifunctional landscaping options (Tuia et al. 2018).

Ex situ conservation refers to the maintenance of coconut genetic diversity outside of its natural habitat. It comprises all coconut germplasm currently maintained as living trees in field genebanks, screen houses, as well as in in vitro conservation, such as the conservation of coconut embryos at sub-zero temperatures and the cryopreservation of pollen, embryos, and embryogenic callus. In addition, researchers can conserve plantlets obtained from embryos and callus in culture vessels for 1–2 years using slow growth approaches; so, this also provides a kind of short-term ex situ conservation approach for coconut.

This chapter presents a range of existing initiatives in coconut conservation. It also emphasizes the biological and sociological constraints jeopardizing this conservation. A prospective approach aims to assess how farmers can better benefit from the advances in coconut biotechnology for germplasm conservation. It argues that the most efficient and effective ways to conserve coconut diversity is by a multifunctional approach, involving the conservation of other crops by the same people on the same lands or in the same laboratories. Even a cryopreserved genebank collection could be multifunctional. To generate economies of scale and to increase their patrimonial value, such genebanks could simultaneously conserve not just coconut but also other appropriate crop and plant species.

4.2 A Conservation Challenged by Strong Biological Constraints

The biology of the coconut plant imposes several unique constraints on its own conservation, including the space and time required for effective conservation, the complexity of its reproductive biology, the lack of vegetative propagation, the need for controlled hand pollination, it's relatively low multiplication rate, and the height to which its palms grow.

4.2.1 Space

Coconut field genebanks require a great deal of space and time to set up and maintain (Bourdeix et al. 2018a). Planting densities generally range between 110 and 250 trees ha^{-1}, according to the type planted and the local cultivation practices. Time from planting to flowering varies from around 1 year for some Dwarf types such as the Indonesian Salak Green Dwarf to 7 years for some Tall types. Tall types require space for at least 90 palms per accession and Dwarfs 45 palms (Santos et al. 1996).

4.2.2 Complexity

Scientists still need to address the complexity of the natural breeding mode of the coconut palm (Fig. 4.1). However, a study conducted in Sri Lanka reported that the principal pollination agents are insects and that wind pollination is negligible. In Sri Lanka, it is the Indian honeybee (*Apis cerana indica*; Fabricius 1798) that plays the major role in pollination (Liyanage et al. 1988). A coconut inflorescence normally develops within the axil of each frond. It contains male and female flowers that may or may not mature at the same time, depending on the type. The male flowers, located in the top portion of the spikelet attached to the peduncle, are more numerous than the female ones that occupy the basal portion of the spikelet. Two mating systems coexist in coconut: the largest group (Tall type and very special compact Dwarf types) is predominantly cross-pollinating, where the female flowers become receptive after the male flowers in the same inflorescence have stopped shedding pollen. However, self-pollination is also possible by pollen from the next succeeding inflorescence. A coconut palm finds a way to reproduce, even if arriving alone on a new island (Bourdeix et al. 2005). Further research is required to assess the rate of inbreeding in Tall types, which varies according to the varieties, the season, the climate, and the conditions of cultivation.[1] In a study by Fernando (1976) in Sri Lanka, the possibilities of inter-spadix pollination were found to be:

- Low for varieties such as the King Coconut, Sri Lanka Green Dwarf, Sri Lanka Red Dwarf, and Bodiri Tall
- Slight for Navasi-thembili and Gon-thembili Tall
- Greater for Sri Lanka Yellow Dwarf and Navasi Tall
- Extremely high for Sri Lanka, Kamandala, and Tan-thembili Tall types, especially when palms are young and vigorous

The potential for self-pollination depends on the length of the male and female flower maturity phases and the chances of overlap within and between spadices. The average pollination distance is a crucial issue for both in situ conservation and seednut production. In the case of Dwarf seed gardens, a 300 m forest barrier was estimated to be enough to provide the necessary isolation distance (De Nucé de Lamothe and Rognon 1975). When Sri Lanka first established their isolated coconut seed gardens, based on experimental evidence, it was considered that a forest barrier of 800 m was required to prevent pollen contamination from outside (Liyanage et al. 1988). The average number of male parental palms naturally contributing to a fruit bunch remains unknown.

[1] Although Tall types are predominantly allogamous, self-pollination can occur when female flowers from one inflorescence are pollinated by the pollen of the next inflorescence on the same palm. Such flower overlapping increases with the speed of inflorescence production, and this depends on the vigour of the tree, growing conditions, and climatic variations. So, selecting well-performing Tall types may result in the choice of palms with a higher than normal selfing rate resulting in a strong inbreeding handicap.

Fig. 4.1 Reproductive biology of the coconut palm. (**1**) Spikelets of a Tahitian Red Dwarf from the opening of the inflorescence to 3 months old. (**2**) Coconut inflorescences. (**3** and **4**) Receptive female flowers with nectar drops pollinated by bees and ants. (**5**) A bee harvesting coconut pollen

4.2.3 Vegetative Propagation

Many scientific publications state that coconut palms cannot naturally propagate through sucker formation. However, recent surveys indicate that such propagation is sometimes possible. In Indonesia, coconut palms able to produce both suckers and fruits have been described (Novarianto and Miftahorrachman 2000). On the Fakarava Atoll in the Tuamotu Archipelago, a young coconut palm able to produce suckers was recently observed (R. Bourdeix, personal communication). These special types have not yet been transferred to any ex situ coconut collection. In Polynesia, farmers dislike and often destroy such coconut palms. The suckers are rare, heavy to carry, and strongly attached to the trunk and induce late flowering of the mother palms. Planting such coconut suckers is much more difficult than planting banana (*Musa* spp.) suckers, while using a coconut seedling is much simpler. A reason why coconut palms do not propagate by natural vegetative means could be that farmers have repeatedly selected against this ability.

4.2.4 Low Multiplication Rate and Controlled Hand Pollination

The low multiplication rate is the main limiting factor for breeding, conservation, and regeneration of coconut. With traditional cultivation methods, most varieties annually produce less than 100 nuts per palm. Seednut production is therefore expensive. Controlled hand pollination requires the covering of inflorescences with a bag that is impermeable to pollen, but permeable to air. When bagged, the 12–16 inflorescences produced per year bears each only 1–3 coconut fruit. Therefore, the annual yield declines to only 20–30 fruit per tree. For only one controlled cross, using a Dwarf as the female, the palms need to be climbed seven or eight times. The production cost for one controlled hand pollination seedling is therefore US$ 8 to 10. Such factors considerably limit the choice between crossing systems and the number of palms per accession to be conserved in the genebanks. After pollination, the female flowers take 10–12 months to mature as viable seednuts. Raising the seedlings in the nursery needs a further 7–12 months before planting is possible. The seed has little

Fig. 4.1 (continued) from male flowers. (**6–11**) Reproductive biology of a young West African Tall. (**6** and **7**) Inflorescence (left side) is open but only male flowers are receptive. (**8** and **9**) Nine days later than in (**6**) and (**7**), female flowers (left side) are receptive, but no male flowers remain so cross-pollination occurs with pollen from another palm. (**10** and **11**). Three days later than in (**8**) and (**9**), some female flowers (left side) are still receptive and the male flowers from the next inflorescence (right side) are now producing pollen so self-pollination can occur between successive inflorescences from the same palm. (**12–15**). Inflorescences of Malayan Red, Niu Leka and Tacunan Green Dwarfs, and Rennell Island Tall

or no dormancy, preventing storage of seednuts. The large size of both seednuts and seedlings makes their transportation expensive, both for farmers (from the seed garden to their fields) and for genebanks (for the international movement of germplasm); it also increases the risk of pest and disease transmission, as their size makes it hard to disinfect for transport. Only about 65–70% of the seednuts give quality selected seedlings that are finally planted in the field (Santos et al. 1996).

4.2.5 Palm Height

In coconut field genebanks, palm climbing has a significant human and economic impact. At about 25 years after planting, Tall types reach a height of about 15 m from the ground to the base of the frond petioles. However, the challenge is not the height itself but the climbing method for making controlled hand pollinations. Staying under the leaf crown is quite simple. It is much more difficult to climb into the coconut leaf crown and reach the young inflorescences for undertaking the pollinations.

4.3 In Situ, On-Farm, and "Beyond Genebanks" Conservation

The 12 million ha of coconut plantations and the millions of additional coconut palms in home gardens are obviously the main places where coconut diversity is conserved. Many farmers and gardeners are still not fully aware of the crucial importance of what they do, i.e. maintaining coconut genetic diversity in agricultural and human landscapes. Increasing their commitment to conservation needs further communication and awareness education.

4.3.1 Farmers' Roles and Rights

Many farmers' organizations around the world are complaining about the present systems for crop conservation. For instance, GRAIN (a small international non-profit organization that works to support small farmers and social movements in their struggles for community-controlled and biodiversity-based food systems) states that: "if Governments were truly interested in conserving biodiversity for food and agriculture, […] they would, as a central priority, focus their efforts on supporting diversity in their countries' farms and markets rather than only betting on big centralized genebanks. This means leaving seeds or planting material in the hands of local farmers, with their active and innovative farming practices, respecting and promoting the rights of communities to conserve, produce, breed, exchange and sell seeds" (GRAIN 2008). The Action Group on Erosion, Technology and Concentration (ETC) also states: "many vital ex situ genebanks are in desperate straits. As much as half of the world's crop diversity may still be in farmers' fields protected only by the family and the community. [.../...] the Governing Body for the Food and Agriculture Organization (FAO) Treaty should devote special attention to the issue of in situ conservation and the urgent need for a financial facility to support this conservation" (ETC 2008). In the Third World Network (an organization formed to strengthen cooperation among development and environment groups in the South), Lim Li Ching (2008) wrote "this focus on ex situ collections in genebanks is highly flawed if it creates the delusion that this is the ultimate approach to seed conservation. In situ or on-farm conservation, where farmers have for generations been preserving seed and crop diversity, is a much more important priority, and yet this form of diversity maintenance for food and agriculture is being increasingly eroded and destroyed".

Article 9 of the International Treaty on Plant Genetic Resources for Food and Agriculture emphasizes the crucial contribution of the local and indigenous communities and farmers for the conservation of plant genetic resources (FAO 2009). In line with national legislation, this article encourages the protection and promotion of farmers' rights regarding plant genetic resources for food and agriculture. This includes (a) protection of relevant traditional knowledge, (b) the right to participate equitably in sharing benefits arising from germplasm use, and (c) the right to participate in making decisions, at the national level, on related matters.

In 2017 in Côte d'Ivoire, West Africa, the first author of this chapter launched a new initiative called "InnoDiv". This initiative aimed to release to local farmers seedlings from all the accessions of the International Genebank for African and Indian Ocean and seedlings from the many experimental crosses obtained in the breeding program derived from this genebank. A local donor provided financial support for this initiative and agreed to devote further significant funds to support further work. This initiative would have been a world first, the first time for an entire international plant collection to be given back to the farmers. Despite the full support of the local coconut research team, higher-level reservations have currently suspended its progress. This reluctance demonstrates some of the challenges, at least for some scientists and policymakers, to sharing with farmers the germplasm conserved ex situ in international genebanks.

4.3.2 Facilitating On-Farm Conservation

There are an estimated 500 million smallholder farmers in the world. Only 7% have access to financing for purchasing improved seeds, fertilizer, or equipment. According to the FAO (2018), the lack of financial access is primarily due to the absence of a formal credit profile of farmers and the cost of developing one. However, new data sets and tools, such as satellite data and machine learning, have helped create credit profiles and financial inclusion for smallholder farmers.

For example, real-time monitoring of the location and condition of croplands can now enable an accurate and quick assessment of the farm's performance and viability to be made. Such tools based on integrated geographical information systems may also help to understand how farmers conserve coconut diversity and help to support them in this activity.

4.3.3 Farmers' Knowledge of Plant Reproduction

Understanding farmer's agricultural practices and knowledge is crucial for successful germplasm conservation. In 2015, the Agropolis Foundation (a granting body that supports and promotes agricultural research and sustainable development) funded a project that examined Côte d'Ivoire farmers' traditional and botanical knowledge concerning the mode of reproduction of five of their crops, including coconut. Overall, 40% of farmers did not know how plants reproduce or referred only to the action of God. A further 19% suggested crop reproduction relates firstly to natural or climatic phenomena, such as sunlight, rainfall, and nutrient-rich soil (19%). Another 22% of farmers described a reproduction mechanism like the scientific version, but in about 14% of these cases, although known, this explanation was not convincing (Bourdeix et al. 2016). The responses also showed a strong difference between region, ethnic group, sex, and age. In French Polynesia, coconut farmers ascribe either a "male or female" gender to their coconut palms and seednuts. From the scientific point of view, the inflorescences have both male and female flowers, but more than 80% of the interviewed farmers did not know this. Although the farmer's descriptions did not align with botanical knowledge, their traditional knowledge is useful from a pragmatic point of view for selecting and breeding their crop. In this situation, imposing scientific knowledge on farmers without due consideration may adversely affect some of their useful traditional practices. Thus, to avoid such negative impact, there is a need to define an appropriate communication strategy from scientists to farmers.

4.3.4 Pollination Isolation

The Polymotu concept (*poly* meaning many, *motu* meaning island) uses the geographical isolation of sites for conservation and reproduction of individual varieties of plants and animals. For instance, when a small island is planted with only a well-chosen set of coconut varieties, breeding occurs only within these varieties and certified seednuts can be naturally produced. Not only islands can be used for this purpose; various kinds of sites inland are suitable if they are protected from pollen contamination, through a combination of barrier planting and geographical isolation.

Instead of climbing palms making controlled pollinations, people can wait for the coconut fruit to fall naturally to the ground. In Polymotu circumstances, open pollination will provide true-to-type and cheap seednuts. Varieties can be conserved if enough palms can be kept alive in the field. In most cases, the duration of a coconut accession will then be increased from 20/30 to 75/100 years. Even if some of the trees die, there is no need to remove the remainder, as is generally done in a classical genebank. Dead palms can be replaced without removing the trees that are still alive.

Three examples of the Polymotu concept have been developed (Bourdeix et al. 2012a, b):

- **Inland:** Côte d'Ivoire is starting to duplicate some accessions of its international genebank in the middle of other tree crop plantations [such as rubber (*Hevea brasiliensis* (HBK) Muell Arg) or oil palm (*Elaeis guineensis* Jacq.)].
- **Ecotourism on Islands:** This approach is presently implemented by replanting three small islets, two of which are in Samoa (as described in Fig. 4.2) and one in French Polynesia.
- **Urban:** This third approach can be illustrated by a proposal submitted to the University of South Pacific (Bourdeix 2014). It involves the integration of a coconut conservation design on the Suva campus in Fiji. The conserved coconut palms would be scattered between all the campus buildings.

In the framework of the CIDP project, the Ministry of Agriculture of the Cook Islands started to implement in 2018 a concept of "a delocalized community-based coconut genebank", like the Polymotu urban approach. Coconut palms, from diverse

Fig. 4.2 A Polymotu design as initially planted on the Nuusafe'e Island, Samoa. This design was conceived to conserve three varieties and produces seedlings of Tall (germinating with green sprouts), Dwarf (red sprouting), and Hybrids (brown sprouting). Unfortunately, land tenure problems have caused the disappearance of most of the coconut palms that were planted in 2012

and easily identifiable varieties, will be planted in many public places. After planting, agricultural officers will record the palms' identity and their location (latitude, longitude, and date of planting), and, only when the palms start to fruit, will data become available in an online database. If the Cook Islands succeeds in implementing this concept, this small country will have in 10 years the largest coconut genebank in the world – without devoting any dedicated land to this activity. Such a genebank will be accessible to all citizens who can gain access to the online database.

From 2009, the Polymotu concept was included as a component of the conservation strategy developed by the International Coconut Genetic Resources Network (COGENT) and the Global Crop Diversity Trust (Crop Trust). Further implementing this concept could strengthen the links between people, landscape, and biodiversity, in the framework of a multifunctional land management approach.

4.4 Ex Situ Field Genebanks

Ex situ collections play a crucial role in the conservation of coconut varieties that are disappearing from home gardens and farmers' fields. They are also useful for germplasm characterization, comparing varieties grown in homogeneous environments, releasing germplasm to farmers or other users, and using germplasm in breeding programs for creating improved cultivars. Such collections form an essential buffer between users and the fast-evolving in situ genetic diversity (Konan et al. 2018a).

4.4.1 Key Threats

"Many vital ex situ genebanks are in desperate straits. As much as half of the world's crop diversity may still be in farmers' fields protected only by the family and the community". This alert message from ETC (2008) fits well with the case of the coconut palm. Indeed, many ex situ coconut field genebanks have experienced genetic erosion due to most particularly land tenure issues and pest and disease attack (Hegde et al. 2018).

Because of land tenure problems, a portion of conserved germplasm has been lost over the past 10 years. For instance, in the international genebank located in Manado, Indonesia, the local mayor had 15 ha of coconut accessions destroyed to build a horseracing track. In Vietnam, local policymakers recently took back about one-fifth of the land initially devoted to the national coconut genebank. In Côte d'Ivoire, the government is presently trying to recover all the land of the present coconut genebank for new real estate and wants to relocate the genebank to another site. A few years earlier, researchers wanted to replant an 8-ha block in the genebank. They felled the old coconut palms but villagers from the neighbourhood came and claimed the land as their own, forbidding replanting. This forced the curator to change the design of the whole collection. To avoid similar problems elsewhere, he planted the new accessions between the rows of the old living palms. Old accessions are now removed only after 3–4 years, when new accessions are well established. Such land tenure problems have resulted in a global loss of at least 54 cultivars in the past 10 years, representing 13% of existing global holdings. A mechanism needs to be established to protect such germplasm, which should be regarded as global public goods.

In many countries, diseases and pests threaten the conservation of coconut genetic resources. A group of phytoplasmas are causing severe damage to palm around the globe. These diseases are called "lethal yellowing-like syndromes" or "lethal yellowing diseases". In Papua New Guinea (PNG), a phytoplasma called Bogia coconut disease is now found less than 5 km from the international coconut genebank in Madang. In Côte d'Ivoire, a lethal yellowing disease present for a long time in Ghana has now been identified in the Grand Lahou region, which is less than 200 km from the international genebank (Arocha-Rosete et al. 2014).

4.4.2 Germplasm Data Management

The International Coconut Genetic Resources Database (CGRD) was created in 1991 by the French Agricultural Research Centre for International Development (CIRAD) and is managed by COGENT (Hamelin et al. 2005; Ruas et al. 2018). In 2013, it contained the data for 24 coconut field genebanks located in 23 countries. Altogether, it represents about 90,000 living palms, equivalent to 530 ha of coconut plantations. The accessions recorded in the CGRD are ranked into three categories: (a) introductions from farmers' fields, (b) accession transfers between ex situ collections, and (c) accessions regenerated within the genebanks (Fig. 4.3). The total number of registered accessions reached 1,680 in 2013. Details on the 416 cultivars, 855 populations, and 1,680 accessions are available on the COGENT website (www.cogentnetwork.org). Five genebanks have an international legal status. They are in Brazil, Côte d'Ivoire, India, Indonesia, and PNG. Data within the CGRD is not regularly updated and must be considered with care so the first author of this chapter has proposed improving the database with an additional data field, not initially planned, the "date of last inventory/counting of palms" for each accession.

4.4.3 Controlled Hand Pollination

The project "upgrading international coconut genebanks and evaluating accessions" was funded by the Crop Trust (Bourdeix et al. 2012a, b). It discovered that ex situ coconut conservation is facing a crisis. Presently 24 genebanks are conserving 725 unique populations with 1,374 living accessions. Some 447 of these accessions, collected during the 1980s, have become very tall without being rejuvenated. It is becoming increasingly dangerous and costly to make the controlled pollinations

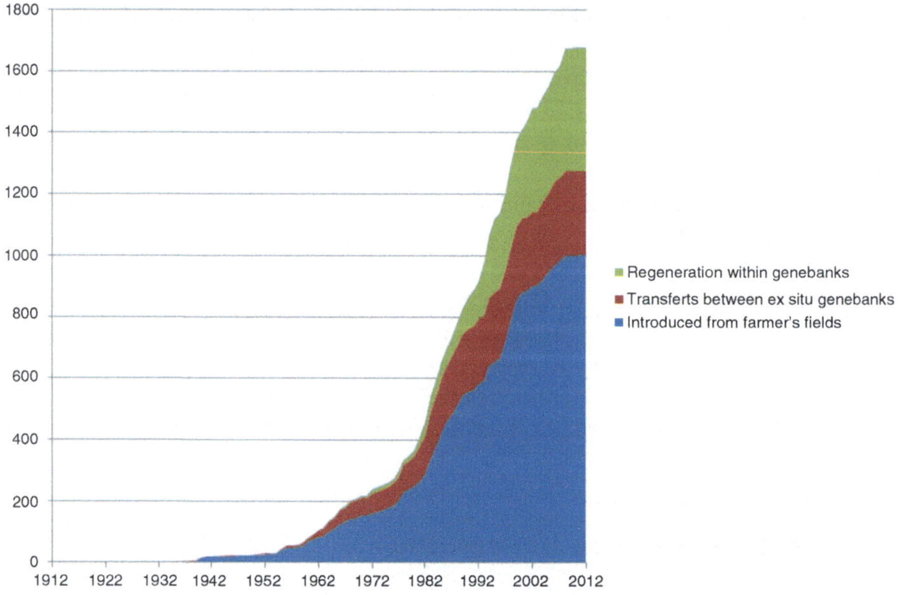

Fig. 4.3 Evolution of accessions recorded in CGRD. (From Bourdeix et al. 2018c)

required for their regeneration. At least 16 genebanks, including 3 out of the 5 international genebanks (Brazil, Indonesia, and PNG), do not have enough capability, laboratory space, equipment, staff, and/or a budget needed to make reliable controlled pollinations. Managers of the collections are just starting to identify and remove duplicated accessions (Bourdeix 2018). There is a huge need for capacity building. The Asia and Pacific Coconut Community (APCC) in 2018 successfully conducted and completed the first international training program on coconut cultivation and management for officers in charge of research and development coconut programs in producing countries at the Coconut Research Institute (CRI) of Sri Lanka. The second training session should include education in the comprehensive theoretical and practical aspects of controlled pollination.

In classical ex situ field genebanks, coconut varieties and populations are conserved as accessions, generally planted close together in the same fields. Each accession generally consists of between 45 and 150 palms. For reproducing accessions in ex situ genebanks, the technique of controlled pollination with bagging of the inflorescence is used. The lifespan of such accessions is about 60 years; however, after 25–30 years, most coconut varieties (except for varieties of the Dwarf type) reach 15 m high or more, and it becomes difficult to climb the palms. The accessions must be regenerated before the inflorescences become inaccessible.

At the Côte d'Ivoire, workers climb costly triple ladders that can reach a height of 14 m only. In Sri Lanka, the pollination workers tightly bind coconut husk with coconut ropes to the trunk to make the trunk itself a ladder, which lasts about 1.5 years. In Thailand, climbers tightly fix bamboo to the trunk and use it as a ladder. In many other genebanks, workers climb manually, which is risky, or dedicated manual machines are used. Rejuvenation programs require the climbing of roughly 75 palms, each about 15–20 times. Production of the 200 seednuts for the duplication of only 1 accession requires 1.5 years of preparation and activity. Another initiative launched in 2012 aimed to estimate the labour needed for characterizing accessions using international standard descriptors. The total labour time needed ranged from 1,409 h (Côte d'Ivoire) to 2,395 h (Indonesia) per accession. It requires collecting about 19,000 data per accession. If accessions stay 30 years in the field, the time needed for characterization is an average of 47–80 h per accession per year. Although the labour cost varies widely between countries, the cost of characterization is high everywhere, even if it is not the same.

Ex situ coconut collections are releasing on-demand conserved germplasm to breeders, farmers, and other users. This system works well for Dwarf-type varieties that reproduce naturally by selfing. Simple genetic markers, such as emerging sprout colour, allow for the discarding of the seedlings from outcrossing at the nursery stage, at least for red and yellow Dwarfs. In Sri Lanka, the early splitting of leaves at the late nursery stage helps to cull illegitimate Dwarf greens from open-pollinated seedling lots. The release of allogamous Tall-type varieties is more challenging. In the genebanks, the only way to reproduce true-to-type Tall populations is making costly controlled pollinations. Producing sufficient seednuts (200) needed to plant 1 ha demands a period of 1.5 years and costs about US$ 2000: "Only scientists with healthy research budgets can afford to order Tall-type varieties from classical coconut genebanks. Most farmers and other users cannot afford it" (Omuru et al. 2018).

4.4.4 Support for Conservation

Although public international resources have partially supported the establishment of international collections, this support has not yet been secured for the long term. The lack of adequate long-term funding for coconut research and ex situ genebanks is threatening coconut genetic resources. Why is it so difficult to sustainably fund coconut genebanks and research? "The ambivalent and multifaceted symbolisms associated with the coconut palm sometimes make stakeholders and even decision-makers forget that coconut cultivation strongly influences the livelihoods of millions of poor farmers" (Konan et al. 2018b). Recently an ethnological approach to coconut symbolisms and their consequences was developed in the Pacific region (Bourdeix et al. 2009a, b). The modern representation of the coconut palm by Pacific Islanders often appears ambivalent. In the collective Western imagination, the coconut palm has become the ubiquitous and anonymous symbol of exoticism and tropical beaches. The image of the coconut palm is widely used to promote tourism and numerous associated products ranging from fashion accessories to financial investments: "The combination of coconut with 'hammocks' or 'monkeys' sometimes reinforces the stereotype of peaceful paradise, far from the stresses of everyday life, an image that does not reflect the true situation of the Pacific Islands.

Islanders become disengaged when confronted with such counterfeit representations that standardize the tropics and deny their cultural identities" (Hedge et al. 2018). Fortunately, this old dynamic is fading because the image of the coconut palm has changed significantly over the last decade. The Western view of the coconut palm is evolving from a "holiday and amusing icon" to a "healthy and natural" palm. This new representation of a single plant is affecting the Western perception of tropical island countries. In some other tropical regions, symbolisms associated to the coconut palm appear to be less ambivalent.

4.5 Cryopreservation and In Vitro Genebanks

In vitro tissue culture approaches provide alternative means for producing coconut planting material for farmers, for safeguarding germplasm and increasing its collecting and exchange using excised embryo transfer, and for long-term conservation using cryopreservation (Oropeza et al. 2018). Nguyen et al. (2015) have provided an extensive review of all tissue culture and associated biotechnologies for coconut conservation and breeding.

Coconut is one of the most recalcitrant plant species to regenerate in in vitro culture. Methods for in vitro culture of coconut embryos started being developed and used in the1950s (Cutter and Wilson 1954; see Chap. 1), but the production of plantlets from embryos wasn't successfully achieved until the 1970s. Problems include the production of plantlets with a balanced development of roots and aerial parts and microorganism contamination. However, using a new ex vitro rooting method, this constraint can be managed (Sisunandar et al. 2010a, b, c, 2018). In most laboratories, the best recovery rates (from embryos in culture tubes to palms in the field) are between 60% and 85%, roughly like that achieved in traditional nurseries, where 65% of seednuts give rise to palms planted in the field. The in vitro culture approach for coconut embryos is the only method for producing the elite type coconuts such as Kopyor (Indonesia) or Makapuno (the Philippines). The main challenge for many laboratories is to obtain replicable results, in a wide range of situations for a wide range of varieties. Conditions do not always allow strict adherence to the rigorous protocols, which can result in lower regeneration rates through bacterial contamination or tissue damage.

To ensure long-term conservation, accessions conserved in vitro can also be "cryopreserved". Cryopreservation stops both the growth of plant cells and all processes of biological deterioration. It preserves the frozen materials in liquid nitrogen at −196 °C for an extended period (in theory for several centuries), whereby such tissues can be subsequently used to regenerate identical and fully viable plants.

Using cryopreservation techniques with either intact zygotic embryos, isolated plumules, or pollen carries several disadvantages over field collections. For example, it does not allow reproducing and thus multiplying a genotype, but only the progeny of this genotype. In addition, there are other costs too, such as the costly hand pollination needed to create the Tall-type material needed for cryopreservation.

Another option is to go back to the sites were the germplasm was initially collected to obtain the tissues for cryopreservation; however, this is often not possible or is expensive to undertake.

4.5.1 Frozen Zygotic Embryos and Plumules

The first attempt to cryopreserve coconut tissues was using immature zygotic embryos treated with a classical chemical dehydration pretreatment and with a slow freezing step (Bajaj 1984). However, the quality of tissue recovery was very low, and no plants were recovered. Later, Lim Li Ching (2008) also using immature zygotic embryos increased the rate of tissue recovery but still were unable to form plants. In 1992, zygotic embryos were successfully cryopreserved using a pre-growth desiccation technique with glucose (600 g L^{-1}) and sorbitol (15%) and rapid freezing to give up to 93% survival, with only one forming a rooted plantlet (Assy-Bah and Engelmann 1992).

Later cryopreservation work took place using the coconut plumule (a single leaf primordium with the shoot meristematic zone). The plumule has advantages and disadvantages over the use of embryos for cryopreservation. The advantages include its small size (ca. 1 mm) and having a higher density of meristematic cells, both features enabling more effective freezing (Malaurie et al. 2002). The main disadvantages include the difficulty in isolating the plumular tissue and the extra care needed to encapsulate the tissues, due to their small size, prior to cryopreservation. In an early study, plumules of Laguna Tall were encapsulated successfully in calcium alginate (3%) and then desiccated using a nutrient medium supplemented with 0.75 M sucrose, followed by further drying over silica gel. The recovered plumules were able to produce a callus upon recovery, but no plants were formed from the callus (Hornung et al. 2001). A more recent protocol for coconut plumule cryopreservation using the encapsulation-dehydration technique gave 60% survival and 20% leafy shoot development from the cryopreserved explants (N'Nan et al. 2012), but again no plants were recovered to soil.

Up until 2010, the cryopreservation protocols developed for coconut were unreliable: The success of tissue recovery following cryopreservation was low and highly variable between varieties. However, in 2010 Sisunandar et al. (2010a, b, c) published a protocol based on the physical dehydration of mature zygotic embryos that could give up to 80% survival rate and 40% production of plants in soil. This study also showed no measurable difference in the physiology and morphology of the plants recovered to those produced without cryopreservation. One year later, Sajini et al. (2011) used a chemical dehydration approach on zygotic embryos that could give 70–80% survival and 20–25% of plant recovery with some plants in soil. Researchers are now working to optimize these two recently developed techniques and to evolve from laboratory protocols to standardized techniques providing regular and consistent results in a wide range of situations and at a larger scale. Further work is underway at the University of Queensland and by a team led by Dr. Bart Panis in Bioversity International, University of Leuven, evaluating the droplet vitrification technique for coconut meristematic tissue cryopreservation (Wilms et al. 2019).

For conserving a single accession in a coconut cryogenebank, the number of embryos required is still to be determined. Some researchers have proposed conserving about 600 embryos per accession: 200 embryos on a recurrent basis, 200 more as safety duplication, and 200 more for responding to requests for germplasm supply. In such a situation, the cost of producing embryos by controlled pollination in an ex situ genebank would reach about US$ 6000 per Tall or Hybrid accession.

A drawback of the controlled pollination process in field genebanks could surprisingly benefit the constitution of a cryogenebank of frozen embryos. To ensure producing enough progeny to duplicate an accession, curators and breeders are forced to undertake high numbers of controlled pollinations: One controlled pollination is normally required to establish one palm planted in the field. Sometimes, unexpected large numbers of seednuts are obtained, more than twice the projected yield. In this situation, surplus embryos may be cryopreserved with limited additional cost. So, an affordable cryogenebank could be gradually realized with surplus embryos obtained during field regenerations and germplasm exchanges.

4.5.2 *Frozen Pollen*

Pollen, available in quite large quantities, can also be easily cryopreserved (Karun et al. 2006, 2014). Coconut genebanks are generally conserving the pollen at −18 °C for only 4–6 months. If desiccated pollen, prepared in a glass tube for controlled hand pollination, is kept for 1 h in liquid nitrogen and then thawed out, the pollen will germinate normally (Haeng-hoon and Engelman 2018).

Technicians can easily collect pollen in ex situ field genebanks when the accessions start to flower. Pollen cryopreservation could be used for palms in farmers' fields or from the original accessions in the genebanks. Such pollen would be used for successive rejuvenations of accession in the genebank. This technique insures that at least half of the genes are fully conserved – those coming for the male parents from which the pollen was cryopreserved. This will reduce the genetic drift caused by successive regenerations. Pollen could be harvested easily when the accessions are 6–8 years old, immediately cryopreserved, and 25–60 years later, when regeneration time comes, this old pollen will be used for hand pollination.

Frozen pollen may serve for exchange, breeding, and regeneration. Exchange of germplasm through pollen seems to pose fewer quarantine problems than is the case for seednuts or other propagules, but this needs further checking. Very small insects, mites, or bacteria may survive in desiccated and frozen pollen.

4.5.3 *Cryopreserving of Embryogenic Callus*

Cryopreservation of the embryogenic callus obtained from plumules has been proposed as an alternative conservation technique. A callus, be it from embryo or from vegetative tissues, represents only one genotype. Thus, to represent an accession, researchers must manage the same number of samples as for embryo conservation. The process seems even more costly than the cryopreservation of embryos, because for each individual plant, you must collect the biological sample, make it grow "in vitro" as a callus, check that this callus can provide new plantlets, and then cryopreserve it.

Nevertheless, this technique could allow conserving a piece of tissue with the potential for regenerating thousands of plants. This process seems adapted to the conservation of a few high value coconut palms to be distributed as cloned planting material.

4.5.4 A Global, Multispecies Cryogenebank

The Global Seed Vault, in Svalbard, Norway, has the largest backup collection of crop seeds originating from most countries around the world. However, some important crops, including coconut, cannot be conserved through seeds. Coconut conservation in field or in vitro collections, as for other crops such as bananas, is labour-intensive and costly. Although cryopreservation setup is initially costly, over the longer term, it is more cost-effective than conservation in field or other in vitro genebanks. With an estimated annual global production of more than 1 billion t, worth at least US$ 100 billion, these crops need a global backup system – a global cryobank. A recent study, commissioned by Bioversity International, the International Potato Center (CIP), the Global Crop Diversity Trust, and the Australian Centre for International Agricultural Research, completed in 2017, investigated the feasibility of establishing a multi-crop species safety backup facility for cryopreserved collections of vegetatively propagated crops or those with recalcitrant seeds (Acker et al. 2017).

The study concluded that a major global initiative is urgently needed to accelerate the development and implementation of multi-crop cryopreservation. The study recommended a collaborative effort to overcome the specific technical and practical constraints posed by cryopreservation protocols to at-risk collections. The study findings indicate that 100,000 unique accessions of the vegetatively propagated and recalcitrant seed crops in Annex 1 and in Article 15 of the International Treaty on Plant Genetic Resources for Food and Agriculture (Plant Treaty) are held in high-maintenance, costly, and potentially vulnerable field and in vitro genebanks. The study recommends that a safety backup cryopreservation facility is set up to accommodate the estimated 5,000–10,000 accessions arising from current, ongoing cryopreservation activities within the Consultative Group on International Agricultural Research (CGIAR) and other genebanks. The facility should follow the principles and policies that govern the Svalbard Global Seed Vault, and its infrastructure and operations should adhere to established technical standards and practices for low-temperature biorepositories (Acker et al. 2017).

4.6 Towards "Networked" and Multifunctional Genebanks

Many coconut genebanks are located close to expanding cities. Land pressure on these genebanks is mounting. Some of those genebanks are now the main green space remaining in their local areas. The status of the coconut genebank should evolve towards a higher benefit for neighbouring citizens: "Genebanks should not be exclusive spaces reserved for researchers. They could evolve towards a kind of botanical garden, public park or green space, open to citizens and where researchers work as well" (Perera et al. 2018a, b). Land pressure will likely decrease if citizens benefit from sharing these spaces.

First, self-funding could be achieved by selling the coconuts produced in the genebanks. A factor limiting fruit yield comes from the organization of host institutions (Perera et al. 2018a, b). Curators often do not engage in increasing the level of production of their genebanks. In many cases, the management and the sale of agricultural production fall under separate administrative services. For example, curators can spend half their annual budget buying fertilizers, yet the genebank doesn't benefit from the resulting yield increases, as all the income is pooled by another service. Some genebanks are recording only 60 fruit per palm per year, when many farmers succeed in obtaining more than the triple of that under comparable conditions (Fig. 4.4). For both demonstration purposes and income generation, most genebanks should have at least one field of Dwarf varieties managed in an intensive way.

In 2012, the COGENT Steering Committee recommended that pilot units for developing new high-value coconut products (HVCPs) and integrated coconut processing centres should preferably be within the coconut genebanks. This would allow the genebank visitors to learn about genetic resources, available planting material, and the new processes for producing HVCPs. Researchers would benefit from a wide range of germplasm for testing their new techniques and equipment. Every site could have a small unit to produce coconut oil and biodiesel. Producing biodiesel and coconut oil would provide energy for the laboratory, offices, and guesthouse, reducing the carbon footprint of the genebank.

For landscaping purposes, genebanks could provide adult palms from certified varieties. When sold to city landscapers, the price of ordinary unidentified coconut palms often reaches about US\$100 m^{-1} of trunk. Genebanks could give value to other coconut products, such as coconut timber and palm hearts, obtained when felling old accession trees. Producing and processing toddy is often more profitable than selling fruit.

Fig. 4.4 (continued) to 350–420 nuts per palm per year. (**3** and **4**). Same variety planted in an international genebank. Curators are sometimes reluctant to apply fertilizers. They would have to spend a large part of their budget to buy fertilizers, but when fruits are sold, the genebank does not benefit the money that goes directly to a centralized sale division. (**5**) Cameroon Red Dwarfs well cultivated in a small Brazilian farm. (**6**) Same variety planted in an international genebank with average management. (**7**) Same variety planted in an international genebank with low management during the dry season. (**8**) Pemba Orange Dwarf cultivated in a garden in Tanzania. This variety is very close to the Cameroon Red Dwarf. (**9**) Huge nursery of Dwarf seedlings in Brazil. The use of polybags is strongly recommended in the case of planting such Dwarf varieties

Fig. 4.4 (**1** and **2**) Brazilian Green Dwarfs well fertilized and irrigated in a small Brazilian farm. Such farmers regularly obtain at least 250 tendernuts per palm per year, so about 55,000 nuts ha^{-1}, for an average gross income of US$ 17,500 per hectare and per year. Best farmers even obtain up

Diversification of coconut genebanks provides another way to increase self-funding. A lucrative option could be to include plots devoted to the conservation of other palm species. There is a huge and very profitable market for adult palms for landscaping of public places and tourist areas. Therefore, the genebank could also sell adult trees from palm species and replace them in the genebank if this operation remains profitable.

Another way for genebanks to increase their resources could be to develop joint ventures with the tourism industry at both international and local levels. Many coconut research centres are small paradises from the aesthetic and environmental perspectives. There is great potential for the development of ecotourism activities.

"A networked collection, also called a virtual collection, is located at more than one geographical/institutional site; it spans the genetic diversity of a given species (genepool) and gathers stakeholders having a mutual interest for rationally conserving and exchanging germplasm" (Bourdeix et al. 2009a, b). Pushing this concept to its extreme, each plant population could be conserved in a separate location, as envisioned in the CNRA project in Côte d'Ivoire (Sect. 4.5). Intermediate strategies are also conceivable. For instance, in Samoa, two small islands have recently been replanted with three varieties each.

However, as many coconut varieties are allogamous, the main limiting factor for classical ex situ conservation is the regeneration of true-to-type accessions via controlled hand pollination. In the case of coconut, this regeneration technique is costly, requiring a well-equipped laboratory, well-trained technicians, field workers able to climb palms, and considerable manpower. Not all genebanks can yet afford this. To overcome this limiting factor, the Polymotu concept is being proposed as a new approach, although it will have to be fully evaluated. Several coconut accessions could be planted, each in a distinct, isolated site. These sites could be islets near bigger inhabited islands, insulated valleys, large plantations of a unique variety, large urban facilities such as a university campus or golf course, or any other designs using a pollen barrier. Reproductive isolation will help ensure cost-effective true-to-type breeding of the coconut varieties through natural pollination.

Establishing a networked collection would require the involvement of more stakeholders, sites, and countries. The criteria for conservation design or accessions to be included in such a networked collection are germplasm uniqueness, genetic representativeness, ability to reproduce true-to-type progeny, and workable policy considerations. The challenge will be to gather in the same framework the many accessions held in very different locations, from international genebanks, public parks, botanical gardens, university campuses, and private farms to islets owned by municipalities, islanders' clans, or tourism enterprises.

The 2018–2028 COGENT global strategy does not yet plan to push the global system of coconut conservation into a networked/virtual collection. However, this novel approach could contribute in the future to the mitigation of the current delineation between in situ and ex situ conservation.

4.7 Conclusion

In the case of the coconut palm, the constraints linked both to the biology of the plant and social issues make conservation and breeding particularly challenging. Ex situ genebanks, be they field, cryopreserved, or in vitro, should be from the beginning conceived as multifunctional. Creating a new ex situ field genebank for conserving only coconut seems less attractive than creating a multifunctional/multispecies genebank. In addition, the patrimonial value of such a single-species genebank may not be enough. If local policymakers decide to recover the land for other purposes, they may also decide to destroy the genebank at the same time. Some researchers believe that such a "pure coconut" field genebank is "already dead even before being planted". To avoid this, ideally the COGENT-International Treaty on Plant Genetic Resources for Food and Agriculture mechanism of a tripartite agreement between the genebank host government and these other two parties needs ratifying to protect the collection in the event of relocation due to threats or policy decisions. We recommend that donors should only fund the construction of new research stations or the installation of new field genebanks where the land on which these facilities will be located is legally recognized as being of public utility by the host government; this should be done prior to the project or at the very beginning of the implementation phase.

To secure conservation of coconut genetic resources, an international socioeconomic study could help coconut field genebanks to increase their profitability: costing conservation activities, increasing self-funding, integrating coconut conservation in landscaping of both public places and tourism locations, and using multifunctional land management approaches.

The authors also do not advocate a situation where coconut farmers will become only diversity users and where national institutions or large private companies will implement all conservation and breeding activities, as illustrated by the example of maize in many countries. Of course, when needed, farmers should have the opportunity to buy good and affordable planting materials, just as they buy other agricultural inputs, and to benefit from the latest biotechnology advances. Maintaining and breeding many varieties by the farmers brings significant benefits, and not only monetary ones, at both local and global levels. As illustrated by the designs and analyses provided in this chapter, multifunctional landscape approaches could also help farmers to improve in situ conservation and the local production of good planting material. As recently stated in a study on incentives for boosting coconut production (Bourdeix et al. 2018d), strengthening communication between farmers, researchers, and policymakers seems crucial to achieve these objectives.

Future effective efforts in coconut germplasm conservation need to integrate "networked" and multifunctional genebanks and harness the latest developments in tissue culture and cryopreservation. Such approaches, often presented as competing or contradictory alternatives, are in fact complementary (Schmitz et al. 2015). The Crop Trust, other conservation groups, and the authors of this chapter now believe that any crop needs to be conserved by using at least two different approaches in at least two different locations. In the case of the conservation of coconut palm, thought in a framework of the COGENT network gathering at least 38 producing countries, a triplication system has even recently been proposed (Bourdeix et al. 2018e).

Acknowledgements Some of the concepts and ideas presented in this chapter came from previous papers published by the authors and from the numerous exchanges conducted during the compilation of the Global Strategy for Conservation and Use of Coconut Genetic Resources. So, the first author would like to thank all the researchers who participated at different levels to these exchanges or who took their time to read and improve the draft document (by alphabetic order of names): Bambridge, Tamatoa, Baudouin, Luc, Billot, Claire, Bretin-Chabrol, Marine, Chomchallow, Narong, Dali, Serge Lida, Donafologo Danièle, Elias, Marlène, Engels, Jan, Hussein, Naheed, George, Maria Luz, Goirand, Renault, Guarino, Luigi, Issali, Emmanuel Auguste, Kanniah, Jayashree, Konan, Jean Louis, Konelio, Misa, Kumar, Vijendra, Labouisse, Jean-Pierre, Louafi, Selim, Lusty, Charlotte, Mirisola, Luis, Ngereza, Andrew Jacob, Planes, Serge, Roux, Nicolas, Ruf, François, Sebastian, Leocadio, Sileye, Tiata, Thampan, Palakasseril Kumaran, Thomas, George and Weise, and Stephan.

References

Acker JP, Adkins S, Alves A et al (2017) Feasibility study for a safety back-up cryopreservation facility. Independent expert report: July 2017. Bioversity International, Rome, 100 p

Arocha-Rosete Y, Konan Konan JL, Diallo A et al (2014) Identification and molecular characterization of the phytoplasma associated with a lethal yellowing-type disease of coconut in Côte d'Ivoire. Can J Plant Pathol 36(2):141–150

Assy-Bah B, Engelmann F (1992) Cryopreservation of immature embryos of coconut (*Cocos nucifera* L.). Cryo-Letters 13:67–74

Bajaj YS (1984) Induction of growth in frozen embryos of coconut and ovules of citrus. Current science (Bangalore), 53(22):1215–1216

Bourdeix R (2014) A proposal for integrating conservation of the coconut palm into the USP campus, Suva Fiji. Available at the URL: http://polymotu.blogspot.fr/2014/06/a-proposal-for-integrating-conservation.html

Bourdeix R (2018) 2.3 The current global ex situ conservation system – chapter 2. Where we are today. In: Bourdeix R, Prades A (eds) A global strategy for the conservation and use of coconut genetic resources 2018–2028. Bioversity International, Montpellier, pp 53–68

Bourdeix R, Konan JL, N'Cho YP (2005) Coconut: a guide to traditional and improved varieties. Editions Diversiflora, Montpellier, 94 p. Available at the URL: http://diversiflora.blogspot.com. Seen 12 Nov 2018

Bourdeix R, George ML, Baudouin L et al (2009a) The concept of "Networked collection" or "Virtual collection": new developments and their applications to the conservation of the coconut palm. Paper presented at the 2nd European Congress of conservation biology, 1–5 September 2009, Prague, Czech Republic. Available at the URL: http://www.eccb2009.org/uploads/book_of_abstracts_errata.pdf

Bourdeix R, Baudouin L, Bambridge T et al (2009b) Dynamics and conservation of the coconut palm *Cocos nucifera* L. In: The Pacific region: towards a new conservation approach. Paper presented at the 11th Pacific science inter-congress, March 2–6, Tahiti, French Polynesia

Bourdeix R, Ruas M, Hamelin C et al (2012a) Upgrading international coconut genebanks and evaluating accessions. Terminal report. (01/11/2011 – 30/05/2012). Bioversity International, Montpellier. Available at the URL: http://www.cogentnetwork.org/images/projects/TR-C60018%20final%20technical%20and% 20financial%20report.pdf

Bourdeix R, Johnson V, Tuia VS et al (2012b) Three declinations of the Polymotu concept: "Inland ex situ", "Ecotourism on islands", "Urban" and their possible applications in Brazil, Côte d'Ivoire, Indonesia, French Polynesia and Samoa. Paper presented at the 45th APCC COCOTECH meeting, 2nd – 6th July 2012, Kochi, India. Available at the URL: http://www.cogentnetwork.org/scientific-publications

Bourdeix R, Perera L, Rivera RL et al (2016) Global coconut communities – status and strategies in in situ diversity management and utilization. In: Coconut – global status and perspectives. Central Plantation Crop Research Institute, Kasaragod

Bourdeix R, Issali AE, Tuia VS (2018a) 1.1.4 Constraints linked to the biology of the plant – chapter 1. Introduction to the global coconut strategy. In: Bourdeix R, Prades A (eds) A global strategy for the conservation and use of coconut genetic resources 2018–2028. Bioversity International, Montpellier, pp 14–19

Bourdeix R, Hamelin C, Ruas M (2018c) 2.3.1 Content of ex situ collections – chapter 2. Where we are today. In: Bourdeix R, Prades A (eds) A global strategy for the conservation and use of coconut genetic resources 2018–2028. Bioversity International, Montpellier, pp 54–55

Bourdeix R, Tuivavalagi N, Mataora V et al (2018d) Germplasm and incentives for boosting coconut production: case studies from the Pacific region and some other countries. Communication at the 48th APCC COCOTECH conference and exhibition, 20–24 August 2018, The Berkeley Hotel Pratunam, Bangkok, Thailand

Bourdeix R, Allou K, Omuru E (2018e) n3.3.3 Triplication of germplasm in distinct geographical sites – chapter 3. Where we need to be to secure diversity and promote use. In: Bourdeix R, Prades A (eds) A global strategy for the conservation and use of coconut genetic resources 2018–2028. Bioversity International, Montpellier, pp 133–138

Cutter VM Jr, Wilson KS (1954) Effect of coconut endosperm and other growth stimulants upon the development in vitro of embryos of *Cocos nucifera*. Bot Gaz 115(3):234–240

De Nucé de Lamothe M, Rognon F (1975) Pollinisation assistée et contamination par des pollens indésirables. Oléagineux 30(8–9):359–364

ETC (2008) Communiqué. Svalbard's doomsday vault the global seed vault raises political/conservation debate. Issue # 98, February 2008

FAO (2009) International treaty on plant genetic resources for food and agriculture. Retrieved from: http://www.fao.org/3/a-i0510e.pdf

Fernando WB (1976) Some observations on the the reproductive patterns of varieties and forms of the coconut (*Cocos nucifera* L.) in Sri Lanka

FAO (2018) Driving financial inclusion for smallholder farmers by leveraging satellite data and machine learning. [Video Webinar]. Retrieved from: http://www.fao.org/e-agriculture/news/e-agriculture-webinardriving-financial-inclusion-smallholder-farmers-leveraging satellite data

GRAIN (2008) Faults in the vault: not everyone is celebrating Svalbard. Retrieved from: https://www.grain.org/en/article/181-faults-in-the-vault-not-everyone-is-celebrating-svalbard

Haeng-hoon K, Engelman F (2018) 3.3.4 Cryogenebanking – chapter 3. Where we need to be to secure diversity and promote use. In: Bourdeix R, Prades A (eds) A global strategy for the conservation and use of coconut genetic resources 2018–2028. Bioversity International, Montpellier, pp 139–142

Hamelin C, Bourdeix R, Baudouin L (2005) The international coconut genetic resources database. In: Coconut genetic resources. IPGRI, Serdang, p 427

Hegde V, Pilet F, Omuru E (2018) 1.1.5 Major threats to coconut genetic resources – chapter 1. Introduction to the global coconut strategy. In: Bourdeix R, Prades A (eds) A global strategy for the conservation and use of coconut genetic resources 2018-2028. Bioversity International, Montpellier, pp 19–22

Hornung R, Domas R, Lynch PT (2001) Cryopreservation of plumular explants of coconut (*Cocos nucifera* L.) to support programmes for mass clonal propagation through somatic embryogenesis. Cryoletters, 22(4):211–220

Karun A, Sajini KK, Nair M et al (2006) Cryopreservation of coconut (*Cocos nucifera* L.) pollen. J Plant Crop 34(3):568

Karun A, Sajini KK, Niral V et al (2014) Coconut (*Cocos nucifera*) pollen cryopreservation. CryoLetters 35(5):407–417

Konan JL, Rivera RI, Bourdeix R (2018a) 2.2.1 Ex situ conservation methods – chapter 2. Where we are today. In: Bourdeix R, Prades A (eds) A global strategy for the conservation and use of coconut genetic resources 2018–2028. Bioversity International, Montpellier, pp 41–44

Konan JL, Omuru E, Bourdeix R (2018b) 2.8 Facing emergency situations: an overview – chapter 2. Where we are today. In: Bourdeix R, Prades A (eds) A global strategy for the conservation and use of coconut genetic resources 2018–2028. Bioversity International, Montpellier, pp 104–112

Lim Li Ching (2008) Global Seed Vault cannot replace on-farm conservation. Retrieved from: http://www.twn.my/title2/susagri/susagri028.html

Liyanage DV, Wickramarathne MRT, Jayasekera C (1988) Coconut breeding in Sri Lanka: a review, CRI. COCOS 6:1–26

Malaurie B, N'Nan O, Hocher V, et al (2002) State of zygotic coconut embryo culture and cryopreservation research at IRD/CIRAD, France. In Coconut embryo in vitro culture: part II. Proceedings of second international on embryo culture workshop, Mérida, Yucatán, Mexico, 14–17 March 2000. International Plant Genetic Resources Institute (IPGRI)

N'Nan O, Borges M, Konan JL et al (2012) A simple protocol for cryopreservation of zygotic embryos of ten accessions of coconut (*Cocos nucifera* L.). In Vitro Cell Dev Biol Plant 48(2):160–166

Nguyen QT, Bandupriya HD, López-Villalobos A et al (2015) Tissue culture and associated biotechnological interventions for the improvement of coconut (*Cocos nucifera* L.): a review. Planta 242(5):1059–1076

Novarianto H, Miftahorrachman (2000) Unique coconuts of Indonesia. COGENT Bulletin: 13–15. Available from the URL: http://www.cogentnetwork.org/images/publications/newsletters/Newsletter4 _nov2000.pdf

Omuru E, Perera L, Lobo D et al (2018) 2.5.1 Planting material for farmers – chapter 2. Where we are today. In: Bourdeix R, Prades A (eds) A global strategy for the conservation and use of coconut genetic resources 2018–2028. Bioversity International, Montpellier, pp 75–80

Oropeza C, Engelman F, Cueto CA et al (2018) 2.2.4 In vitro culture and cryopreservation – chapter 2. Where we are today. In: Bourdeix R, Prades A (eds) A global strategy for the conservation and use of coconut genetic resources 2018–2028. Bioversity International, Montpellier, pp 50–53

Perera L, Konan JL, Tulalo M (2018a) 3.3.1 Business plans for genebanks – chapter 3. Where we need to be to secure diversity and promote use. In: Bourdeix R, Prades A (eds) A global strategy for the conservation and use of coconut genetic resources 2018–2028. Bioversity International, Montpellier, pp 130–132

Perera SACN, Gunn B, Rivera RI (2018b) 3.9.4 DNA analysis in farmer's fields – chapter 3. Where we need to be to secure diversity and promote use. In: Bourdeix R, Prades A (eds) A global strategy for the conservation and use of coconut genetic resources 2018–2028. Bioversity International, Montpellier, pp 171–172

Ruas M, Hamelin C, Bourdeix R (2018) 3.8.2 International databases on ex situ conservation – chapter 3. Where we need to be to secure diversity and promote use. In: Bourdeix R, Prades A (eds) A global strategy for the conservation and use of coconut genetic resources 2018–2028. Bioversity International, Montpellier, pp 166–167

Santos GA, Batugal PA, Othman A (1996) Manual on standardized research techniques in coconut breeding (STANTECH). IPGRI–APO, Singapore, p 46

Sajini KK, Karun A, Amarnath CH et al (2011) Cryopreservation of coconut (*Cocos nucifera* L.) zygotic embryos by vitrification. CryoLetters 32(4):317–328

Schmitz OJ, Lawler JJ, Beier P et al (2015) Conserving biodiversity : practical guidance about climate change adaptation approaches in support of land-use planning. Nat Areas J 35(1):190–204

Sisunandar, Rival A, Turquay P et al (2010a) Cryopreservation of coconut (*Cocos nucifera* L.) zygotic embryos does not induce morphological, cytological or molecular changes in recovered seedlings. Planta 232(2):435–447

Sisunandar, Rival A, Samosir Y et al (2010b) Cryopreservation of coconut (*Cocos nucifera* L.): the influence of embryo maturity upon rate of recovery and fidelity of seedlings. In Vitro Cell Dev Biol Anim 46:2–3

Sisunandar, Sopade PA, Samosir Y et al (2010c) Dehydration improves cryopreservation of coconut (*Cocos nucifera* L.). Cryobiology 61(3):289–296

Sisunandar, Alkhikmah, Husin A et al (2018) Ex vitro rooting using a mini growth chamber increases root induction and accelerates acclimatization of Kopyor coconut (*Cocos nucifera* L.) embryo culture-derived seedlings. In Vitro Cell Dev Biol Plant 54(5):508–517

Tuia VS, Sebastian L, Bourdeix R (2018) 2.2.2 In situ and on-farm conservation – chapter 2. Where we are today. In: Bourdeix R, Prades A (eds) A global strategy for the conservation and use of coconut genetic resources 2018–2028. Bioversity International, Montpellier, pp 45–46

Wilms H, Rhee JH, Rivera RL et al (2019) Developing coconut cryopreservation protocols and establishing cryo-genebank at RDA; a collaborative project between RDA and Bioversity International. III international symposium on plant cryopreservation 1234, pp 343–348

Chapter 5
Collecting Coconut Germplasm for Disease Resistance and Other Traits

Roland Bourdeix, Gilles Coppens d'Eeckenbrugge, Jean Louis Konan, Hengky Novarianto, Chandrika Perera, and Valentin Luis Fredrik Wolf

5.1 Introduction

The description of the extent and distribution of coconut genetic diversity is essential for determining what to conserve as well as where and how to achieve it (Ramanatha Rao et al. 2005). Genetic variation arises from mutation and recombination. Artificial selection, genetic drift, and gene flow further modify its repartition (Fig. 5.1).

Many factors influence crop evolution and the distribution of genetic diversity temporally and spatially, such as past and present climatic, edaphic, and biotic constraints and biological traits such as population size and breeding systems. They also include agricultural and social habits of human groups who have cultivated the crop (Leclerc and Coppens d'Eeckenbrugge 2011).

Knowing what has already been collected is a prerequisite for launching new coconut collecting programmes. The International Coconut Genetic Resources Network (COGENT) aims to promote collaboration for the conservation and use of coconut genetic resources. It currently comprises 39 member countries that, all together, represent more than 90% of the world coconut production. In 1991,

---

R. Bourdeix (✉) · G. Coppens d'Eeckenbrugge · V. L. F. Wolf
CIRAD – UMR AGAP, CIRAD (Agricultural Research for Development), Montpellier, France

AGAP, University Montpellier, CIRAD, INRA, Montpellier SupAgro, Montpellier, France
e-mail: Roland.bourdeix@cirad.fr

J. L. Konan
National Centre for Agricultural Research, Abidjan, Ivory Coast

H. Novarianto
Indonesian Coconut and Palm Research Institute (IPRI), Manado, North Sulawesi, Indonesia

C. Perera
Faculty of Agriculture, University of Peradeniya, Peradeniya, Sri Lanka

S. Adkins et al. (eds.), *Coconut Biotechnology: Towards the Sustainability of the 'Tree of Life'*, https://doi.org/10.1007/978-3-030-44988-9_5

Fig. 5.1 (1), (2), and (3). The Niu Leka Dwarf from Taveuni Island in Fiji was the first Compact Dwarf to be described by scientists, but this is not the most productive. (4), (5), and (6). During **Fig. 5.1** (continued) surveys conducted in French Polynesia, the first Compact Dwarfs with

COGENT requested CIRAD to develop an International Coconut Genetic Resources Database (Hamelin et al. 2005). This database, known as CGRD, presently contains data collected in 23 countries from 24 coconut field genebanks (Ruas et al. 2018). In collaboration with 90 key experts from 20 countries, COGENT recently published a Global Strategy for Conservation and Use of Coconut Genetic Resources, with the objective of optimizing the conservation and use of coconut diversity (Bourdeix and Prades 2018).

Starting with a presentation of the main challenges related to coconut germplasm collecting, this chapter illustrates the importance of developing ethnological and historical approaches to optimize collecting activities. Then it discusses collecting strategies with emphasis on tolerance to pests and diseases, varieties with special traits such as Compact Dwarfs, and collecting using geographical and molecular approaches. In link with the COGENT strategy, we tried to estimate what should be collected globally during the next decade. Special attention is paid on how to strengthen stakeholders' involvement in the collecting activities.

5.2 Challenges in Collecting Coconut Germplasm

5.2.1 Biological Constraints

The height of the palms and the large size of seednuts make their collection and transportation difficult and expensive, for farmers as well as collectors. Furthermore, such large propagules are very difficult to disinfect, which increases the risk of disease or pest diffusion related to germplasm movements.

Tall coconut populations are allogamous and heterozygous, which often hampers and complicates the identification and collection of palms with favourable traits as well as their later use in breeding programs (Bourdeix et al. 2018a). When researchers find such rare palms, they collect only 1 to 20 embryos or seednuts per palm. It is worth noting that germination will not occur in all collected fruit. For those that mature, the targeted characteristic in the ex situ collection often comes from selfing, leading to an associated risk of inbreeding depression. Given the long coconut life cycle, a few targeted progenies will be available 6–7 years later in the collection, and breeders will need another generation to fix and to multiply this progeny.

Fig. 5.1 (continued) orange-red fruits were identified in Moorea Island. (7). Some of the best Compact Dwarfs from Moorea Island produce big round fruits with a tasty coconut water and a specific nice pink colour in the immature husk. (8). Around Suva in Fiji, more than 20 new varieties of Compact Dwarfs were observed, such as this yellow form with pointed fruits. Some may derive from crosses between Compact Dwarfs and the Malayan Red Dwarf. (9) and (10). Two more Compact Dwarf varieties from Tahiti Island. (11) and (12). A Compact Dwarf variety from Tubuai Island, Australes archipelago, in French Polynesia, showing very unusual terminal flowers at the end of the spikelets

5.2.2 Collecting Seednuts, Embryos, and Pollen

From 2009 to 2012, the Crop Trust funded an international project involving COGENT member countries to optimize, validate, and apply a standard embryo culture protocol (Cueto et al. 2012a). This project published improved guidelines available on the COGENT website (Cueto et al. 2012b). However, when collecting in farmer's fields, cryopreservation of embryos may again require a refinement of these techniques. In cassava (*Manihot esculenta* Crantz; Dumet et al. 2013), it was shown that the efficiency of cryopreservation with tissues directly taken from farmers' fields, when compared to sampling from ex situ genebanks, was considerably reduced (70%). The challenge remains to define methods providing regular and consistent results when being applied on a large scale and in a wide range of situations.

Even though pollen can be collected from the field, collectors are unlikely to do so as it requires hiring climbers to reach the inflorescences, spending more time at each collecting site, and the need for pollen sieving and drying facilities, and the lifespan of pollen is no more than 5 days in natural conditions (Karun et al. 2014). So, pollen collected in a farmers' fields must be desiccated and, if possible, cryopreserved during the survey. Despite the difficulties in collecting and storing pollen, the length of breeding programmes can be reduced by one generation if the pollen of intended male parents is collected with the seednuts. Thus, recent germplasm exchanges have involved both seednuts and pollen, allowing direct pollination of Sri Lankan varieties by materials from the Côte d'Ivoire coconut collection. The resulting hybrids are being evaluated in multilocational trials, while the exotic coconut accessions of approximately the same age are growing in the field genebanks.

5.2.3 How Ethnological and Historical Approaches Impact Collecting

Optimizing coconut collecting and conservation requires collectors to understand the biology but also social and historical dynamics shaping its genetic diversity and uses. Coconut palms were naturally distributed in the tropics of the Indo-Pacific basin, in a coastal habitat dominated by small islands, where ocean drift played a major role in their long-distance dissemination. Indeed, supposedly wild populations are only found in small islands and peninsulas in the Indian and Pacific Ocean, confirming the importance of such sites in the natural distribution of the species, as underlined by Beccari (1916), Hill (1929), and Sauer (1967, 1971). On continents and large islands, naturally dominated by biodiverse tropical forests, spontaneous coconut populations are limited to the shoreline by competition and predation. But the recognition of wild populations is not an easy task. A thick husk favouring buoyancy and delaying germination has been associated with primitive forms (Harries 1978); however, it may also be associated with artificial selection for coconut fibre.

Finally, only the Dwarf varieties appear to be domesticated, while the difference between the wild Tall-type coconut and the human-selected Tall types is more complicated (Coppens d'Eeckenbrugge et al. 2018a).

Both linguistic and ethnobotanical evidence indicates that coconut domestication began in two different regions in the Pacific and Indian Ocean, respectively. From then on, Austronesian seafarers and, later, Arab traders have been largely credited for its diffusion (Gunn et al. 2018). However, archaeobotanical data are too scarce to disentangle the relative contributions of nature and prehistoric man.

From 1870, copra and coconut oil has become a colonial business: "Commercial plantations run by European settlers or companies had a ripple effect on the establishment of small plantations by Pacific islanders, geared to copra production. The number of coconut palms in the Pacific region greatly increased, probably 50 to 100-fold" (Bourdeix et al. 2009). In most cases, planting techniques on the atolls consisted in bringing importing seednuts from larger islands. The natural vegetation was cleared, let it dry for a month, and burned. Such planting techniques were harmful to the diversity of both existing coconut palms and endemic species. Furthermore, in many Pacific islands, measles and influenza brought by European mariners decimated local populations. These events altered the social representations of the coconut palm for the Pacific islanders and exacerbated the mix of coconut biological resources and the erosion of traditional knowledge. More than half of the coconut landraces created by the islanders over several millennia are estimated to have been lost. Traditional varieties were diluted in the mass of coconut palms planted to produce copra (Bourdeix et al. 2009)

Finally, the populations observed by modern collectors may result from rare founding events as well as repeated introductions, either natural or artificial, the massive substitution of local materials, or any combination of all these processes. This complicates tremendously the interpretation of coconut diversity observations and thence the prioritization and choice of collected germplasm.

Nowadays, traditional coconut plantations in the Pacific region show a high level of variability, especially in fruit shape and weight. Generally, a few palms of traditional Polynesian varieties, or mix of these varieties, may survive in the middle of copra coconut plantations. Some of them produce long coconuts with a thick husk that were used for making ropes, some produce large shells with flat bottoms that were used as containers, some have soft and/or sweet kernels or husks, and some produce sweet and tasty coconut milk. Probably half of the Pacific coconut varieties presently conserved in ex situ genebanks were collected from such mixed populations.

The situation has changed since, in 1993, the STANTECH manual argued: "ideally, it is best to identify and collect as much genetic variability as possible". Recent surveys (Perera 2018) have shown that many farmers would like to plant discrete, recognized cultivars, i.e. "often intentionally bred and selected subset of a species that will behave uniformly and predictably when grown in an environment to which it is adapted". They are no more interested in growing populations with a mix of high- and low-yielding palms, producing small and big fruits with heterogeneous composition. Farmers generally appreciate a diversity of planting material, but they want to choose the diversity to be used.

Approaches to collecting have evolved, at least in the Pacific region. Because of the many mixed populations, and because breeding operations take so long, the collecting process should be selective. Instead of sampling a population mixed for historic reasons, we advise to collect, within this population, palms presenting specific phenotypes. The surveys should focus on collecting structured genetic diversity that can be used "in the short term" for breeding programs. This concept of "short term" remains relative, since a single generation of field evaluation already requires about 12 years. Such a survey may aim at a kind of "reconstruction" of traditional coconut varieties already half-lost because of historical mixing.

5.2.4 Legal Issues

In many countries, laws restricting access to plant genetic resources have been recently adopted. Intellectual property rights (e.g. UPOV variety protection), the recognition of national sovereignty and restrictions to germplasm access, have made the genetic diversity much less available. From a legal and practical perspective, it seems that the simplest way for collectors is to buy seednuts or seedlings from stakeholders and get a receipt. Many farmers want to provide the planting material for free; however, more legal problems can occur in this case. Some countries legally forbid their farmers to sell planting material, adding another layer of complexity to the collector's task.

5.3 Collecting for Tolerance While Avoiding Contamination

Pest and diseases seriously threaten coconut cultivation and conservation of genetic resources. A group of insect-vectored phytoplasmas cause lethal, fast-spreading diseases in coconut plantations. They are responsible for various forms of coconut lethal yellowing-like diseases (LYLDs). Numerous outbreaks of LYLDs in coconut have been recorded in Central America, the Caribbean, and Africa since the late nineteenth century and have caused the death of millions of coconut palms. The only coconut LYLD found in the Pacific region is the Bogia coconut disease in Papua New Guinea (PNG).

These LYLDs are caused by several different, but related, phytoplasmas in different places around the world. As obligate parasites, phytoplasmas only survive inside the plants that host them and the insects that spread them from plant to plant. They cannot be readily cultured and can only be studied with the tools of molecular biology. Therefore, they have complex grouping and naming conventions.

5.3.1 Difficulty of Collecting for Tolerance to Phytoplasmas

Phytoplasma epidemics cause the greatest devastation. Their dynamics make the collecting of tolerant plant material difficult. When LYLD strongly attacks a coconut plantation, generally all palms die rapidly even though the cause of their death is not LYLD. The death of half or more coconut palms leaves crownless rotting trunks that constitute breeding sites for insects (such as *Oryctes* and *Rynchophorus*) that will kill all coconut trees, even those that are tolerant to LYLD.

As the epidemic unfolds, it can spread rapidly and kills thousands or millions of coconut palms before stopping abruptly. It may enter a latency phase and disappear completely for 20–50 years or even longer. For instance, in Tanzania, from 1960 to 1992, it killed eight million palms, about 38% of the mainland palm population (Schuiling et al. 1992). Despite the disease and the severity of the dry season, new plantings have continued, to the extent that there was a 40% increase in the planting area between 2000 and 2012 (Liberty et al. 2018).

Collectors often harvest and plant seednuts from the few remaining living palms that escaped LYLD. When LYLD does not attack these progenies, they sometimes pretend that they have succeeded in selecting disease-tolerant coconut palms. In many cases, this is not true. Palms may persist because the epidemic has stopped. One of the authors (R. Bourdeix) observed this in Tanzania, when visiting LYLD field tests. Most of the palms on a livestock breeding station, which appeared tolerant in their place of origin, died from LYLD when planted in a region where the disease was still active (Mpunami et al. 2013).

Furthermore, what is generally collected (either seednuts or plumules of the embryo for cloning) is not the genotype of the surviving palms, but its seed progeny, the pollen parent being unknown, whether tolerant or not. True-to-type tissues such as inflorescence or pollen, for cloning or breeding purposes, would provide more certainty.

In Jamaica, local and introduced coconut varieties have been field tested for LYD tolerance (Whitehead 1968). According to Romney (1972), no other variety or hybrid than Malayan Dwarfs was found to be sufficiently resistant for farm planting. In 1974, the Jamaican breeding program diffused the Maypan hybrid as a new tolerant planting material (Harries and Romney 1974). This hybrid was successfully used for about 30 years, then it started to die from the disease. In 2008, Baudouin et al. studied the genetic contamination of "Maypan" hybrids by LYLD-sensitive Jamaica Tall and found it was insufficient to be the cause of the disease outbreak. They concluded that neither "Maypan" nor its parents can be deemed resistant in the present context.

Exploration of resistant germplasm for the Weligama coconut leaf wilt disease (WCLWD, a phytoplasma) in Sri Lanka resulted in the identification of the Sri Lanka Green Dwarf to be the only variety with high levels (98%) of resistance. A few susceptible palms (2%) were old, weak palms, indicating a possible breakdown of the resistance mechanism as the Green Dwarf palms age. The commercially grown Sri Lanka Tall palms were highly susceptible to WCLWD, but some apparently tolerant genotypes have been hybridized with the Sri Lanka Green Dwarf palms. The resulting hybrids are being evaluated in comparison with susceptible varieties (Perera and Dissanayaka 2013; Perera et al. 2014).

Even in the best cases, resistance or tolerance to a pathogen is not likely to be permanent. Over time pathogens can naturally evolve mechanisms to overcome genetic tolerance or resistance of their host. For annual crops, it seems that the resistance of a variety to a pathogen has, on average, a lifespan of only 10–15 years. After this delay, the variety is no longer resistant because the pathogen has evolved. There is good reason to think that if, today, we plant a resistant palm, 50 years later and before its natural death, this coconut palm will become sensitive to newly emerging forms of the pathogen. Therefore, even though good and tolerant varieties are available, research to collect and breed the next good variety should never stop.

In the Dominican Republic, the LYLD never became epidemic on the island. The local genetic diversity of the local Tall, known as "Criollo", was studied by using 13 microsatellite (SSR) markers from the COGENT kit (Martinez et al. 2010). The "Criollo" coconut proved to be a typical Indo-Atlantic variety and is probably highly susceptible to the usual LYLD pathogen. Local conditions and the nature of the local phytoplasma strain may explain this special epidemiology.

Mexico has conducted two-step collecting programs. A first review of historical knowledge enabled researchers to select key sites for collecting coconut samples. They conducted a pre-survey at 41 sites and studied the pattern of morphological variation of the nut (Zizumbo-Villarreal et al. 1993). Collecting was then conducted at 18 sites. The collected populations were grouped into five ecotypes: Malayan Yellow Dwarf (MYD), Mexican Atlantic Tall (MXAT), and three distinct Mexican Pacific Tall varieties (Zizumbo-Villarreal and Colunga-GarcíaMarín 2001; Zizumbo-Villarreal et al. 2008). Fruit of these selected populations were collected, taken to Yucatan, established in the field, and exposed to LYLD for 15 years. Accumulated mortality was highest in MXAT (64%) and lowest in MYD (6%). The disease killed about 20% of the palms of the most tolerant Tall population (MXPT2), which was selected for further breeding purposes and seed production. Mexico also launched the search for molecular markers associated with phytoplasma tolerance (Cardena et al. 2003), followed by a team from Côte d'Ivoire (Konan et al. 2007).

These results from Mexico illustrate well the difficulty to find tolerant material ready to be used, or ready to be cloned, directly from farmers' fields. Delivering good planting material requires collecting but also breeding and seedling production programs. After collecting in farmers' fields, at least one more generation is needed. If seednuts are to be used, the best palms from the most tolerant variety should be intercrossed using controlled pollination, their progeny (i.e. generation 2) should be planted in isolation, and from this fruiting progeny, mass seednut production can start. If clones are to be disclosed, true-to-type tissues such as inflorescences must be used; or the best palms from the most tolerant variety should be intercrossed using controlled pollination, and the embryos (zygotic tissues) from this progeny can be converted into clones. Then those clones should preferably be tested again in the fields to identify the best ones. As discussed below, the absence of transmission of LYLD through seednuts or clones is to be verified. Planning an additional generation of field tests helps in this process.

5.3.2 *Transmission of Phytoplasmas by Seednuts or Seedlings*

Two research teams have tested the transmission of phytoplasmas through seeds. In Mexico, phytoplasmas were detected in coconut plantlets obtained through in vitro germination of zygotic embryos from the seeds of infected palms (Oropeza et al. 2017). In Côte d'Ivoire, similar experiments gave negative results (Daramcoum et al. 2018).

5.3.3 *Collecting for Tolerance to Other Pests and Diseases*

The Cadang-Cadang viroid found in the Philippines, a diverse species of *Phytophthora* killing palms and/or causing the fall of immature fruits, and the foliar decay virus in Vanuatu are some other important coconut diseases. For Cadang-Cadang, no genetic resistance or tolerance has been discovered. A test of 93 coconut populations showed that all were susceptible to inoculation (Orolfo et al. 2000); however, genetic resistance to natural infection may be available in some populations (Randles and Rodriguez 2003). *Phytophthora heveae* and *P. palmivora* cause bud rot and nut fall leading to either death of the coconut palm or harvest losses that can exceed 30%. Many varieties and hybrids are sensitive to at least one form of disease expression, while others present good tolerance (Franqueville et al. 1989). The variation existing within populations allowed selecting tolerant half-sib hybrid lines (Bourdeix et al. 1992).

The main pests of coconut are: the beetles (*Oryctes rhinoceros* L., *Scapanes australis*), which feed on young leaves; the red palm weevil (*Rynchophorus sp.*) who use the tissues injured by the beetles to penetrate to the heart and kill the palm; xylophageous nematodes (called Red Ring Disease); the coconut mite (eriophyid, *Aceria guerreronis* Keifer), which damages the epidermis of the fruits, reducing their size; leaf-eating caterpillars; the coconut white grub (*Leucopholis coneophora* Burmeister); whiteflies (genera *Aleurotrachelus* and *Aleurodicus*); and the coconut hispine beetle (*Brontispa longissima* Gestro). As vectors, leafhoppers (family Cicadellidae) often transmit lethal diseases. Various degrees of susceptibility (more often than tolerance) have been observed for many pests: *O. rhinoceros* and *S. australis* (Foale 1987), *B. longissima* (Foale 1987), *Rhynchophorus* sp. (Chowdappa et al. 2018), and mites (Aratchige et al. 2016; Chowdappa et al. 2018). As further examples, two coconut accessions, Sri Lanka Yellow Dwarf and Gon Thembili Tall, were identified to be less susceptible to *A. guerreronis* infestation in Sri Lanka (Perera et al. 2015).

For the decade (2018–2027), COGENT suggested its member countries to collect about 100 populations with putative tolerance to biotic and abiotic stresses (Perera et al. 2018). Collecting actions were to focus on areas where tolerance was to be expected to be found. The "Sri Lanka Green Dwarf" and "Vanuatu Tall" are the most tolerant varieties to LYLD in Ghana, while LYLD is present neither in Sri Lanka nor in Vanuatu. Thus, collecting areas should not be only those under present disease pressure. The expectation of collecting LYLD-tolerant palms in diseased zones is not completely hopeless but, as already initiated in Mexico, implies a minimum 10-year phase of validation by field testing before any conclusion can be reached. Progenies of putative tolerant palms should be planted in comparison with a sensitive control, using adapted experimental design, preferably in multisite experiments. It is difficult to know where the disease will stay active and where it will strike next. Once they have obtained results, breeders must take care not to transmit the phytoplasma in infected seedlings. Efficient disease indexing methods and quarantine centres will be necessary for safely releasing the collected germplasm to ex situ genebanks.

5.3.4 Collecting Compact Dwarf and Other Special Varieties

Among the COGENT recommendations in 2012, the fourth one deals with collecting activities: "Strengthening coconut genetic research, coconut conservation and specific uses of traditional coconut varieties in the Pacific Region." Because of the coconut water boom and other socio-economic factors, the cultivation of Dwarf varieties is now expanding. Over the last 10 years, it has become increasingly difficult to find coconut climbers. In Indonesia and in many South Pacific countries, young workers in rural areas are less interested in coconut harvesting and especially in harvesting sap from very tall trees. Indonesian farmers started to use some Dwarf varieties instead of Tall types to produce neera (sap) and coconut sugar (Novarianto et al. 2017).

Except in the Pacific region, there is a narrow genetic basis for Dwarf expansion, which relies mainly on the "Malayan Dwarf" type. This type is characterized by early flowering, precocity, slow vertical growth, thin stem easily damaged by cyclones, relatively small fruit with rubbery copra, low resilience, and susceptibility to drought. The most cultivated one is the "Brazil Green Dwarf", originating from the Philippines and appreciated for its sweet water. The second most cultivated is the Aromatic Green Dwarf whose genetic history seems different. According to recent surveys conducted in 2018 in Thailand, Thai farmers created this variety from a natural hybrid between a Yellow Dwarf (very probably of Malayan type) and a Green Tall with aromatic characteristics. From this unique palm, they selected the "Aromatic Green Dwarf" but also a few, less known, Aromatic Yellow Dwarf varieties (Dumhai et al. 2019).

In Indonesia, the Kopyor coconut is a naturally occurring mutant having fluffy solid endosperm instead of the typical solid endosperm. It is an Indonesian delicacy that sells as much as ten times higher prices than an ordinary coconut. Similar mutants have been found in other countries, mostly among Tall populations: Macapuno (Philippines), Makhrao Kathi (Thailand), Dikiri Pol (Sri Lanka), and Thairu Thengai (India). In Indonesia, researchers recently described four Kopyor Dwarf varieties, named by their colours – Green, Yellow, Brown, and Red (Novarianto et al. 2014), but there are probably more types to discover, differing by other traits, mainly fruit characteristics.

In 2016, the exploration of Indonesian coconut germplasm led researchers to identify and characterize a new kind of coconut tree with limited vertical growth in Bido Village, Morotai Island. This "Bido Tall" is genetically distinct from other Dwarf and Tall varieties conserved in the Indonesian International Genebank. It has been crossed with eight Dwarf varieties for progeny testing, established on a 7-ha seed garden for mass release to farmers (Novarianto et al. 2018).

The preferentially cross-pollinating Compact Dwarf varieties are generally called Compact Dwarfs or Niu Leka-type Dwarfs, because the "Niu Leka Dwarf" from Fiji was the first described of this type. Harries (1978) has described the Niu Leka as follows: "Except for the very short internodes, which reduce trunk height and produce a dense leaf canopy, all other characteristics resemble those of tall varieties. The well-developed bole, the trunk girth, the predominantly cross-pollinating flowering pattern, the lack of bright red or yellow fruit colours, the large fruit size and an exclusive distribution (until very recently) within certain Pacific islands all point to a completely distinct selection process".

At least Fijian, Tongan, and Samoan people already knew Compact Dwarf varieties in the 1850's. In Fiji, the first mention of the Niu Leka coconut variety was found in the Fijian-English dictionary of Hazlewood (1850), in the definition of the verb valueka: "to fruit whilst very short, or young, as a Niu Leka". Leka mean short. Niu or better Ni'u is presented as a contraction of Ni au or Ni ka'u. In Tonga, Niu Leka was also cited in an early dictionary, as a coconut variety (Rabone 1845). In Samoa, it was cited as Niu Le'a (Pratt 1862).

Marechal (1928) reported the first coconut-controlled pollination, involving the Malayan Red (MRD) and the Niu Leka (NLAD) Dwarfs (Marechal 1928). The Niu Leka Dwarf and its hybrid with the MRD have been disseminated worldwide. Before leaving Fiji in the 1930s, Marechal foresightedly gave the MRD x NLAD hybrid and its progeny to many Fijian gardeners and farmers. We do not know whether the amazing Compact Red and Yellow Dwarf varieties presently found in Pacific gardens are ancient varieties that have existed for hundreds of years or progenies of Marechal's hybrids, carefully selected in the last 80 years by hundreds of Pacific gardeners and farmers. DNA analysis of these Compact Dwarf varieties should soon help to better understand the full story. Some of these varieties are mainly autogamous. There is still no official name for these new kinds of varieties, but Vijendra Kumar (Fiji) suggested calling them "Super Dwarfs", as they may bear dwarfism genes from both Malayan-type and Compact Dwarfs.

Recent surveys conducted in Fiji, the Cook Islands, and Samoa have shown that the most interesting diversity is more commonly found in gardens than in farms (R. Bourdeix and V. Kumar, personal communication). In many Pacific countries, coconut plantations serve mainly to get cash when needed; but the few coconut palms kept in gardens are for daily food: People know and select these very carefully, especially for slow vertical growth, fruit quality, and taste.

In Florida, genetic diversity and population structure of coconut germplasm were analysed with microsatellite DNA markers (Meerow et al. 2003). The study gave interesting results on a variety called "Fiji Malayan", resulting from Marechal hybridization and described as "Compact" or "Super" Dwarfs in this chapter. As in Fiji, Florida gardeners seem to disseminate those varieties more efficiently than scientists. Some of these Dwarfs have shown tolerance to the local LYLD.

COGENT should promote the cultivation of Compact Dwarf varieties and the testing and use of hybrids varieties between Malayan and Compact Dwarf types. These varieties have slower vertical growth, thicker stems, and better resilience than the Malayan types; some of them have better fruits characteristics, allowing to produce not only water but also other products such as coconut milk, cream, and virgin oil. Some red allogamous forms could be used to design new seed gardens (mixed with green Tall types) for producing hybrids without manual emasculation. Coconut breeders working in COGENT country members could first receive pollen of Compact Dwarfs and use it rapidly to create new hybrids. Together with the understanding of dwarfism inheritance (from both Malayan and Compact types), this could generate the most important advance in coconut breeding for the next 30 years.

Coconut palms producing both fruits and suckers have been described. This germplasm urgently needs to be studied further and conserved. Hormonal balance in these rare plants may bring crucial information for in vitro propagation of the coconut palm.

For the next decade, the COGENT strategy envisions the collection of about 50 populations or varieties of Compact Dwarfs (or hybrids between Compact and Malayan-type Dwarfs) and 50 Tall-type populations or varieties with special traits such as new texture and taste of kernel, sweet husk, different kind of aromatic coconut water, or other rare characteristics (Kumar et al. 2018).

In the framework of the Coconut Industry Development for the Pacific (CIDP) project, CIRAD launched a reflection on how to organize a regional contest for coconut varieties (Bourdeix and Leroy 2018). This new approach will help to locate crucial germplasm and increase public awareness about coconut diversity. The multiple objectives would be to:

- Strengthen local interactions among coconut farmers.
- Encourage farmers to develop private initiative in coconut seed production.
- Strengthen interactions between farmers and national services in charge of extension and research.
- Locate coconut germplasm, assess its diversity, and facilitate the collecting process.
- Help farmers preserve disappearing traditional varieties.
- Boost global communication in the coconut value chain in the Pacific region.

Communication using social media may bring substantial benefits. Advertising campaigns can be launched with limited budgets and target a specific audience using keywords such as coconut varieties, coconut farming, diversity, traditional varieties, virgin coconut oil, farmers' association, etc. Strengthening the communication on coconut diversity should concern not only coconut farmers but also gardeners. Most Compact Dwarfs planted in gardens are overused: It is extremely difficult to get mature seednuts. In the framework of a recent Darwin Project (Upgrading and Broadening the South-Pacific International Coconut Genebank), CIRAD and the Ministry of Agriculture of Fiji started to pay in advance for seednuts or seedlings from gardeners and farmers. This encourages stakeholders to take more care to their palms and, may be, to continue marketing seednuts and seedlings.

5.4 Balance and Plans for Global Collection Programmes

5.4.1 Nomenclature and Recording of Coconut Accessions

During the first COGENT meeting held on May 1992, country representatives discussed the status of existing collections and outlined what would become the CGRD (Bourdeix and Prades 2018). On the advice of the first author of this chapter, researchers were proposed to systematically provide an international name to any accessions they collect. Researchers from the Philippines (Geraldo Santos' team) were the most reluctant to agree to this new rule; they were possibly right but they finally joined the consensus of other countries.

In 2018, 26 years later, it remains unclear if this systematic naming policy was beneficial or not; in any case, it has strongly impacted the conservation of coconut germplasm and users' representations of coconut diversity at the global level. In a COGENT ex situ genebanks, each new accession is registered under an international cultivar or population name and an abbreviation. COGENT also requested national researchers not to create a new cultivar name for each and very sample collected in farmers' fields. For that purpose, the new notion of "populations within a cultivar" was introduced. In the coconut nomenclature, the term *population* denotes "minor phenotypic and/or geographical differentiation within a cultivar". This contributed to reduce the number of cultivar names, to limit unnecessary and costly conservation of similar samples under different names.

Genebank curators and national researchers are those who decide to give the international names to the germplasm found in their country. To achieve international standardization, however, they are prompted to endorse specific guidelines. Those guidelines' "useful definitions of terms and nomenclature" have been released in the book *Coconut Genetic Resources* (Batugal et al. 2005), as well as in the "catalogue of conserved germplasm". An extensive list of international names of existing coconut cultivars was also released (Bourdeix et al. 2005, 2010) and recently updated in the COGENT strategy.

International coconut names may evolve with time. For instance, Vanuatu researchers have attributed population names to the accessions they collected as

"Vanuatu Tall": Nipeka, Pélé, Walanaro, Waluembue, etc. These researchers are characterizing these populations, based on productivity, morphology, and molecular traits. If the population of "Vanuatu Tall Walanaro" shows well-established distinct traits, then it will have to be renamed as a new cultivar, possibly "Walanaro Tall". On the contrary, if the populations of "Vanuatu Tall Waluembue" and "Vanuatu Tall Nipeka" seem very similar, these accessions (or hybrids between selected palms of these accessions) should be merged under one name only. When the time will come to regenerate those accessions, they will be merged, because it is useless to conserve two separate accessions for the same germplasm.

The last major update of the CGRD dates to 2012, so data is becoming obsolete. According to this database, 1,005 accessions have been successfully collected in farmers' fields and transferred ex situ. Around 419 cultivars are kept by COGENT ex situ genebanks. A fifth to a third of these conserved cultivars may be duplicates, mixed populations, or represented by an insufficient number of palms (Perera et al. 2018). Most genebanks have started to detect and remove duplicated accessions.

5.4.2 Combining Geographical and Molecular Approaches

An approach developed since the 1990s consisted of using geographic information systems and related databases to assess the repartition of coconut diversity and to identify gaps for conducting novel expeditions. In the 1990s, COGENT launched a global study on coconut collecting strategies. From 1995 to 2012, CIRAD experts trained coconut researchers from 13 COGENT countries on inputting data into the Coconut Genetic Resources Database (CGRD) (Bourdeix 1997a, b; Labouisse and Bourdeix 2003). Most of the geolocalization of the collecting sites was then done manually on paper maps. Two important next steps were a first report on collecting strategies (Bourdeix et al. 2005) and geographical studies for mapping and modelling the distribution of coconut and its genetic diversity (Coppens d'Eeckenbrugge et al. 2018a, b).

COGENT ex situ collections conserve accessions from 44 countries and territories. According to the FAO, 92 countries and territories are producing coconut. Thus, 48 of them (51%) are not yet represented in the COGENT ex situ collections. We calculated ratios between the number of accessions conserved ex situ and the area under coconut palms: This index is 84 accessions per million hectares on average. At regional levels, it varies from 64 (in Africa) to 282 (in the Pacific Ocean) (Bourdeix et al. 2018a). In 2013/2014, Komba Mayossa and Coppens d'Eeckenbrugge inferred the global coconut potential distribution for current climatic conditions, using Maxent, on a sample of 1,875 coconut palm observation sites (Coppens d'Eeckenbrugge et al. 2018a) (Fig. 5.2). From this work, Coppens d'Eeckenbrugge and Ullivari derived a map of collecting gaps at the global scale (Fig. 5.3) and were included in COGENT Strategy (Coppens d'Eeckenbrugge et al. 2018b).

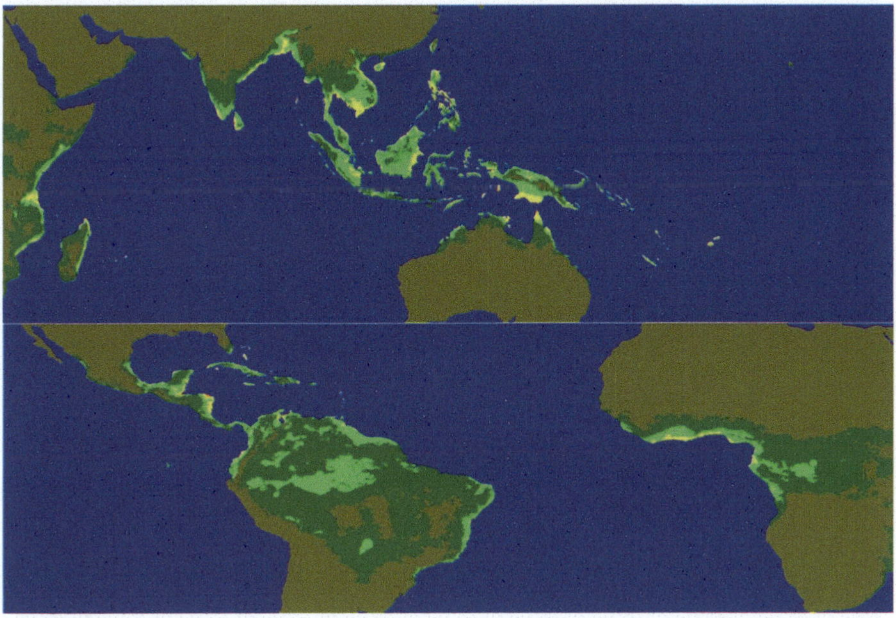

Fig. 5.2 Coconut potential distribution (without irrigation). Dark green areas correspond to marginal climatic conditions; light green to orange areas correspond to increasingly favourable climates. Source: COGENT Strategy (Coppens d'Eeckenbrugge et al. 2018b)

Microsatellite DNA marker studies have shown the existence of two highly differentiated gene pools: the Pacific and Indo-Atlantic. Genetic diversity of neutral markers is mainly found within cultivars and between gene pools (Gunn et al. 2011). Wolf and Coppens d'Eeckenbrugge used a microsatellite dataset from 86 coconut populations (provided by Baudouin), for a first study of the distribution of genetic diversity, using Jost's J1 Index (Fig. 5.4). The Indo-Pacific region presents the highest diversity, with hotspots in Vanuatu, PNG, Southeast Asia, and East Africa. Atlantic shores, where the coconut was introduced in historical times, appear less diverse. However, this representation remains incomplete, as many neighbouring areas correspond to gaps in collections and/or genetic studies. Important gaps seem to persist in the affiliated COGENT germplasm collections – Micronesia, Melanesia, large parts of Polynesia, Australia, Malaysia, Indonesia, Myanmar, India, Madagascar, Somalia, and Kenya – among zones where coconut palm is likely to be native.

The potential impact of climate change on coconut distribution has been anticipated for the year 2050, using three climate models (BCC-CSM1-1, CCSM4, and HadGEM2-ES) and three gas emission/radiative forcing scenarios (Wolf and Coppens d'Eeckenbrugge, unpublished). All results converge in predicting a very strong increase of areas climatically suitable for coconut, thus discarding a global negative impact of climate change per se. This seems reassuring in terms of genetic erosion risk, as well as production. However, a few regions, including

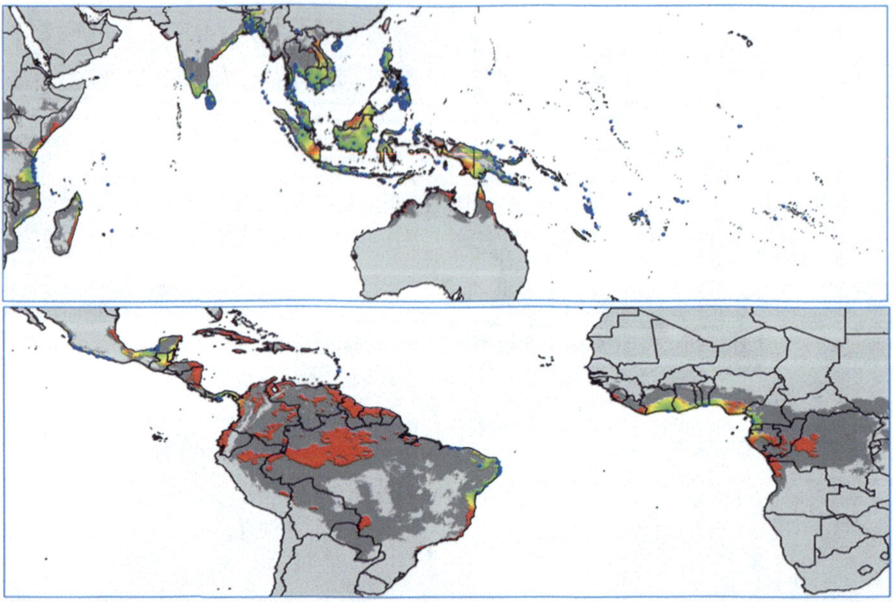

Fig. 5.3 Origins of accessions inventoried in the COGENT CGRD database (blue dots). Coconut suitable areas (see Fig. 5.1) are represented with a green-to-red colour gradient according to their distance from the closest collecting site: yellow between 220 and 320 km and red to deep red beyond 420 km. Dark grey areas correspond to marginal climates for coconut growth and development. Source: COGENT Strategy (Coppens d'Eeckenbrugge et al. 2018b)

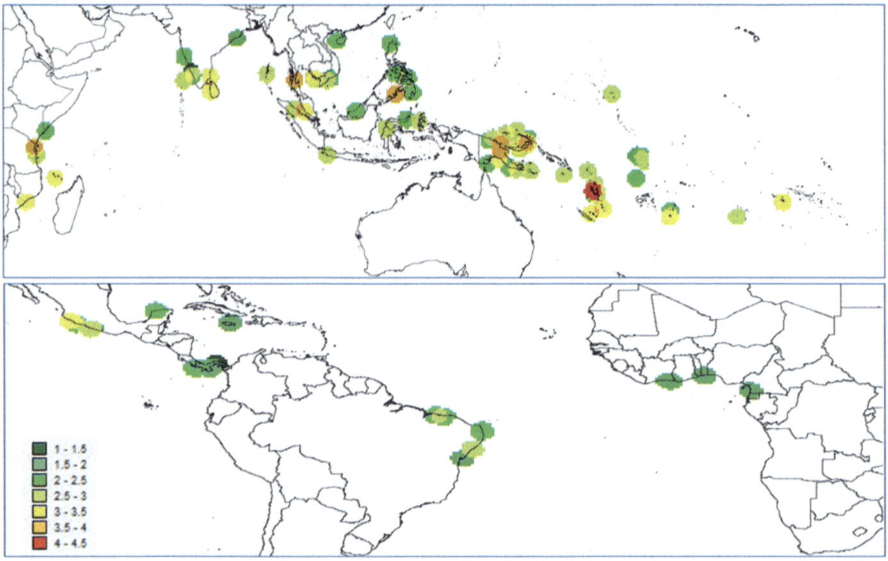

Fig. 5.4 Distribution of coconut genetic diversity (Jost's J1 Index), according to currently available data (30 SSR loci, 778 individuals from 86 populations)

south-eastern India and Bangladesh, would be negatively affected. Due the eastward shift of favourable areas around the Bay of Bengal, Bangladesh could lose about half of its coconut production areas. The 0.2–0.4 m sea-level rise expectation for 2000–2050 might be doubled given that the West Antarctic Ice Sheet has already been set into motion. Even so, its impact on coastal coconut populations should be negligible for continents and large islands, where shoreline retreat should be modest, compared to the extension of climatically suitable areas. The situation would be different for distant, small, and low-lying islands, as well as delta areas, particularly in the Indo-Pacific region, where sea rising rates are up to thrice the global rate. Thus, there are severe threats on the Maldives, Laccadive, Mascarene archipelagos, and the myriad of far Polynesian, Micronesian, and Melanesian islands, all regions hosting an important diversity of coconut germplasm, not well studied so far.

The COGENT Strategy plans to collect 100–200 populations following the approach of filling geographical gaps (Coppens d'Eeckenbrugge et al. 2018b). It also plans to collect embryos and pollen from 100 to 200 additional coconut populations located in the most isolated small islands (Bourdeix et al. 2018a, b, c, d). Many archipelagos, such as the Tuvalu, Tuamotu, Kiribati, Maldives, the many Indonesian atolls, etc., could be concerned by this kind of survey. Tonga is one of the highest priorities, for reasons linked to the historical context and the very high isolation index of some Tongan islands. Those islands will need to be carefully selected based on the following criteria

- Most endangered by sea-level rise and/or climate change
- Most remote and accessible only by boat (no airport) but preferably where people are living or were living less than 100 years ago
- Islands where coconut has been culturally important or where copra was never an important business (as traditional varieties will probably be better conserved)
- Islands having special interest for other crops or animals, as other species than coconut will probably be collected in the same survey

5.4.3 Plans for Global Collecting Programmes

During surveys conducted in 2012 (Bourdeix et al. 2018a), most of the COGENT experts and genebank curators estimated that:

- No more than one-third of the existing useful diversity has been adequately transferred to referenced ex situ collections.
- Diversity is quickly disappearing from farmers' fields.
- Specific diversity, such as Compact Dwarfs, is needed for immediate use and is not yet available from COGENT genebanks.

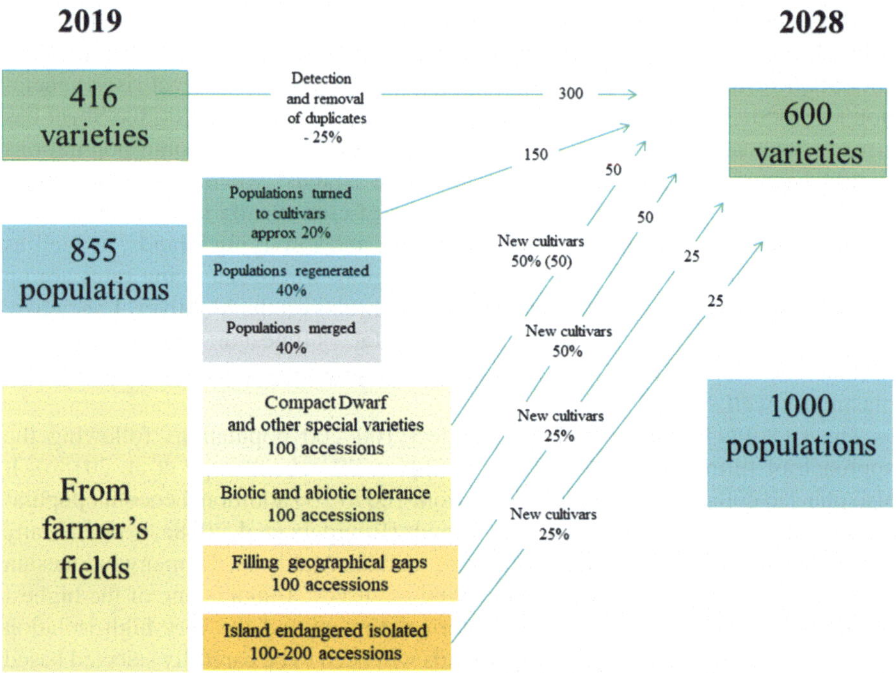

Fig. 5.5 Forecasts for collecting coconut varieties and populations during the next decade (2019–2028)

A tentative representation of what could happen during the next decade in coconut collecting and conservation can be proposed (Fig. 5.5), if enough funds are allocated for these collection activities and if global coordination is effectively ensured. The 416 varieties and 855 populations currently conserved ex situ would be reduced to about 450 varieties and 350 populations, respectively. All the new collections to be carried out will increase total numbers to about 600 varieties and 1,000 populations in 2028. This corresponds to the collection and ex situ transfer of about 50 varieties and populations per year. The assessments presented during the 2014 COGENT Steering Committee meeting indicate that, globally, COGENT countries are presently collecting more, so that this number of 50 per year is probably underestimated.

If global coordination is not ensured, the number of conserved accessions will also increase, but, very probably, a lower number of duplicate populations will be detected and merged. Global coordination also helps to protect coconut genebanks from land tenure issues and from destruction by local decision-makers. Genebank destruction is the greatest risk that could compromise the forecasts presented above (Fig. 5.5).

5.5 Conclusion

Expecting to plan coconut collecting activities at the global level may seem unrealistic. This challenging exercise was attempted in the framework of the strategic foresight carried out by COGENT member countries. In 2012, member countries were collaborating more closely than nowadays (2018) and the CGRD received significant update. In 2014, COGENT members agreed that the global objective within the next decade was to collect up to 500 well-chosen populations or varieties and transfer them to ex situ genebanks. COGENT member countries will probably collect more germplasm in the framework of their national programs; but defining common priorities at the global level helps to coordinate international actions.

Although highly threatened, the accessions already conserved in field genebanks are conserved in relative security when compared to the many farmers' varieties – we could say "elders' varieties" – which are disappearing because of socio-economic, sanitary, and climatic changes. As a first step to face this emergency, surveyors may collect germplasm and cryopreserve it in a dedicated genebank. Thus, when COGENT field genebanks and breeders will have both the interest and the funding to study this germplasm, the cryo-genebank will release part of this frozen material. In exchange, the field genebanks will have to commit to giving back double the quantity of embryos and/or pollen from the next generation. This is probably how cryo-genebanks could be most useful, as duplicating the germplasm already existing in ex situ genebank would be very costly (see Chap. 4).

For collecting on the most isolated small islands, we envisioned a boat, fully equipped with laboratory facilities including cryopreservation, to reach these islands and collect germplasm. For both economy of scale and multifunctional approach, this boat would preferably not only collect coconut germplasm but also other crop germplasm or natural resources. This emblematic project could be developed by a multi-crop research team.

Over the next decade, coconut cloning will develop. In this context, there is a danger that the novelty of the method makes us forget the basics of plant breeding. The genetic value of palms selected as clones in the field needs experimental confirmation under controlled conditions. Advances in technology may lead to neglecting field research and applying highly sophisticated methods to questionable biological samples while leading to false resistance assumptions. It is not enough to collect embryos from apparently resistant palms, to grow embryogenic callus from these embryos, and to regenerate plantlets for having the best clones revolutionizing coconut cultivation. If collecting processes are important, they cannot replace breeding programs. Progenies selected by breeders during decades will very probably give the best clones. At the present stage, an additional coconut generation seems inevitable for comparing the real value of clones from various origins with well-chosen genetic controls. On the other hand, new genomic techniques will soon contribute to shorten and optimize collecting, breeding, and cloning programs.

References

Aratchige NS, Kumara ADNT, Suwandharathne NI (2016) The coconut mite: current global scenario. In: Economic and ecological significance of arthropods in diversified ecosystems. Springer, Singapore, pp 321–342

Batugal P, Ramanatha Rao V, Oliver J (eds) (2005) Coconut genetic resources. International Plant Genetic Resources Institute – Regional Office for Asia, the Pacific and Oceania (IPGRI-APO), Serdang

Beccari O (1916) Il genere Cocos Linn e le palme affini. Istituto agricolo coloniale italiano. Istituto Agricolo Coloniale Italiano, Firenze, p 143

Bourdeix R (1997a) Mission report. Coconut germplasm in Tanzania, Sri Lanka and India, October and November 1996. Paris, CIRAD-CP, p 48

Bourdeix R (1997b) Mission report. Training operations in Africa and Latin America/Caribbean. Training courses for instructors on the use of the STANTECH manual (standardized research techniques for coconut breeding). Ivory Coast, Grand-Bassam, 16th–26th June 1997; Jamaica, Kingston, 14th–25th July 1997 Paris: CIRAD-CP, p 60

Bourdeix R, Leroy T (2018) Preparing the world first regional coconut varietal contest. In: Bourdeix R, Labouisse JP, Mapusua K et al (eds) Coconut planting material for the Pacific region. Available at the URL: https://replantcoconut.blogspot.com

Bourdeix R, Prades A (2018) A global strategy for the conservation and use of coconut genetic resources 2018–2028. Bioversity International, Montpellier. Available at the URL: https://www.bioversityinternational.org/fileadmin/user_upload/Cogent_bourdeix_2018.pdf

Bourdeix R, N'Cho YP, Sangare A et al (1992) The improved PB 121 coconut hybrid, a cross between the Malayan Yellow Dwarf and selected West African Tall parents. Oleagineux (France) 47(11):619

Bourdeix R, Konan JL, N'Cho YP (2005) Coconut : a guide to traditional and improved varieties. Editions Diversiflora, Montpellier, p 94

Bourdeix R, Baudouin L, Bambridge T et al (2009) Dynamics and conservation of the coconut palm *Cocos nucifera* L. In: The Pacific region: towards a new conservation approach. Paper presented at the 11th Pacific science inter-congress, March 2–6, Tahiti, French Polynesia

Bourdeix R, Batugal P, Oliver JT et al (2010) Catalogue of conserved coconut germplasm. International Plant Genetic Resources Institute – Regional Office for Asia, the Pacific and Oceania (IPGRI-APO), Serdang

Bourdeix R, Allou K, Omuru E (2018a) 3.3.3 triplication of germplasm in distinct geographical sites – chapter 3. Where we need to be to secure diversity and promote use. In: Bourdeix R, Prades A (eds) A global strategy for the conservation and use of coconut genetic resources. Biodiversity International, Montpellier, pp 133–138

Bourdeix R, Baudouin L, Santos GA (2018b) 2.1.3 international coconut nomenclature – chapter 2. Where we are today. In: Bourdeix R, Prades A (eds) A global strategy for the conservation and use of coconut genetic resources 2018–2028. Biodiversity International, Montpellier, pp 39–40

Bourdeix R, Chong F, Maskromo I (2018c) 3.5.4 islands most isolated and/or endangered by climate change – chapter 3. Where we need to be to secure diversity and promote use. In: Bourdeix R, Prades A (eds) A global strategy for the conservation and use of coconut genetic resources 2018–2028. Biodiversity International, Montpellier, p 152

Bourdeix R, Devakumar K, Pole F (2018d) 2.3.4 collecting germplasm – chapter 2. Where we are today. In: Bourdeix R, Prades A (eds) A global strategy for the conservation and use of coconut genetic resources 2018–2028. Biodiversity International, Montpellier, pp 60–63

Cardena R, Ashburner GR, Oropeza C (2003) Identification of RAPDs associated with resistance to lethal yellowing of the coconut (*Cocos nucifera* L.) palm. Sci Hortic 98(3):257–263

Chowdappa P, Hegde V, Mohan C et al (2018) Pest and disease-free coconut. Indian Coconut J 60(12):24–25. Available at the URL: https://krishi.icar.gov.in/jspui/bitstream/123456789/13476/1/Pest%20&Disease.pdf

Coppens d'Eeckenbrugge G, Komba P, Ullivarri V (2018a) 3.5.3 filling geographical gaps – chapter 3. Where we need to be to secure diversity and promote use. In: Bourdeix R, Prades A (eds) A global strategy for the conservation and use of coconut genetic resources 2018–2028. Bioversity International, Montpellier, pp 150–152

Coppens d'Eeckenbrugge G, Duong NTK, Ullivarri A (2018b) 2.4.3 geographic information systems – chapter 2. Where we are today. In: Bourdeix R, Prades A (eds) A global strategy for the conservation and use of coconut genetic resources 2018–2028. Bioversity International, Montpellier, pp 71–74

Cueto CA, Johnson VB, Bourdeix R et al (2012a) Validation of a coconut embryo-culture protocol for the international exchange of germplasm, Terminal Report (15/10/2009–28/02/2012). Bioversity International, Rome. Available at the URL: http://www.cogentnetwork.org/images/projects/tr-c60014-terminal%20report-finalv4.pdf

Cueto CA, Johnson VB, Bourdeix R et al (2012b) Technical guidelines for the safe movement and duplication of coconut (*Cocos nucifera* L.) germplasm using embryo culture transfer protocols. COGENT; Bioversity International, Montpellier. Available at the URL: http://www.cogentnetwork.org/images/publications/tg-coconutembryotransfer.pdf

Daramcoum WAMP, Konan Konan JL, Martial YSD et al (2018) Molecular diagnosis of phytoplasma transmission from zygotic embryos to in vitro regenerated plants of coconut palm (*Cocos nucifera* L.). Afr J Biotechnol 17(26):862–869

Dumet D, Diebiru E, Adeyemi A et al (2013) Cryopreservation for the 'in perpetuity' conservation of yam and cassava genetic resources. CryoLetters 34(2):107–118

Dumhai R, Wanchana S, Saensuk C et al (2019) Discovery of a novel CnAMADH2 allele associated with higher levels of 2-acetyl-1-pyrroline (2AP) in yellow dwarf coconut (*Cocos nucifera* L.). Sci Hortic 243:490–497

Foale MA (1987) Coconut germplasm in the South Pacific Islands (No. 113880). Australian Centre for International Agricultural Research, Canberra

Franqueville HD, Taffin GD, Sangare A et al (1989) Detection of phytophthora heveae tolerance characters in coconut in Côte d'Ivoire. Oléagineux (Paris) 44(2):93–103

Gunn BF, Baudouin L, Olsen KM (2011) Independent origins of cultivated coconut (*Cocos nucifera* L.) in the old-world tropics. PLoS One 6(6):e21143

Gunn B, Myrie WW, Baudouin L (2018) 1.1.1 origin, history and dynamics of coconut cultivation – chapter 1. Introduction to the global coconut strategy. In: Bourdeix R, Prades A (eds) A global strategy for the conservation and use of coconut genetic resources 2018–2028. Bioversity International, Montpellier, pp 3–7

Hamelin C, Bourdeix R, Baudouin L (2005) The international coconut genetic resources database. Coconut Genetic Resources, 427

Harries HC (1978) The evolution, dissemination and classification of *Cocos nucifera* L. Bot Rev 44(3):265–319

Harries HC, Romney DH (1974) Maypan: an Fl hybrid coconut variety for commercial production in Famaica. World Crops

Hazlewood D (1850) A Feejeean and English dictionary: with examples of common and peculiar modes of expression and uses of words. Also, containing brief hints on native customs, proverbs, the native names of the natural productions of the islands, notices of the islands of Feejee, and a list of the foreign words introduced. Printed at the Wesleyan Mission Press, Vewa

Hill AW (1929) The original home and mode of dispersal of the coconut. Nature 124(3117):133

Karun A, Sajini KK, Niral V et al (2014) Coconut (*Cocos nucifera* L.) pollen cryopreservation. CryoLetters 35(5):407–417

Konan KJNL, Koffi KE, Konan JL et al (2007) Microsatellite gene diversity in coconut (*Cocos nucifera* L.) accessions resistants to lethal yellowing disease. Afr J Biotechnol 6(4):341–347

Kumar V, Prades A, Perera SACN et al (2018) 3.5.1 compact dwarfs and other special varieties – chapter 3. Where we need to be to secure diversity and promote use. In: Bourdeix R, Prades A (eds) A global strategy for the conservation and use of coconut genetic resources 2018–2028. Bioversity International, Montpellier, pp 148–149

Labouisse JP, Bourdeix R (2003) Coconut germplasm collecting, characterisation and conservation in Cook Islands, Kiribati, Marshall Islands and Tuvalu: Project LOA IPGRI 00/015, March 2000–February 2001. Final report Santo: VARTC, p 126

Leclerc C, Coppens d'Eeckenbrugge G (2011) Social organization of crop genetic diversity. The G× E× S interaction model. Diversity 4(1):1–32

Liberty C, Foale M, Arancon R (2018) 1.1.2 cultivation and current production of coconut – chapter 1. Introduction to the global coconut strategy. In: Bourdeix R, Prades A (eds) A global strategy for the conservation and use of coconut genetic resources 2018–2028. Bioversity International, Montpellier, pp 7–9

Marechal H (1928) Observations and preliminary experiments on the coconut palm with a view to developing improved seed nuts for Fiji. Fiji Agric J 1(2):16–45

Martinez RT, Baudouin L, Berger A et al (2010) Characterization of the genetic diversity of the Tall coconut (Cocos nucifera L.) in the Dominican Republic using microsatellite (SSR) markers. Tree Genet Genomes 6(1):73–81

Meerow AW, Wisser RJ, Brown SJ et al (2003) Analysis of genetic diversity and population structure within Florida coconut (*Cocos nucifera* L.) germplasm using microsatellite DNA, with special emphasis on the Fiji dwarf cultivar. Theor Appl Genet 106(4):715–726

Mpunami A, Mugini J, Tembo P et al (2013) Disease control research. In: Kullaya A, Temu N, Seguni Z, Mpunami A, Chipungahelo G, Masumbuko L, Mkumbo K, MAdulu R (eds) Twenty-five years of coconut research and development in Tanzania. Mikocheni Agricultural Research Institute, Dar es Salaam, pp 99–155

Novarianto H, Maskromo I, Dinarti D et al (2014) Production technology for kopyor coconut seed-nuts and seedlings in Indonesia. CORD Int J Coconut Res Dev 30(2):31–40

Novarianto H, Maskromo I, Mashud N et al (2017) Development of coconut sugar by using Dwarf coconuts. Final report of collaboration research the IAARD and PT. UNILIVER, p 57

Novarianto H, Tulalo M, Mawardi S et al (2018) BidoTall coconut the Dumpy as pollen source to produce high quality coconut hybrids. Paper presented at the 48th APCC COCOTECH Conference & Exhibition, 20–24 August, The Berkeley Hotel Pratunam, Bangkok, Thailand

Orolfo MB, Estioko LP, Rodriguez MJB (2000) Screening of coconut populations for resistance to coconut cadang-cadang viroid (CCCVd). PCA-ARDB Annual Report

Oropeza C, Cordova I, Puch Hau C et al (2017) Detection of lethal yellowing phytoplasma in coconut plantlets obtained through in vitro germination of zygotic embryos from the seeds of infected palms. Ann Appl Biol 171(1):28–36

Perera L (eds) (2018) 3 where we need to be to secure diversity and promote use. In: Bourdeix R, Prades A (eds) A global strategy for the conservation and use of coconut genetic resources 2018–2028. Bioversity International, Montpellier, pp 113–176

Perera SACN, Dissanayaka HDMAC (2013) Management of the Weligama coconut leaf wilt disease: screening and breeding coconuts for resistance/tolerance to WCLWD. In: Gunasena HPM, Gunethilake HAJ, Fernando LCP, Everard JMDT, Appuhamy PAHN (eds) Wiligama coconut leaf Wilt disease – six years after (sds). Coconut Research Institute, p 96–106

Perera SACN, Herath HMNB, Wijesekera HTR et al (2014) Evaluation of coconut germplasm in Weligama and Matara area in the southern province of Sri Lanka for resistance to Weligama coconut leaf wilt disease. COCOS 20(1):15–20

Perera SACN, Kamaral LC, Fernando WBS (2015) Molecular assessment of *Cocos nucifera* L. Var. Sri Lanka yellow dwarf for genetic purity and aceria mite tolerance. Int J Mol Evol Biodivers 5(1):1–5

Perera SACN, Odewale JO, Omuru E et al (2018) 3.5.2 collecting for pest and disease tolerance – chapter 3. Where we need to be to secure diversity and promote use. In: Bourdeix R, Prades A (eds) A global strategy for the conservation and use of coconut genetic resources 2018–2028. Bioversity International, Montpellier, p 150

Pratt G (1862) A Samoan dictionary: English and Samoan, and Samoan and English; with a short grammar of the Samoan dialect. London Missionary Society's Press, London, UK

Rabone S (1845) A vocabulary of the Tongan language, arranged in alphabetical order: to which is annexed a list of idiomatic phrases. Wesleyan Mission Press

Ramanatha Rao V, Hodgkin T, Bourdeix R (2005) Locating coconut genetic diversity, pp 13–31. In: Batugal P, Ramanatha Rao V (eds) Coconut genetic resources. IPGRI-APO, Serdang, p 779. http://www.cogentnetwork.org/index.php?page=books

Randles JW, Rodriguez MJB (2003) Coconut Cadang-Cadang viroid. In: Hadidi A, Flores R, Randles JW, Semancik JS (eds) Viroids, 1st edn. CSIRO Publishing, Collingwood, pp 233–241

Romney DH (1972) Past studies on and present status of lethal yellowing disease of coconuts. PANS Pest Articles News Summ 18(4):386–395

Ruas M, Hamelin C, Bourdeix R (2018) 2.4.2 managing international coconut databases – chapter 2. Where we are today. In: Bourdeix R, Prades A (eds) A global strategy for the conservation and use of coconut genetic resources 2018–2028. Bioversity International, Montpellier, pp 69–71

Sauer JD (1967) Plants and man on the Seychelles coast: a study in historical biogeography. Plants and man on the Seychelles coast: a study in historical biogeography. The University of Wisconsin Press, Madison, Milwaukee, p 132

Sauer JD (1971) A reevaluation of the coconut as an indicator of human dispersal. In: Riley CL, Kelley JC, Campbell WP, Rands RL (eds) Man across the sea. Problems of pre-Columbian contacts. University of Texas Press, Austin, pp 309–319

Schuiling M, Kaiza DA, Mpunami A (1992) Lethal disease of coconut palm in Tanzania. II. History, distribution and epidemiology. Oleagineux 47:516. (No. A-)

Whitehead RA (1968) Selecting and breeding coconut palms (*Cocos nucifera* L.) resistant to lethal yellowing disease. A review of recent work in Jamaica. Euphytica 17(1):81–101

Zizumbo-Villarreal D, Colunga-GarcíaMarín P (2001) Morphophysiological variation and phenotypic plasticity in Mexican populations of coconut (*Cocos nucifera* L). Genet Resour Crop Evol 48:547–554

Zizumbo-Villarreal D, Hernández F, Harries HC (1993) Coconut varieties in Mexico. Econ Bot 47:65–78

Zizumbo-Villarreal D, Colunga-GarcíaMarín P, Fernández-Barrera M et al (2008) Mortality of Mexican coconut germplasm due to lethal yellowing. Bulletin de Ressources Phytogénétiques, p 23

Chapter 6
Diversity Studies Using Molecular Markers

Chandrika Perera, H. D. Dharshani Bandupriya, Regi J. Thomas, and Roland Bourdeix

6.1 Introduction

Molecular markers can directly identify variation within organisms at the nucleic acid level, accurately distinguishing genetic variation from environmental variation. The discovery in the 1980s followed by the application of molecular technology in plants was a benchmark in breeding research in many of the important crop plants worldwide. While the initial applications of deoxyribonucleic acid (DNA) molecular markers were in annual plants, they also provided great potential to be used in perennial and challenging-to-breed plants including coconut. Conventional approaches of phenotypic evaluation of coconut germplasm and coconut breeding have resulted in achieving the objectives of such programmes to a considerable level. However, the use of molecular markers would be highly attractive and effective to overcome the inherent constraints of conventional approaches due to comparatively longer duration of these approaches in coconut. As is the case with many other crop plants, the molecular markers provide accurate and reliable approaches for evaluating the genetic diversity of coconut, as well as in the development of genetic linkage maps for marker-assisted selection (MAS). The application of molecular markers in breeding research of coconut began in the 1990s. This chapter

C. Perera (✉)
Faculty of Agriculture, University of Peradeniya, Peradeniya, Sri Lanka

H. D. D. Bandupriya
Department of Plant Sciences, University of Colombo, Colombo, Sri Lanka

R. J. Thomas
The Central Plantation Crops Research Institute, Kasargod, Kerala, India

R. Bourdeix
CIRAD – UMR AGAP, CIRAD (Agricultural Research for Development), Montpellier, France

AGAP, University of Montpellier, CIRAD, INRA, Montpellier SupAgro, Montpellier, France

S. Adkins et al. (eds.), *Coconut Biotechnology: Towards the Sustainability of the 'Tree of Life'*, https://doi.org/10.1007/978-3-030-44988-9_6

discusses the use of molecular markers in genetic diversity analysis, approaches to MAS made so far, and the future potential for wider applications in coconut.

6.2 Genetic Resources of Coconut

The coconut (*Cocos nucifera* L.) belongs to the monocotyledon family Arecaceae. The coconut genome is large representing 32 diploid chromosomes (2n = 2x = 32). The coconut palm does not record any closely related species or wild relatives; there are only different types of coconuts believed to be at different evolutionary and domestication stages.

6.2.1 Classification of Coconut Genetic Resources

Coconuts are broadly classified into two groups, Tall (*typica*) and Dwarf (*nana*). There is a high phenotypic variation in coconut, and non-standard names are currently used to identify different coconut populations. This causes a lot of difficulties in developing a standard and formal systematic classification for coconut germplasm in the world. According to Menon and Pandalai (1958), the term "variety" denotes a single or a group of strains which differ from the other groups in structure or function and have the ability of true-to-type reproduction. However, using this definition, many reported types of coconuts cannot be considered as varieties. There are many different forms or types of coconut reported in the literature, but an acceptable and standard classification that describes the world coconut germplasm is yet to be formulated. Currently, different countries adopt different classifications to describe the genetic resources available in each part of the world.

The first classification of global coconut germplasm was recorded by Narayana and John (1949). In this classification, Tall and Dwarf coconuts were identified as distinct types of coconut. Both these types were divided into varieties based on the botanical features. Accordingly, Tall coconuts recorded three varieties, namely, *typica*, *spicata*, and *androgena*, while the Dwarf coconuts were divided into two varieties: *nana* and *javanica*. Of the Tall and Dwarf types, Tall coconut was the most abundant type throughout the world and the type that was planted on a commercial scale. In contrast, the Dwarf coconut type was less common and not grown on commercial scale. The variety *nana* of the Dwarf type was less vigorous but was reported to be early bearing reaching the reproductive phase approximately 3 years after planting, and the second variety *javanica* was more vigorous and comparatively late bearing taking about 5–6 years for floral initiation.

Based on a literature survey conducted by Gangolly et al. (1957), a coconut classification was presented by Menon and Pandalai (1958). This classification was like the previous classification by Narayana and John in identifying the two main types of coconut as Tall and Dwarf. In the same year, Liyanage (1958) reported a

Table 6.1 General characteristics defining Tall, Dwarf, and Intermediate types of coconut

Feature	Tall	Dwarf	Intermediate
Stature	Tall and hardy	Short	Intermediate to tall and Dwarf
Crown	Large, long fronds	Small, shorter fronds	Intermediate in size
Root bole	Present	Absent	Absent
Fruit size	Large	Small	Small
Fruit colour	Green to reddish brown	Green, orange, yellow, or brown	Mostly yellow to brownish colour
Pollination behaviour	Out-breeding	In-breeding	In-breeding

classification for Sri Lankan coconut germplasm, proposing three varieties, namely, *typica* (Tall), *nana* (Dwarf), and a new variety *aurantiaca*. The new coconut variety *aurantiaca* included coconuts of intermediate height between the Tall and Dwarf types (Table 6.1).

A subsequent coconut classification was reported based on the pollination or breeding behaviour of coconut (Fremond et al. 1966). As per this classification, Tall and Dwarf coconuts, which were previously categorized as types, were named as two distinct coconut varieties: Tall characterized by cross-pollinating breeding behaviour, resulting in allogamous populations, and the Dwarf by self-pollinating breeding behaviour and autogamous populations. However, the Tall coconut variety is known to display low levels of self-pollination via inter-spadix pollination during favourable time periods for inflorescence emission. In addition, the Dwarf coconut variety naturally displays some degrees of cross-pollination when they are present in the vicinity of Tall coconuts (Whitehead 1976).

In 1978, Harries proposed a different classification, identifying two types of coconuts as "*Niu kafa* type" and "*Niu vai* type". The fruits of *Niu kafa* type were large with a thick husk and a low volume of water. They were long and angular in shape and were slow to germinate. In contrast, the fruits of '*Niu vai* type' were more spherical shaped with a higher volume of endosperm and less husk content. They also were comparatively early germinating and were recorded as the more domesticated type of coconut that were selected and cultivated. Harries (1978) further suggested that the widely spread pan tropical coconut populations of today resulted from the introgression of the *Niu kafa*- and *Niu vai*-type coconuts and the selection and dissemination of them by humans (Harries 1978).

The identification of two main types of coconut is a common feature of the majority of coconut classifications to date. Each of these two groups displays certain phenotypic variations within the main type. In Tall coconuts, the size and shape of the fruit and shell and kernel thickness contribute mostly to this variation, Sri Lanka Tall Bodiri and Laccadive Micro Tall being examples. For Dwarf coconuts, the main feature contributing to within-type variation is the fruit colour which is indicated in the standard nomenclature as Sri Lanka Yellow Dwarf, Chowghat Green Dwarf, Cameroon Red Dwarf, etc. Such phenotypic variations within the type were referred to as forms within each variety by Liyanage (1958) and as phenotypically distinct variants by Bourdeix et al. (2005).

World coconut classifications do not include some of the different phenotypes present in certain countries and are showing semi-Tall or semi-Dwarf stature, Gangabondam in India and Niu Leka Dwarf in Fiji being examples. Therefore, there is an important need for an extensive evaluation and characterization of different types/forms of coconut grown worldwide in order to develop a standardized classification. A comprehensive classification should include a standard international nomenclature, local or vernacular names, and important specific descriptors for each coconut accession at the entry of them into a genebank for better utilization of coconut germplasm in breeding programmes.

6.2.2 Domestication of Coconut

Gunn et al. (2011) in the most extensive investigation to date reported the domestication history and the population structure of global coconut germplasm by analysing 1,322 palms spanning the geographical and phenotypic diversity of coconut at 10 microsatellite marker loci. The team reported two genetically distinct subpopulations relating them to Pacific and Indo-Atlantic Ocean basins and, accordingly, suggested two independent origins in the two regions; the islands in Southeast Asia and southern regions of the Indian subcontinent with the same population structure maintained to date despite the long-term mediation by humans in cultivation and dispersion. The selection by humans favoured the *Niu vai* type: more round-shaped nuts described by Harries (1978), Dwarf growth habit, and self-pollinating breeding behaviour. Gunn et al. (2011) located coconuts with admixture between the Pacific group and the Indo-Atlantic group in the south-western Indian Ocean and further revealed substructure within the Pacific coconuts. Further, data reveals that the original coconut gene pools in the South Asia and Atlantic oceans consist only of Tall-type coconuts. In contrast to the origin and domestication of Tall-type coconuts, the Dwarf coconuts have been found to be evolved from the cross-pollinating Tall coconuts in Southeast Asia. Dwarf coconuts have acquired self-pollinating breeding behaviour, but evolution from the cross-pollinating Talls and subsequent mutations have resulted in a considerable level of distinguishable phenotypic diversity in Dwarf coconuts (Coppens et al. 2018).

A study by Gunn et al. (2018) reports the distribution of coconut dating back to 28–44 million years ago, during the Eocene to Oligocene era, but that the distribution was limited to Southeast Asia and Oceania. The South Indian subcontinent coconuts came into being only 10,000 years ago when agriculture started, followed by domestication through cultivation. Based on archaeological studies, Summerhayes (2018) stated that the evidence from DNA data on the domestication of coconut may have the omission of 'equifinality theory' which states that similar results can be arrived at by two different processes. The author reveals that the coconuts in South Asia date back only to the mid-Holocene era, i.e. about 6,000 years ago, while coconuts were reported to be present in the Western Pacific region 20,000 years ago. Accordingly, combining DNA evidence with archaeological and palynological data, Summerhayes states that the coconut domestication has taken place in the Pacific followed by subsequent events of admixture elsewhere in the world.

6.3 Conservation of Coconut Genetic Resources

Globally, coconut genetic resources are endangered due to a relatively high rate of genetic erosion caused by several factors including urbanization and industrialization, infrastructure development, a shift from coconut cultivation to high-value cash crops, natural disasters and biotic stresses, and pests and diseases. In addition, the wide-scale replanting of coconuts from relatively few numbers of high-yielding improved cultivars, especially in areas of natural coconut stands where the genetic diversity is high, results in the dwindling of the genetic diversity of coconut germplasm. Consequently, a few decades ago, the identification, collection, and conservation of coconut genetic resources were recognized as an important objective in individual coconut-growing countries and later expanded to be a phenomenon of international interest, mainly because the future of coconut breeding would be based on the availability of diverse germplasm. Accordingly, in 1992, the international agency Coconut Genetic Resources Network (COGENT) was founded, under the umbrella of the International Plant Genetic Resources Institute (IPGRI, now renamed as Bioversity International). Strengthening and standardizing the individual country programmes and bringing the operations under an international forum for knowledge and material sharing were the main objectives of COGENT (http://www.cogentnetwork.org). In the year 1997, the COGENT initiated a systematic programme for collection and conservation of coconut genetic resources, which was funded by the Asian Development Bank.

6.3.1 Nomenclature of Accessions Conserved in Genebanks

In order to standardize and internationalize the coconut germplasm collection and conservation, the COGENT initiated the international Coconut Genetic Resources Database (CGRD) for the computerized cataloguing of the global coconut germplasm by facilitating the uploading of the data of national germplasm repositories by individual countries. Each conserved coconut accession, in the CGRD, is identified with a unique international name, comprising of the type (whether Tall or Dwarf), the geographical reference, and in the case of a Dwarf the respective colour form. West Coast Tall, Sri Lanka Tall, East Coast Tall, Malayan Yellow Dwarf, and Brazilian Green Dwarf are a few examples for this nomenclature in CGRD. In addition to this information, in the CGRD, accessions are assigned a more specific geographic location within each country in the naming of accessions conserved in the international field genebanks. Examples are San Ramon Tall in the Philippines and Sri Lanka Tall Ambakelle. Further studies will be essential to determine whether there is actual genetic variation among these "ecotypes" as they are referred to.

The standard data recording process of CGRD is aimed at enhancing the awareness and knowledge of global coconut germplasm for exchange of germplasm among countries for enriching the national coconut breeding programmes as required (Hamelin et al. 2005). The CGRD database facilitates the compilation of passport, characterization, and evaluation data for the accessions recorded in it. Each individual country was to provide the data on collected accessions, the passport data, and to proceed with the collection of characterization and evaluation data to be fed into the CGRD at each stage of characterization.

6.4 Evaluation of Coconut Genetic Diversity

The coconut germplasm collection and conservation programme resulted in the initiation of the ex situ field genebanks of coconuts as national germplasm repositories. Those countries which had conserved their limited germplasm collections used this opportunity to expand the collections and establish new ex situ field genebanks in a systematic manner. This programme further resulted in several international ex situ field genebanks, in addition to the national ex situ field genebanks of coconut. The value from conservation was to be the characterization of the conserved material to facilitate the utilization in coconut breeding programmes. However, to date, the conserved germplasm in these ex situ field genebanks has not been systematically characterized, apart from several scattered studies intended either to update the CGRD or to find desirable parental material for genetic improvement programmes indicating gaps in all aspects of characterization: morphological, molecular, and biochemical assays.

6.5 Morphological Assessment of Coconut Genetic Diversity

The assessment of genetic diversity of species was entirely dependent on the morphological assays during the early period of the nineteenth century. Scoring for morphological markers is less technology-demanding than molecular markers yet suffers from several drawbacks. The limited number of morphological markers, the manifestation of the environmental effects on the phenotype masking the genetic variation, and the growth stage-dependent expression of morphologies are some examples for such drawbacks of morphological markers. Despite the said disadvantages, morphological characterization, if properly designed to assess for the environmental effects, reveals the overall manifestation of the genetic potential of the plant, and accordingly, several studies have been carried out for assessing the genetic diversity of coconut by morphological means.

The availability of a comprehensive and well-defined descriptor list is a prerequisite for the process of morphological characterization. Bioversity International has coordinated the preparation of descriptor lists for important crop species in the world including coconut. These descriptor lists describe each accession of a crop in four basic data categories: passport, characterization, preliminary evaluation, and further characterization and evaluation.

Passport data are the most preliminary data to be collected of a germplasm recorded at the same time the collection is made and at the site of collection. This preliminary description is important for the identification and standard nomenclature of a germplasm accession. The site of collection (recorded as village/state/country), location in terms of longitude and latitude, collector's number, date the collection was made, botanical and vernacular names, sample type (whether wild/weedy/landrace/cultivar, etc.), source (field/farm store/institute, etc.), and the site environmental characteristics are some examples for passport data parameters. Currently ethnobotanical information and digitalized location identifying, global positioning system (GPS), and the management of data via advanced databases have been introduced with the extensive adoption of novel technology.

Characterization data consists of the characters found to be highly heritable and can be easily distinguishable. These are qualitative traits which are stable across environments and are governed by one or a few major genes. Fruit shape and fruit colour are examples for such traits in coconut. During preliminary evaluation, a limited number of additional traits, which are useful to identify the specific germplasm, are scored. The scoring for evaluation data is complex because they involve the traits which are of quantitative nature and are influenced by the environment. Information on stem, leaf, inflorescence and flower, fruit, and seeds are the evaluation data of the germplasm. The use of adequate sample sizes and proper experimental designs are important criteria for accurate assessment of the genetic diversity for quantitative traits. Further characterization and evaluation of germplasm include the scoring for potential agronomic characters which are useful for crop improvement: physiology, pathology, entomology, cytogenetics, biochemistry, and recently molecular information. Internationally accepted norms are followed in scoring, coding, and recording of each of the descriptor states.

6.5.1 Descriptors for Morphological Characterization

Several research teams have published on the general morphology of the coconut palm in different geographic locations and reported a high phenotypic variation within as well as among populations for fruit traits. Accordingly, variations in fruit size, fruit shape, colour of the epicarp of the fruit, and proportional weight of various fruit components such as husk, shell, kernel (the solid endosperm), and water (the liquid endosperm variations) have been highlighted. A survey of the diversity of fruit components of coconuts in the South Pacific revealed a high diversity among populations displaying both wild-type characters and more domesticated traits and a range of mixed phenotypes of the two extremes (Ashburner et al. 1997a). The results of a study by Foale (1991) revealed different shapes, varying from angular to pear or near-spherical shapes (Figs. 6.1, 6.2, and 6.3) and from shorter to longer fruits of coconuts. Furthermore, higher morphological diversity has been observed in coconuts in Southeast Asia compared to those of South Asia, Africa, and South America (Benbadis 1992; Whitehead 1976).

Zizumbo-Villarreal (2005) studied the patterns of variations of coconut in Mexico using fruit morphological descriptors and reported three main Tall-type groups originating from different historical introductions. Several studies on the systematic characterization of coconut using standard descriptor lists have also been reported. Morphological diversity related to stem, leaf, inflorescence, and fruit morphology of Sri Lankan indigenous coconut varieties was studied (Perera and Ekanayake 2008), and the results revealed high levels of variation among the varieties examined. Further, Perera et al. (2009) characterized 24 germplasm accessions conserved in the ex situ field genebanks of Sri Lanka using morphological descriptors and recorded higher variations in Tall coconuts than in Dwarf accessions. Despite these attempts, comprehensive analysis of conserved coconut accessions for all the morphologies – stem, leaf, inflorescence, and fruit diversity – using morphological descriptors remains scarce due to various constraints and difficulties. However, attempts in the use of molecular markers for assessing the genetic diversity of coconut have been on the rise for the last two decades.

Fig. 6.1 Crown of a Tall-type coconut from South Asia having elongated fruit

Fig. 6.2 Crown of a Tall-type coconut palm from Southeast Asia and the Pacific region having large spherical-shaped fruit

Fig. 6.3 Crown of a Dwarf-type (yellow) coconut palm having smaller fruit

6.6 Molecular Approaches for Assessing Coconut Genetic Diversity

The use of molecular marker technology in coconut research was initiated in the mid-1990s and has continued to advance. Molecular markers have been applied in coconut for elucidating the genetic diversity of coconut germplasm, developing linkage maps, detection of intracellular pathogens, DNA finger printing of coconut, and hybridity testing. The initial attempts were on diversity analysis of coconut germplasm mainly with restriction fragment length polymorphism (RFLP), randomly amplified polymorphic DNA (RAPD), and amplified fragment length polymorphism (AFLP) markers, with the current trend of increased use of simple sequence repeats (SSR).

6.6.1 Early Studies of Molecular Characterization of Coconut

One of the preliminary studies using molecular markers in assessing the genetic diversity of coconut has been reported in 1995. In this study, 17 varieties of coconut, representing different geographical regions, were evaluated through polymerase chain reaction (PCR) amplification of *copia*-like 16R1 repetitive elements (Rohde et al. 1995). The results of this study provided molecular evidence for the genetic similarity of Tall coconuts from East Africa to coconuts from the Indian Ocean and the grouping of the West Coastal Panama Tall coconuts with coconuts from the Pacific and the Southeast Asia.

In another preliminary research study, utilizing molecular markers in assessing the genetic diversity of coconut was the use of RFLP markers to study a pool of coconut palms representing different geographical zones (Lebrun et al. 1998). This research was carried out by hybridizing 9 cDNA clones and a mitochondrial DNA clone developed from rice and 1 ribosomal DNA clone derived from wheat with the DNA extracted from a coconut palm pool of 100 individuals of 10 Tall coconuts and 7 Dwarf coconuts. The results provided evidence for two groups of coconut representing the geographical regions: the first being the coconuts from Far East and the Pacific and the second representing coconuts in the South Indian subcontinent – India, Sri Lanka, and East Africa. In addition, a comparatively high level of diversity was observed in the coconuts from the Far East and the Pacific compared to the rest. Furthermore, the coconut variety Panama Tall clustered together with coconut varieties representing the Pacific region, while the African variety, West African Tall, grouped with the coconuts from South Asia.

In yet another earlier study, 17 coconut populations from the South Pacific region were subjected to RAPD analysis, and the results reported a within-population diversity of about 60% which is a considerably high value (Ashburner et al. 1997b). The results from this study further illustrated the occurrence of two geographic groups in addition to distinguishing two single populations in the 17 populations studied. The study thus revealed a low rate of gene migration among the coconut populations in the said region, with the possibility of founder effects, followed by selection during the coconut domestication in the Pacific. The team further recommended the germplasm collection programmes in the region focus on populations and not the individual palms to capture the high genetic variation that was observed among populations.

Other research teams reported diversity studies on coconut in respective local germplasm, using RAPD. Accordingly, the conserved coconut germplasm in Sri Lanka was analysed using RAPD markers, and the results indicated a rather narrow genetic base (Dasanayaka et al. 2003; Everard 1996). Duran et al. (1997) studied 48 East African Tall (EAT) coconut genotypes using 22 RAPD primers. A total of 238 amplicons were produced out of which 204 (86%) were polymorphic and others were monomorphic. The dendrogram constructed, based on the RAPD analysis of 48 coconut accessions, clustered into two major groups recording an average dissimilarity index of 0.35. Upadhyay et al. (2002) analysed 14 coconut accessions of indigenous and exotic origin, including Tall-, Dwarf-, and Intermediate-type coconuts, conserved ex situ in the Coconut Germplasm Centre in Kerala, India. The genetic relationships among the tested coconut accessions revealed that the Intermediate coconuts were genetically closer to Dwarf coconuts than the Tall coconuts.

Ratnambal et al. (2001) used RAPD markers to screen the marker polymorphism to identify a set of informative markers and to characterize coconut germplasm in India. Hundreds of primers were screened in this study to detect polymorphism and allelic diversity in coconut. Among these, only 34% primers were polymorphic recording 1–16 polymorphic bands per RAPD primer. Daher et al. (2002) assessed the genetic divergence among 19 coconut populations by RAPD. The markers used permitted the identification of each of the populations, showing that they were genetically different revealing the lack of duplicates in the collection. The genetic diversity of 30 ex situ conserved coconut accessions representing the Pacific and Nicobar Islands was genotyped with RAPD markers (Sankaran et al. 2012). The resultant dendrogram from cluster analysis revealed two main clusters and the distinct genetic variation among the accessions. The informativeness of the tested 13 RAPD primers was also revealed recording an average polymorphic information content (PIC) value of 0.29, with a range of PIC values from 0.46 to 0.17, for primers OPF-19 and OPH-25, respectively (Sankaran et al. 2012).

Jayalekshmy and SreeRangasamy (2002) reported that intervarietal variation could be detected by RAPD markers. Out of the ten primers used, seven gave amplification products showing polymorphism between Tall and Dwarf varieties. They also reported that RAPD markers appeared to be of high value for characterizing the genetic resources. The use of AFLP markers in the diversity analysis of coconut was reported by Perera et al. (1998) and Teulat et al. (2000), and during the same period, diversity studies using inverse sequence-tagged repeat (ISTR) markers were reported by Duran et al. (1997) and Rohde et al. (2000).

6.6.2 Development and Use of SSR Markers in Diversity Analysis of Coconut

The need for the use of coconut-specific molecular markers was later highlighted to expand and enhance the molecular marker applications in coconut. Accordingly, two groups (Perera et al. 1999; Rivera et al. 1999) successfully developed simple sequence repeat (SSR) or microsatellite markers in 1999, using Sri Lanka Tall and Tagnanan Tall, respectively, as the base genetic material. These SSR markers are co-dominant and thus have been immensely useful in the evaluation of the genetic diversity of coconuts. The use of SSR markers was highly advantageous to genotype coconuts, due to the highly heterozygous nature resulting from out-breeding pollination behaviour. Accordingly, the use of SSR markers was effective in providing clear information on genetic relationships and identification of representative collections (Dasanayaka et al. 2003; Perera et al. 2000, 2001, 2003; Meerow et al. 2003; Teulat et al. 2000), hybridity testing (Perera et al. 2004), identifying somaclonal variations in tissue-cultured coconuts, and developing linkage maps of coconut (Baudouin et al. 2006; Herran et al. 2000; Lebrun et al. 2001). Standardization of the techniques is a must for facilitating the comparison of results derived across laboratories in different countries. In order to fulfil this requirement, an SSR marker kit, comprising of markers at 14 microsatellite loci, was developed and relevant software for the analysis of data was introduced (Baudouin and Lebrun 2002).

The use of SSR markers has been the most common marker system used to date, for the evaluation of the coconut germplasm. The genetic diversity of coconut was evaluated at 8 SSR marker loci, by analysing 130 individuals representing 75 Tall and 55 Dwarf coconut ecotypes spanning the coconut-growing areas in the world (Perera 1999). The results revealed higher allele richness in Tall coconuts compared to Dwarf coconuts, thus indicating the comparatively low genetic diversity in the Dwarfs. Simultaneous studies including 20 coconut varieties from the southern coast of Asia and the Pacific and 31 varieties from the same region (Rivera et al. 1999; Teulat et al. 2000) reported results which agree with that of Perera (1999). Thirty-three Sri Lankan Tall coconut populations collected across the coconut-growing areas in the country were subjected to SSR analysis (Perera et al. 2001), and the results provided evidence for lack of population differentiation in Sri Lanka Tall coconuts.

Microsatellite marker analysis of a collection of 179 global coconut accessions, displayed comparatively higher genetic variation and a heterozygosity value of 30% in the naturally cross-pollinating Tall coconuts, compared to low genetic variation and a low heterozygosity value of 2.5% in the naturally in-breeding Dwarf coconuts (Perera et al. 2000, 2001, 2003). Further studies revealed very low (5%) population differentiation in Tall coconuts from Sri Lanka (Perera et al. 2001), contrary to findings by Ashburner et al. (1997b), who reported higher genetic diversity between populations in Southeast Asia and Pacific coconut germplasm. Both studies reiterate the need to adjust conservation strategies accordingly. These findings have assisted in changing the strategies in the collection of genetic resources in different parts of the world and in addition provided insights into the genetic base of coconut. As an example, the revealing of the narrow genetic base of Sri Lankan coconuts led to the revision of breeding programmes by incorporating imported exotic material as parents in recent crossing projects. Molecular analysis of global coconut germplasm, represented by 51 Tall accessions and 49 Dwarf accessions, revealed the presence of 2 major groups of Tall coconuts: the first being the Tall coconuts from Southeast Asia and the Pacific including Panama West Coastal Region Talls and the second group comprising of Tall coconuts from South Asia and East and West Africa. The entirety of the global Dwarf coconut germplasm grouped into a sub-cluster within the main cluster of coconuts from the Southeast Asia and the Pacific (Perera et al. 2003).

Information derived from microsatellite markers on the genetic variation in coconut germplasm agree with the findings of other molecular marker techniques ISTR (Rohde et al. 1995) and RFLP (Lebrun et al. 1998). The combined results of several of these studies revealed the presence of two main groups of coconut: the first being the Southeast Asian and the Pacific island coconuts and the second being the Indo-Atlantic coconuts. As per the molecular evidence, derived from the studies described so far, Dwarf coconuts in the world are classified as a subgroup within the main group of Tall coconuts from the Southeast Asia and the Pacific. In addition, a reduction of allelic diversity is revealed among the Dwarf coconut accessions, indicating the evolution of the Dwarf group from within the Tall coconuts from the Southeast Asia and the Pacific, supporting the findings of Teulat et al. (2000), indicating a common origin for global Dwarf coconut germplasm.

Diversity of the coconut chloroplast genome has been studied by the polymerase chain reaction (PCR) amplification of chloroplast DNA of a sample of 130 individual palms of global coconut germplasm, with the objective of identifying genetic lineages in coconut (Perera 1999, 2002). These studies revealed the lack of chloroplast variation of the tested samples, providing evidence for close ancestry within the tested coconut germplasm. Later, moving towards high-throughput genotyping of coconut germplasm, the high-throughput marker system Diversity Arrays Technology (DArT) was validated for coconut, elucidating the genetic diversity of a collection of Sri Lankan coconut germplasm (Perera and Kilian 2008).

6.6.3 Use of Molecular Markers to Screen National-Level Repositories of Coconut Germplasm

Since early studies on the development of basic theories and preliminary screenings, many countries and different research groups have attempted diversity analysis of coconut using molecular markers in their respective genebanks/countries for different purposes. The findings of such investigations are being used for better understanding of the true genetic diversity among the genetic resources conserved in in situ and ex situ field genebanks. Below are some examples of such studies.

Simple sequence repeat markers have been used to study the tolerance/susceptibility of coconut accessions of West African Tall, to lethal yellowing (Konan et al. 2007a, b). The analysis of data at 12 SSR marker loci revealed the clustering patterns of the susceptible accessions from that of the tolerant accessions. The results further indicated that the two groups were genetically distant from each other, as determined by the specific alleles and the frequency variation of shared alleles of the accessions. The same genotypic profiles were used to determine the genetic diversity of the accessions using the 58 alleles that resulted recording an average of 4.8 alleles per SSR marker locus. This study demonstrated the feasibility of large-scale molecular screening of coconut accessions for desirable traits, in selecting of parents for breeding programmes, and for developing mapping populations for tagging genes for lethal yellowing.

A total of 30 coconut samples comprising of 12 accessions from China and 18 accessions from Southeast Asia were evaluated using 30 new microsatellite markers developed with the sequence data generated from an Illumina transcriptome profile (Yong et al. 2013). The results displayed variable levels of allelic polymorphism among the accessions. The analysis of population structure with the same genotypic profiles revealed the Chinese accessions to be a subset of the Southeast Asian coconut accessions. Accordingly, the authors suggested that the evolution of the Chinese accessions had not been independent from that of the tested accessions in Southeast Asia. With the combined results of the population structure analysis and historical evidence, it was concluded that the dissemination of coconuts to the Hainan Province of China occurred along the sea currents and human-mediated dispersal was responsible for coconuts moving from Southeast Asia to the Yunnan Province in China.

Investigations were carried out to elucidate the genetic diversity of coconut accessions using the PCR-based molecular marker systems RAPD, inter-simple sequence repeat (ISSR), and SSR (Kandoliya et al. 2018), with a total of 45 markers representing 15 markers each from the above marker systems. The three marker systems, RAPD, ISSR, and SSR, generated 82, 82, and 28 bands, respectively. All the marker systems recorded higher percentages of polymorphic alleles while recording a few unique bands as well. In addition, the values of similarity coefficient of clusters of the marker systems recorded values ranging from 22 to 83% for RAPD, 26 to 86% for ISSR, and 50 to 97% for SSR. Accordingly, it was concluded that the performance of the three molecular marker systems was comparable and equally reliable in assessing the genetic diversity of coconut genotypes.

The SSR marker technology was used to evaluate the genetic diversity among 48 individual coconut palms collected from the lowland coastal belt in Kenya (Oyoo et al. 2016). The information derived from the 15 SSR marker loci, analysed with Popgene version 1.31, revealed genetic diversities ranging from 0.0408 for marker locus CAC68 to 0.4861 for marker locus CAC23, recording a mean of 0.2839. The investigation further revealed the marker loci which are polymorphic for Kenyan germplasm and more importantly revealed a high within-population variation of 28%, compared to a low variation of only 2% recorded between populations, suggesting the non-dependence of molecular variation in the region they are cultivated.

The genetic diversity of 14 coconut accessions was evaluated at 8 SSR marker loci (Pradeepkumar et al. 2011). The accessions grouped into three clusters in the dendrogram. Cluster 1 consisted of five accessions from New Guinea, Cluster 2 of four from French Polynesia, and Cluster 3 of five accessions from the South Pacific, giving evidence for the geographical variations. The effect of controlled pollination on the maintenance of the levels of genetic diversity upon was investigated in three Tall coconut accessions, namely, Mozambique Tall (MZT), Gazelle Peninsula Tall (GPT), and Tahitian Tall (THT) (Yao et al. 2013). The genotypic analysis of the parents (G0) and the progeny derived via controlled pollination (G1) at 15 SSR loci revealed a slight reduction of gene diversity, varying from 0.69 to 0.587, low values of Jaccard dissimilarity index varying from 0.072 to 0.133, and low levels of genetic diversity ranging from 0.005 to 0.007 between the parental and regenerated populations. Accordingly, it was concluded that the genetic integrity of the original accessions conserved in field genebanks can be maintained satisfactorily by controlled pollination in the rejuvenation process.

The genetic diversity within and between populations of Brazilian Tall coconuts, as represented by 195 palms belonging to 10 populations, was studied at 13 SSR marker loci (Ribeiro et al. 2010). The results revealed 68 alleles, averaging 5.23 alleles across populations, varying from 2 to 13 per locus and mean gene diversity (He) and observed heterozygosity (Ho) values of 0.459 and 0.443, respectively. The among population genetic distances varied from 0.034 to 0.390, and the results provided molecular evidence for the presence of two groups, the first comprising of the Baía Formosa, Georgino Avelino, and São José do Mipibu populations and the second consisting of the Japoatã, Pacatuba, and Praia do Forte populations. The comprehensive analysis of data indicated spatial genetic structuring of populations, by geographically close populations displaying higher genetic similarities.

SSR markers were utilized to evaluate the genetic variation of coconuts in the Andaman and Nicobar Islands of India (Rajesh et al. 2008). The results of this study revealed 7.35 alleles and an average heterozygosity of 0.29. A mean fixation index (FST) of 0.49 indicated a high level of population differentiation among the tested coconut accessions. The highest heterozygosity was observed in Tall coconut accessions, as expected, and most of the rare alleles were recorded in Tall coconuts sampled in the Nicobar Islands. Tall coconut ecotypes were reported to display greater heterozygosity values ranging from 0.18 to 0.37 in comparison with Dwarf coconuts, the values for which ranged from 0.03 to 0.07 (Thomas et al. 2013). In this study, a total of 90 sample trees, originating from 6 ecotypes, were evaluated, at 14 SSR marker loci. The differences in observed and expected heterozygosity in Tall ecotypes were an indication of the genetic basis of resistance to diseases, by the combined analysis of SSR marker data with the morphological data scored.

Studies have been conducted to determine the genetic variability among Tall coconut accessions conserved at the International Coconut Genebank for Latin America and the Caribbean using SSR markers (Loiola et al. 2016). The study revealed information for decision-making regarding the conservation of coconut germplasm and for the higher accuracy of selecting diverse parents to be utilized in crossing programmes, including selecting varieties for resistance to lethal yellowing. The molecular genetic diversity among 14 coconut accessions from India was determined at 8 SSR marker loci (Pradeepkumar et al. 2011). The eight markers produced a high level of polymorphism, with an average of 4.166 alleles per locus.

The dendrogram developed, based on the SSR analysis, separated the 14 coconut accessions into 3 major clusters. The smaller similarity coefficient value indicated the absence of similarity between the genotypes.

Rajesh et al. (2012) studied the genetic purity among coconut hybrids at 50 SSR marker loci. Chowghat Green Dwarf (CGD) and West Coast Tall (WCT) were used as parents for hybrid production. Among 50 SSR markers, 17 displayed the complementary banding patterns of both the parents, and the selfed progenies showed the banding patterns of only the mother palm. The study also revealed the importance of SSR markers in hybridity testing. Kriswiyanti et al. (2013) determined the genetic variation of Tall coconuts in Bali, based on the analysis at six SSR marker loci. In total, 80 alleles were identified with an average of 13.33 alleles per locus. The mean values of gene diversity and observed heterozygosity were 0.883 and 0.542, respectively. Gene diversity ranged from 0.85 to 0.92, with a mean of 0.88; the overall results explained a high genetic diversity among the Tall coconuts in Bali.

6.6.4 *Expansion of Molecular Diversity Studies for Specific Findings*

The early attempts on the use of molecular markers for diversity analysis only served the intended purpose. Subsequently, the molecular marker research was planned in such a way as to reveal additional information on the studied genetic resources. Accordingly, Kamaral et al. (2014) conducted a research study to characterize 15 Sri Lanka Yellow Dwarf (SLYD) coconut palms using 10 SSR markers. All 10 microsatellite primers produced polymorphic amplicons resulting in 34 alleles, scored in the 15 individuals of SLYD palms. A total of 22 heterozygous loci were identified with the results further revealing the existence of high genetic diversity within the SLYD coconut palms.

Shalini et al. (2007) conducted a study using 3 coconut populations with varied yield traits (high, medium, and low) at 32 SSR marker loci to determine the genetic variation. High- and medium-yielding populations showed maximum heterozygosity, which indicated that they are not undergoing a population expansion. But the low-yielding population exhibited a significant deficiency in genetic diversity, which indicated that they are undergoing a rapid population expansion. Further advancing the genetic diversity studies, the molecular marker technology was adopted for the determination of the population structure of coconut. The studies on determining the population structure and genetic diversity revealed relevant genetic information on Florida coconuts, with special reference to Fiji Dwarf coconut cultivars (Meerow et al. 2003).

Gunn et al. (2011) reported one of the most comprehensive studies on the domestication of coconut through the analysis of the population structure of 1,322 accessions, representing wide geographical and phenotypic diversity, using 10 microsatellite markers. The study concluded independent origins and persistent population structure of coconuts in the two main regions, despite the long-term cultivation of coconut and human-mediated dispersal. Genetic diversity and the population structure were studied in Sri Lankan Yellow Dwarf coconut phenotypes by Kamaral et al. (2016). In this experiment, the Yellow Dwarf coconut variety was purified from a mixture of phenotypes, and a novel semi-Tall self-pollinating coconut phenotype, termed Sri Lanka Yellow Semi Tall, was identified.

6.7 Genetic Linkage Mapping in Coconut

In addition to being used extensively in genetic diversity studies, molecular markers have been used in linkage and QTL mapping of coconut. Selection of or construction of a suitable mapping population is a crucial step in linkage mapping, while it is a must to have a dense coverage of molecular markers. A population for linkage mapping should be segregating for traits to be mapped. Basic segregating populations, such as the early filial generations and backcrosses or the advanced segregating populations including recombinant inbred lines (RILs) or doubled haploid lines (DHLs), are being used for linkage mapping for self-pollinated crops. The development of suitable mapping populations is a highly challenging task in coconut, hindering the process and resulting in linkage maps lacking the power for fine location of gene/QTL.

Accordingly, the information generation in QTL mapping in an outbred species, including coconut, is concentrated on the available pedigree populations. However, in coconut, the family sizes are limited in numbers, preventing the formation of a sufficiently large mapping population. Formation of a large number of families, followed by the analysis of data to model the genic inheritance of multiple pedigree populations with appropriate statistical software, would be a solution to this problem (Kearsey and Luo 2003). The naturally outbreeding nature of Tall coconuts, resulting in heterozygous individuals, and the inherently inbreeding characteristic of Dwarf coconuts are a specific feature in coconut, facilitating the development of a segregating population by crossing the Tall with Dwarf coconuts (Bandaranayake 2006). Yet, for the resulting population to possess enough levels of segregation, the selected Tall parent should possess heterozygosity and polymorphism at important loci (Perera 2010).

The success of fruit setting upon artificial hand pollination of coconut is generally low, resulting in low numbers of seeds and progeny from a single mother tree. Due to this, it takes a considerable period to produce a population of reasonable size, inducing a long age gap among the progeny. This limitation can be overcome by the combination of progeny from several half sib families of a short age gap, in a single mapping population. A practical approach to develop such a mapping population would be to select a highly heterozygous male parent and pollinate enough Dwarf female parents using the pollen of the selected Tall male parent (Perera 2010).

6.7.1 Linkage and QTL Maps of Coconut for Marker-Assisted Selection

Despite the above-mentioned difficulties, several research groups have developed linkage maps for coconut. The first genome map of coconut was constructed with an F_1 population of a cross between East African Tall and Laguna Tall (Rohde et al. 1999). The genotypic data for this map was derived with ISTR markers. A second genome map of coconut was produced in the Philippines, using a mapping population developed by crossing Malayan Yellow Dwarf with Laguna Tall. The molecular markers – AFLP, ISTR, RAPD, and ISSR – were utilized to derive the genotypic data for this linkage framework map, which positioned 382 makers covering the 16 linkage groups of coconut. The QTL map of the second mapping population was successful in identifying six QTLs governing early germination (Herran et al. 2000), providing the opportunity for marker-assisted selection in coconut. The second QTL map was expanded to include QTL governing vegetative traits leaf production and girth (Ritter et al. 2000). A subsequent mapping population constructed in the Ivory Coast, using Cameroon Red Dwarf and Rennell Island Tall as parents, resulted in a framework map anchoring 280 markers, in addition to identifying QTL for yield traits: numbers of nuts, bunches, and fruit components (Baudouin et al. 2006; Lebrun et al. 2001).

6.7.2 Improvements for Genetic Linkage Mapping in Coconut

Any mapping population should be developed using phenotypically and genotypically segregating parents, and the population itself should comprise of enough individuals. In coconut, a simulation study revealed the optimum size of a linkage mapping population to be about 400 individuals.

Evaluation of global coconut germplasm has resulted in the identification of two main groups of coconut: Southeast Asia and the Pacific group being the first and the South Asian and Atlantic group being the second. Accordingly, it is expected that maximum segregation would be the result in a cross between these two groups, making a highly informative population for linkage mapping. However, the earlier-mentioned mapping populations were derived from the crossing of varieties included in the same group, resulting in uninformative non-segregating loci in mapping. Given that about 84% of DNA loci generated in the Malayan Yellow Dwarf and Laguna Tall mapping population is monomorphic provides an example, revealing identical alleles at many loci between parents.

The availability of polymorphic markers for a mapping population is essential for gene mapping in coconut for a dense marker coverage to saturate the 16 linkage groups of coconut. Currently this requirement can be fulfilled with the availability of a large collection of SSR markers and the possibility to move forward with high-throughput marker systems.

6.8 Conclusions

The availability of many polymorphic markers and the facilities for high-throughput genotyping and sequencing enables the creation of accurate and reliable information from research on diversity analysis and on population structure. The density of molecular marker coverage of the coconut genome is a crucial consideration in fine-scale characterization and evaluation of germplasm and genome mapping in coconut. The molecular marker kit developed in 2002 for the analysis of coconut comprises of 14 genomic SSR marker loci (Perera et al. 2018; Pokou et al. 2018). However, with the development of sequencing projects, which have become faster and more economical, research needs to be directed towards identifying a comprehensive set of more targeted loci representing the coconut genome. The representative marker loci should cover functional loci of important phenotypic and agronomic traits for molecular marker studies to be more effective and practically useful. Here again concerted efforts of the coconut molecular biologist are needed to decide the molecular marker system or combined systems to be used to develop high-throughput systems, which are affordable and feasible, to be used even by resource-poor laboratories in coconut-growing countries. However, such research should be coordinated at an international level to develop standard methodologies for the molecular characterization of coconut. The information thus derived will be more targeted and useful for formulating and refining further collection and conservation of coconut germplasm, management of genebanks, identification of duplicates, and determining the strategies for rejuvenation of the existing field genebanks. It could also enhance the utilization of genebank material by assisting in parental selection in coconut breeding programmes aimed at combining the desirable characters from diverse parents into novel cultivars. Such measures will bring the much-awaited benefits of molecular marker techniques to farmers and other stakeholders of the coconut value chain.

References

Ashburner GR, Thompson WK, Halloran GM et al (1997a) Fruit component analysis of south Pacific coconut palm populations. Genet Resour Crop Evol 44:327–355

Ashburner GR, Thompson WK, Halloran GM (1997b) RAPD analysis of South Pacific coconut palm populations. Crop Sci 37:992–997

Bandaranayake CK (2006) An effective population size for reliable map resolution of Coconut (*Cocos nucifera* L.). CORD 22(2):33–40

Baudouin L, Lebrun P (2002) The development of a microsatellite kit and dedicated software use with coconuts. Burotrop Bull 17:16–20

Baudouin L, Lebrun P, Konan JL et al (2006) QTL analysis of fruit components in the progeny of a Rennell Island Tall coconut (*Cocos nucifera* L.) individual. Theor Appl Genet 112:258–268

Benbadis BK (1992) Coconut and date palm. In: Hammerschlag FA, Litz RW (eds) Biotechnology of perennial fruit crops. CAB International, Wallingford, pp 383–400

Bourdeix R, Santos G, Labouisse JP et al (2005) Useful definition of terms and nomenclature. In: Batugal P, Ramanatha Rao V, Oliver J (eds) Coconut genetic resources. IPGRI, Rome, pp 9–10

Coppens G, Gunn B, Baudouin L (2018) Coconut domestication - Chapter 2. Where we are today. In R. Bourdeix & A. Prades (Eds.), A Global Strategy for the Conservation and Use of Coconut Genetic Resources 2018–2028 (pp 36–38) Montpellier, France. Bioversity International
Daher RF, Pereira MG, Tupinamba EA et al (2002) Assessment of coconut tree genetic divergence by compound sample RAPD marker analysis. Crop Breed Appl Biotechnol 3(2):431–438
Dasanayaka PN, Everard JMDT, Karunanayake EH et al (2003) Characterization of coconut germplasm by microsatellite markers. Trop Agric Res 15:51–61
Duran Y, Rohde W, Kullaya A et al (1997) Molecular analysis of east African Tall coconut genotypes by DNA marker technology. J Genet Breed 51:279–288
Everard JMDT (1996) Use of molecular markers for breeding of the coconut palm (*Cocos nucifera* L.). MSc. Thesis, University of New England, Armidale, Australia
Foale MA (1991) Coconut genetic diversity: present knowledge and future research needs. In: Papers of the IBPGR workshop on coconut genetic resources held in Cipanas, Indonesia, 8–11 October 1999. IBPGR, Rome, pp 46–53
Fremond Y, Ziller R, de Nuce de Lamothe M (1966) Le cocotier. Maisonneuve and Larose, Paris
Gangolly SR, Satyabalan K, Pandalai KM (1957) Varieties of coconut. Indian Cocon J 10:3–28
Gunn BF, Baudouin L, Olsen KM (2011) Independent origins of cultivated coconut (Cocos nucifera L.) in the old-world tropics. Plos One 6(6):e21143
Gunn B, Myrie WW, Baudouin L (2018) Origin, history and dynamics of coconut cultivation - Chapter 1. Introduction to the Global Coconut Strategy In R Bourdeix & A Prades (Eds.), A Global Strategy for the Conservation and Use of Coconut Genetic Resources 2018–2028 (pp 3–7). Montpellier, France. Bioversity International
Glenn R. Summerhayes (2018) Coconuts on the Move: Archaeology of Western Pacific. The Journal of Pacific History 53(4):375–396
Hamelin C, Bourdeix R, Baudouin L (2005) The international coconut genetic resources database. In: Batugal P, Ramanatha Rao V, Oliver J (eds) Coconut genetic resources. IPGRI, Rome, pp 282–301
Harries HC (1978) The evolution, dissemination and classification of *Cocos nucifera* L. Bot Rev 44:205–317
Herran A, Estioko L, Becker D et al (2000) Linkage mapping and QTL analysis in coconut (*Cocos nucifera* L.). Theor Appl Genet 101:292–300
Jayalekshmy VG, Sree Rangasamy SR (2002) Cluster analysis in coconut (*Cocos nucifera* L.). J Plant Crop 30(2):18–22
Kamaral LCJ, Perera SACN, Perera KLNS et al (2014) Genetic diversity of the Sri Lanka yellow dwarf form as revealed by microsatellite markers. Trop Agric Res 26(1):131–139
Kamaral LCJ, Dassanayaka PN, Perera KLNS et al (2016) SSR markers reveal the population structure of yellow (dwarf) coconuts in Sri Lanka. Tree Genet Genomes 12(6):116
Kandoliya UK, Joshi AK, Mori DS et al (2018) Genetic diversity analysis of coconut (Cocos nucifera L.) genotypes and hybrids using molecular marker. Indian J Agric Biochem 31(1):25–32
Kearsey MJ, Luo ZW (2003) Mapping, characterization and deployment of quantitative trait loci. In: Plant molecular breeding. Blackwell Publishing Ltd, Oxford, pp 1–29
Konan KJN, Koffi KE, Konan KJL et al (2007a) Microsatellite gene diversity in coconut (*Cocos nucifera* L.) accessions resistant to lethal yellowing disease. Afr J Biotechnol 6(4):341–347
Konan KJL, Konan KJN, Koffi KE et al (2007b) Coconut microsatellite gene diversity analysis technology transfer to Cote d' Ivoire. Biotechnology 6(3):383–387
Kriswiyanti E, Temaja GRM, Sudana IM et al (2013) Genetic variation of coconut Tall (*Cocos nucifera* L.) in Bali, Indonesia based on microsatellite DNA. J Biol Agric Healthc 13:208–224
Lebrun P, N'Cho YP, Seguin M et al (1998) Genetic diversity in coconut (*Cocos nucifera* L.) revealed by restriction fragment length polymorphism (RFLP) markers. Euphytica 101:103–108
Lebrun P, Baudouin L, Bourdeix R et al (2001) Construction of a linkage map of the Rennell Island Tall coconut type (*Cocos nucifera* L.) and QTL analysis for yield characters. Genome 44:962–970
Liyanage DV (1958) Varieties and forms of coconut palms grown in Ceylon. Ceylon Cocon Quart 9:1–10

Loiola CM, Azevedo AON, Diniz LEC et al (2016) Genetic relationships among tall coconut palm (Cocos nucifera L.) accessions of the International Coconut Genebank for Latin America and the Caribbean (ICG-LAC), evaluated using microsatellite markers (SSRs). Plos One 11(3):e0151309

Meerow AW, Wisser RJ, Brown JS et al (2003) Analysis of genetic diversity and population structure within Florida coconut (*Cocos nucifera* L.) germplasm using microsatellite DNA, with special emphasis on the Fiji Dwarf cultivar. Theor Appl Genet 106:715–726

Menon KPV, Pandalai KM (1958) The coconut, a monograph. Indian Central Coconut Committee, Ernakulam

Narayana GV, John CM (1949) Varieties and forms of coconut. Madras Agric J 36:349–366

Oyoo ME, Muhammed N, Cyrus KN et al (2016) Assessment of the genetic diversity of Kenyan coconut germplasm using simple sequence repeat (SSR) markers. Afr J Biotechnol 15(40):2215–2223

Perera L (1999) Assessing genetic diversity in coconut using molecular markers. Ph.D Thesis, University of Dundee, Scotland

Perera L (2002) Chloroplast DNA variation of coconut is opposite to its nuclear DNA variation. CORD 18(2):56–73

Perera SACN (2010) QTL analysis in coconut via genome mapping: principles, requirements and prospects. Cocos 20:57–65

Perera SACN, Ekanayake GK (2008) Characterization of Sri Lankan indigenous coconut (*Cocos nucifera* L.) varieties for diversity in quantitative morphology. Trop Agric 157:25–42

Perera SACN, Kilian A (2008) Diversity arrays technology: a high throughput molecular marker system for coconut. Pragna (IFS Newsl Sri Lanks) xix (1 Special issue), pp 60–64

Perera L, Russell JR, Provan J (1998) Evaluating genetic relationships between indigenous coconut (*Cocos nucifera* L.) accessions from Sri Lanka by means of AFLP profiling. Theor Appl Genet 96:545–550

Perera L, Russell JR, Provan J (1999) Identification and characterization of microsatellites in coconut (*Cocos nucifera* L.) and the analysis of coconut populations in Sri Lanka. Mol Ecol 8:344–346

Perera L, Russell JR, Provan J et al (2000) Use of microsatellite DNA markers to investigate the level of genetic diversity and population genetic structure of coconut (*Cocos nucifera* L.). Genome 43:15–21

Perera L, Russell JR, Provan J et al (2001) Levels and distribution of genetic diversity of coconut (*Cocos nucifera* L., var. Typica form typica) from Sri Lanka assessed by microsatellite markers. Euphytica 122:381–389

Perera L, Russell JR, Provan J et al (2003) Studying genetic relationships among coconut varieties/populations using microsatellite markers. Euphytica 132:121–123

Perera L, Fernando WBSF, Hearth N et al (2004) Use of microsatellite DNA markers for population analysis, variety identification and for hybridity testing of coconut in Sri Lanka. In: Peiris TSG, Ranasinghe CS (eds) Proceedings of the international conference to mark the 75th anniversary of Coconut Research Institute, Sri Lanka, Part II. Ceylon Printers, Colombo, pp 3–15

Perera SACN, Ekanayake GK, Attanayake RB (2009) Characterization of conserved coconut germplasm in Sri Lanka with morphological descriptors. CORD 25(1):46–53

Perera SACN, Gunn B, Rivera RI (2018) 3.9.4 DNA analysis in farmer's fields – chapter 3. Where we need to be to secure diversity and promote use. In: Bourdeix R, Prades A (eds) A global strategy for the conservation and use of coconut genetic resources 2018–2028, Montpellier, pp 171–172

Pokou DN, Manimekelai R, Fan HK (2018) 3.9.3 DNA analysis in ex situ genebanks – chapter 3. Where we need to be to secure diversity and promote use. In: Bourdeix R, Prades A (eds) A global strategy for the conservation and use of coconut genetic resources 2018–2028, Montpellier, p 171

Pradeepkumar S, Manimekalai R, Ranjithakumari BD (2011) Microsatellite marker-based characterization of south pacific coconut (*Cocos nucifera* L.) accessions. Int J Plant Breed Genet 5(1):34–43

Rajesh MK, Nagarajan P, Jerard BA et al (2008) Microsatellite variability of coconut accessions (*Cocos nucifera* L.) from Andaman and Nicobar Islands. Curr Sci 94(12):1627–1631
Rajesh MK, Thomas RJ, Rijith J et al (2012) Genetic purity assessment of D X T hybrids in coconut with SSR markers. Indian J Genet Plant Breed 72(4):472–474
Ratnambal MJ, Kumaran PM, Arunachala V et al (2001) Coconut genetic resources and molecular approaches. Indian J Plant Genet Resour 14(2):182–184
Ribeiro FE, Baudouin L, Lebrun P et al (2010) Population structures of Brazilian Tall coconut (*Cocos nucifera* L.) by microsatellite markers. Genet Mol Biol 33(4):696–702
Ritter E, Rodriguez MJB, Herran A et al (2000) Analysis of quantitative trait loci (QTL) based on linkage maps in coconut (*Cocos nucifera* L.). In: Arencibia A (ed) Plant genetic engineering towards the third millennium. Elsevier, Amsterdam, pp 42–48
Rivera R, Edwards KJ, Barker JHA et al (1999) Isolation and characterization of polymorphic microsatellites in *Cocos nucifera* L. Genome 42:668–675
Rohde W, Kullaya A, Rodriguez J et al (1995) Genome analysis of Cocos nucifera L. by PCR amplification of spacer sequences separating a subset of copia-like 16RI repetitive elements. J Genet Breed 49:179–186
Rohde W, Becker D, Kullaya A et al (1999) Analysis of coconut germplasm biodiversity by DNA marker technologies and construction of a first genetic linkage map. In: Oropeza C, Verdeil JL, Ashburner GR, Cardena R, Santamaria JM (eds) Current advances in coconut biotechnology. Kluwer, Dordrecht, pp 99–120
Rohde W, Herran A, Estioko L et al (2000) Mapping of DNA markers, homeotic genes and QTLs in coconut (*Cocos nucifera* L.) and synteny studies with oil palm. In: Proceedings of the international symposium on oil palm genetic resources and utilization, Kuala Lumpur, pp pAC1–AC21
Sankaran M, Damodaran V, Singh DR et al (2012) Characterization and diversity assessment in coconut collections of Pacific Ocean Islands and Nicobar Islands. Afr J Biotechnol 11(97):16320–16329
Shalini KMS, Lebrun PBA, Baudouin L et al (2007) Identification of molecular markers associated with mite resistance in coconut (*Cocos nucifera* L.). Genome 50:35–42
Teulat B, Aldam C, Trehin R (2000) An analysis of genetic diversity in coconut (*Cocos nucifera*) populations from across the geographic range using sequence-tagged microsatellites (SSRs) and AFLPs. Theor Appl Genet 100:764–771
Thomas RJ, Rajesh MK, Kalavathi S et al (2013) Analysis of genetic diversity in coconut and its conservation in root (wilt) disease affected areas of Kerala: a community participatory approach. Indian J Genet Plant Breed 73(3):295–301
Upadhyay A, Jose J, Manimekalai R et al (2002) Managing plant genetic diversity. Proceedings of international conference, Kula lumpur, Malaysia, p 61–66
Whitehead RA (1976) Coconut. In: Simmonds NW (ed) Evolution of crop plants. Longman, London, pp 221–225
Yong X, Luo Y, Yang Y et al (2013) Development of microsatellite markers in Cocos nucifera and their application in evaluating the level of genetic diversity of *Cocos nucifera*. Plant Omics 6(3):193–200
Yao SDM, Konan KJL, Pokou ND et al (2013) Assessment of the genetic diversity conservation in three tall coconut (*Cocos nucifera* L.) accessions regenerated by controlled pollination, using microsatellite markers. Afr J Biotechnol 12(20): 2808–2815
Zizumbo-Villarreal D, Fernandez-Barrera M, Torres-Hernandez N et al (2005) Morphological variation of fruit in Mexican populations of *Cocos nucifera* L. (Arecaceae) under in situ and ex situ conditions. Genet Resour Crop Evol 52:421–434

Chapter 7
Genome Studies for Effective Management and Utilization of Coconut Genetic Resources

Luc Baudouin

7.1 Introduction

Coconut (*Cocos nucifera* L.) is the emblematic tree of the tropical landscape. Most coconuts grow close to sea level but can be found in the high valley of Markham River (450 m above sea level) and even by Lake Tanganyika (768 m). It has been, for more than a century, the main oil crop on the international market, before being superseded by oil palm (*Elaeis guineensis* Jacq.). Besides oil, coconut has many different uses, involving all parts of the plant, from the root (medicinal use) to the terminal bud (a delicious salad) (Batugal 2005). The huge nut is covered by a thick layer of fibre or "husk". Its hard shell encases a layer of solid endosperm, edible and rich in oil, and a cavity which contains the liquid endosperm or "coconut water". All parts of the plant have food and non-food uses, including the stem and leaves, used in construction and handicrafts.

There are two main types of coconut: the Tall coconut is robust, fast growing and mostly cross-pollinating, while the Dwarf coconut grows slowly and is mostly self-pollinating. All Dwarf coconuts are related and may have emerged from a Southeast Asian Tall population through domestication (Perera et al. 2016). Other, least frequent types are the Compact Dwarf, originating from the South Pacific, which is cross-pollinating and various self-pollinating semi-Talls. All these types are interfertile.

Coconut is mostly cultivated by smallholders, on farms whose area is often less than 1 ha, and copra is often their main source of cash income. The copra is the dried solid endosperm, which can be stored for a certain period before being transported to the mill where the oil is extracted. Today, the coconut oil trade is stagnating,

L. Baudouin (✉)
CIRAD, UMR AGAP, F-34398 Montpellier, France

AGAP, Université de Montpellier, CIRAD, INRA, Montpellier Supagro, Montpellier, France
e-mail: luc.baudouin@cirad.fr

S. Adkins et al. (eds.), *Coconut Biotechnology: Towards the Sustainability of the 'Tree of Life'*, https://doi.org/10.1007/978-3-030-44988-9_7

while other oils grow considerably, especially palm oil. However, markets are expanding rapidly for some coconut products such as virgin coconut oil (extracted from the fresh endosperm) and coconut water, which is now produced on a large scale in Brazil and Thailand.

7.2 Factors Hindering Coconut Breeding

Coconut is a long-lived multipurpose crop tree producing huge fruit, and this is precisely what makes it so attractive. It is also what makes its genetic improvement relatively inefficient (Batugal et al. 2009). A breeding cycle takes about 15 years and requires a large land area. In addition, coconut being a multipurpose crop, multiple selection criteria have to be used, which imposes a limit to the selection intensity that can be applied to each of them. Due to the low number of fruit each palm produces per year, the rate of multiplication is low. Given improved seeds are difficult to produce and expensive, this discourages farmers from replacing their old coconut palms, whose production is slowly declining but may last for many more years. In addition, several epidemic diseases have become a serious subject of concern. Breeding for Lethal Yellowing resistance is especially difficult, because there is no method for inoculating the disease to young plants. As a result, breeders have to rely on field trials planted in affected areas, and results are not available until 10 or more years after initiation (Baudouin et al. 2009).

There are several types of improved varieties. Due to self-pollination, Dwarf varieties can be reproduced easily by collecting seeds from a palm of the desired type. Phenotypical traits, such as sprout colour, are used to discard the rare off-types. Tall coconuts are mostly cross-pollinating, and in the absence of a vegetative multiplication method, it is not possible to reproduce a specific individual. In most cases, improved Tall varieties consist simply of multiplying superior populations by natural pollination. Some progress can be obtained by felling the poorest individuals and collecting seeds from the best mother palms. Hybrid varieties are obtained by spraying a mix of talcum powder and of pollen from the selected male parent onto the receptive female flowers of the other parent, after removing the male flowers. These are most of the time "population hybrids", i.e. derived from a set of individuals from each parental population. They are thus less uniform than, e.g. maize (*Zea mays* L.) hybrids. Most hybrid varieties are Dwarf × Tall hybrids (De Nucé De Lamothe and Rognon 1973), but Tall × Tall and Dwarf × Dwarf (Huang et al. 2014) also exist. The best hybrids are excellent in terms of oil production but are not always considered as adapted to other uses. They can be improved by progeny testing the pollen donors (Bourdeix et al. 1992). This method is tedious and exploits only a small part of the diversity of the species.

Coconut genomics has been an active research domain for 20 years (Arunachalam 2012; Perera et al. 2017) and continues to progress rapidly. Past and current research aimed at answering three broad questions: How did coconut, as a species, evolve and differentiate itself from its closest parents? What is the distribution of coconut genetic diversity and which were the main driving forces? How do genome expression in coconut and its regulation contribute to its phenotype? The pursuit of these objectives inevitably requires a large quantity of resources, including a reference coconut genome sequence. We anticipate that these results will benefit coconut genetic improvement by lifting some of the above-mentioned hindrances.

7.3 Where Does Coconut Come From?

A book by Nayar (2017) was recently published about the origin of coconut. Even with the wealth of information it brings together, the origin issue doesn't seem to be settled. There is a definite advantage in genome studies to compare the species of interest with other species, especially related ones. Close relatives inherited comparable genomes from their common ancestor but evolved in different environments. While they exhibit a high degree of synteny, the environmental conditions where they evolved resulted in differences such as gene family expansions or gene regulatory pathways. Among palms, genomic studies are active in two economically important species: oil palm and date palm (*Phoenix dactylifera* L.). Other, more distant palms, such as nipa and rattans, have also been studied to some extent.

Coconut is a member of the palm family (Arecaceae). It belongs to the subfamily Arecoideae and to the tribe Cocoseae. Among them Bactridinae and Eleaidinae have spiny leaves, while Attaleineae – which includes coconut – are not spiny. However, the exact position of coconut in this subtribe is unclear. Phylogenic studies involving DNA[1] sequences associate *Cocos* to different genera: *Parajubaea* (Baker and Couvreur 2013; Gunn 2004), *Attalea* (Baker and Couvreur 2013; Hahn 2002; Meerow et al. 2014) and *Syagrus* (Meerow et al. 2009). In a cladistics study based on morphologic traits (Noblick et al. 2013), *Cocos*, *Voanioala* and *Attalea* are embedded in a paraphyletic grade of *Syagrus*.

While a vast majority of Attaleineae species are South American, four genera (out of 11) are found in the Indian Ocean region or farther away, *Beccariophoenix*, *Jubaeopsis*, *Voanioala* and *Cocos*, but none in West Africa. This strongly suggests that the ancestors of coconut existed in Antarctica. The diversification of Attaleineae would have begun from 23 to 37 million years (MY) ago (Baker and Couvreur 2013; Gunn 2004), and expansion towards Africa would have taken place 29 MY ago. This is somewhat difficult to reconcile with the presence of fossils attributed to *Cocos* towards the limit of the Cretaceous and Palaeocene periods (60–65 MY) in Colombia (Gomez-Navarro et al. 2009) and in India (Srivastava and Srivastava 2014; Tripathi et al. 1999), given what we know about continental drift. A more

[1] DNA (deoxyribonucleic acid) is a molecule forming a long chain (the double helix) which holds the genetic information of a living organism. It is in the nucleus of the cell. It contains genes which are transcribed into messenger ribonucleic acid (mRNA), which are, in turn, translated into polypeptide chains in ribosomes. Other types of RNA are ribosomal RNA (rRNA), transfer RNA (tRNA), microRNA (miRNA), etc.

satisfying picture of coconut origin might emerge from taking into account a low base substitution rate in chloroplastic as well as nuclear (for synonymous changes only) palm genomes, compared to grasses (Gaut et al. 1996; Wilson et al. 1990). This would also involve the consideration of recently discovered fossils. In the meantime, the origin of coconut will remain an "abominable mystery", and we should be content with two certainties: (1) Coconut descends from ancestors, which grew in South America tens of millions of years ago. (2) At the dawn of humankind, some 10,000 years ago, coconut existed only in the Indian and Pacific Ocean regions (Gunn et al. 2011).

7.4 Marker Development for Genetic Diversity Studies

Understanding the organization of the genetic diversity of coconut is essential to rationalize coconut genetic diversity management and for devising genetic improvement programmes. Morphometric traits can be used in the framework of a research station (N'Cho et al. 1993); however, they are influenced by environmental conditions, and it is difficult to compare populations at several locations. Biochemical markers such as leaf polyphenols may give interesting results (Jay et al. 1991), but repeatability across experiments is not guaranteed. There are few isozyme studies in coconut (Cardena et al. 1998; Parthasarathy et al. 2004), but while they ensure repeatability, they capture only a small portion of the species' diversity.

7.4.1 Dominant Markers

The main advantage of presence-absence markers is to allow the generation of many polymorphic markers at a low cost, using only a small number of marking systems. Their drawback is that identifying alleles is difficult (except in mapping populations). In addition, random amplified polymorphic DNA (RAPD) and, to a lesser extent, amplified fragment length polymorphism (AFLP) are sensitive to experimental conditions, and their repeatability is questionable. They can however provide useful results in "one-shot" experiments. The earliest diversity study published was done using polymerase chain reaction (PCR) and primer pairs tailored to amplify highly repetitive EcoRI elements. This inverse sequence-tagged repeat (ISTR) approach (Rohde et al. 1995) allowed the identification of 36 polymorphic bands in 21 individuals from a wide panel of diversity. The RAPD technique was used in several studies. In the South Pacific, 123 markers were used to differentiate coconut populations collected at 17 sites (Ashburner et al. 1997). In India, 8 primers detected 77 polymorphic markers in 81 palms representing 20 collection accessions

(Upadhyay et al. 2004). Thirty-three accessions from the CPCRI[2] genebank were analysed with 55 primers, yielding 538 markers (Manimekalai and Nagarajan 2006). In Mexico, 33 of 80 tested primers amplified 82 markers in 3 populations (20 individuals per population). Among them, five appear to be linked with resistance to lethal diseases, which are a serious cause of concern in this country. In Tanzania, a RAPD analysis performed on germplasm conserved at Chambezi revealed the presence of two groups, respectively, localized in the Southern and in the Northern parts of the coastal belt. This finding is in agreement with what is known about coconut history in the country (Masumbuko et al. 2014).

The AFLP method was used to study the collection from Sri Lanka (Perera et al. 1998). Forty-two palms belonging to 12 accessions were analysed using 8 primer pairs, yielding 198 polymorphic bands. Palms from the same accession cluster together, and local Talls (cross-pollinating) are easily distinguished from Dwarfs and semi-Talls, which are introduced and self-pollinating. The latter exhibit less diversity than the Talls, which is consistent with their mating systems. The local "Dwarf Yellow" is a notable exception. This accession is considered as representing the Malayan Yellow Dwarf, a widespread cultivar, which was subsequently studied using microsatellites and proved to be very uniform in most countries, except in Sri Lanka and Jamaica (Kamaral et al. 2016; Lebrun et al. 2007). One way of making presence-absence markers more repeatable is to convert them into PCR markers. Out of 200 RAPD-derived sequence characterized amplified region (SCAR) markers, one was found to be specific to Tall coconuts (Rajesh et al. 2013).

7.4.2 Codominant Markers

Unlike the above markers, restriction fragment length polymorphism (RFLP) and microsatellite or SSR (for simple sequence repeat) markers make it possible to identify alleles. At the cost of added complexity, they are more informative and offer a better estimation of gene flow. To study the diversity of the ex situ collection of Côte d'Ivoire, RFLP markers were used (Lebrun et al. 1998, 1999). In the second of these studies, 42 accessions, representing the major coconut cultivation areas, are represented by 289 individuals when analysed with RFLPs. Most complementary DNA (cDNA) probes were heterologous, from 15 from oil palm, 1 from maize, 4 from rice (*Oryza sativa* L.) and 1 from coconut, to which a mitochondrial probe cytochrome oxidase (Cox1) from wheat (*Triticum aestivum* L.) was added. In combination with 4 restriction enzymes, 25 marking systems revealed 60 polymorphic bands. Most individuals fall into two groups: The largest one and the most diverse represents Tall coconuts from the Pacific (Southeast Asia to the South Pacific and Panama), as well as all Dwarfs, irrespective of their origins. The second group represents coconuts from South Asia and West Africa. They represent two clearly

[2] Central Plantation Crops Research Institute, India.

differentiated gene pools, which must have remained in reproductive isolation for a long period and came into contact only recently. In fact, a third group is intermediate and is located at both ends of the Indian Ocean; the Andaman Islands and East Africa represent zones of contact between the first two groups. In the first case, this is due to their geographical situation; in the second case, it results from "Malayo-Polynesian" migrations to Madagascar and the Comoros. This picture of coconut diversity concords broadly with what was observed with leaf polyphenols (see Jay et al. 1991 above).

Microsatellite markers have been extensively used during the last 20 years for coconut. Among the first published marker sets were those produced by researchers from Sri Lanka and the Philippines (Perera et al. 1999; Rivera et al. 1999). The names of the markers from the Philippines begin with "CNZ" or simply "CN", such as in "CN1H2". Thirty-eight markers were used to characterize 20 Tall and Dwarf genebank accessions, mostly from the Philippines. The Dwarf accessions are relatively diverse (the Philippines is in the suspected region of origin of Dwarfs) but cluster together. The exotic accessions, from the Solomon Islands and French Polynesia, are genetically distant from the local accessions. The names of the markers from Sri Lanka begin with "CAC". Eight were used in successive studies at the local (Perera et al. 2001) and global level (Perera et al. 2000, 2003). These studies confirm the pattern of coconut genetic diversity already identified with RFLPs. The accumulation of this data makes it possible to specify the limits of coconut gene pools, especially in Thailand. Combinations of markers developed by Sri Lankan and Filipino teams were used to characterize diversity in Kenya (Maurice et al. 2016) and in Hainan Island (Liu et al. 2011). These markers were also used in Florida (Meerow et al. 2003), showing that the "Fiji Dwarf" (also known as Niu Leka Dwarf, which is cross-pollinating and unrelated to other Dwarfs) differed strongly from the other studied germplasm. In the meantime, another microsatellite set was developed in France (Baudouin and Lebrun 2002) as part of a project dedicated to create a reference microsatellite kit to identify coconut cultivars for the International Coconut Genetic Resources Network (COGENT). This initial set (identified by an initial "CnCir") comprised of 14 markers, whose technical data are given in the CGRD database ("CGRD Version 6.1" 2012), and was later extended further. A large data set is available in TropGeneDB ("Tropgene" 2013; Hamelin et al. 2013), along with technical data on a large number of markers, mainly produced by CIRAD[3]. The COGENT project included training sessions, and the methodology was disseminated among coconut genetic research agencies throughout the world. As a result, the kit markers were used – alone or in combination with other marker types – by different research teams. This allowed to characterize ex situ accessions like at the CNRA[4] genebank (Konan et al. 2007, 2011) and to study diversity within a given country, Brazil (Ribeiro et al. 2010, 2013) and Dominican Republic (Martinez et al. 2010). Several regions were studied: in India, Kerala and

[3] Centre International de Recherche en Agronomie pour le Développement, France

[4] Centre National de Recherche Agronomique, Côte d'Ivoire.

Pondicherry (Rajesh et al. 2008a), the Lakshadweep (Devakumar et al. 2010; Rajesh et al. 2014) and Andaman Islands (Rajesh et al. 2008b).

China too produced its own marker sets. A 30 marker set identified by "WCYZ" was used to distinguish differences between 30 coconut accessions (presumably individuals) from China and Southeast Asia (Yong et al. 2013). Transcriptome sequencing makes it possible to identify many gene-based microsatellite sequences. Based on the 57,304 unigenes identified by Fan et al. (2013), 6,608 SSR were identified, resulting in 191 polymorphic markers. Interestingly, 2 out of a subset of 80 markers were found to be associated with palm height.

7.4.3 Geographical Distribution of Coconut Genetic Diversity

After more than 20 years of research devoted to the polymorphism of coconut molecular markers, we currently have an impressive set of tools available. We also now have a detailed account of the geographical distribution of the species' diversity. Noteworthy, this account is independent of the marker set used as shown by the comparative studies made by Duran et al. (1997) and Teulat et al. (2000). We can thus summarize it according to the most extensive study which involved more than 1,200 palms and 10 SSR markers (Gunn et al. 2011).

Coconut is naturally adapted to seashores as its fruit is disseminated by sea currents. Atolls represent a refuge, which allows the build-up of substantial populations, and serve as stepping stones for long distance dissemination (Harries and Clement 2014). A striking feature of atolls is that corals adapt to variations of the sea level, which plays an important protective role for the coconut environment during the periods of global temperature change.

At the onset of agriculture, coconut was distributed in two distinct areas: The largest one extended from the Sunda archipelago to the Philippines and Papua New Guinea. It certainly extended eastwards into the South Pacific, but it is difficult to determine to what extent. The other area included the South Asian archipelagos, Maldives, Lakshadweep (or Laccadive), Sri Lanka and possibly South India. The resulting populations (Pacific and Indo-Atlantic groups) are genetically well differentiated because they evolved separately during a long period, which could have extended for as much as 3.9 MY (Meerow et al. 2014). This corresponds to the time of the narrowing of the gap between the Sahul (Australia and Papua) and the Sunda tectonic plates. The resulting restriction of the communication between the Ocean was reinforced by the Holocene ice ages (from −650,000 to −12,000 years common era) and the lowering of the sea level to an extent of 150 m, leaving only a narrow channel between Timor and Australia, thus severing the contacts between these populations (Fig. 7.1). As a result, coconut populations from both populations diverged strongly.

After the last glacial episode, the connections were re-established, to some extent, through natural dissemination, but essentially through human movement, which could be tracked by combining molecular studies with historical and

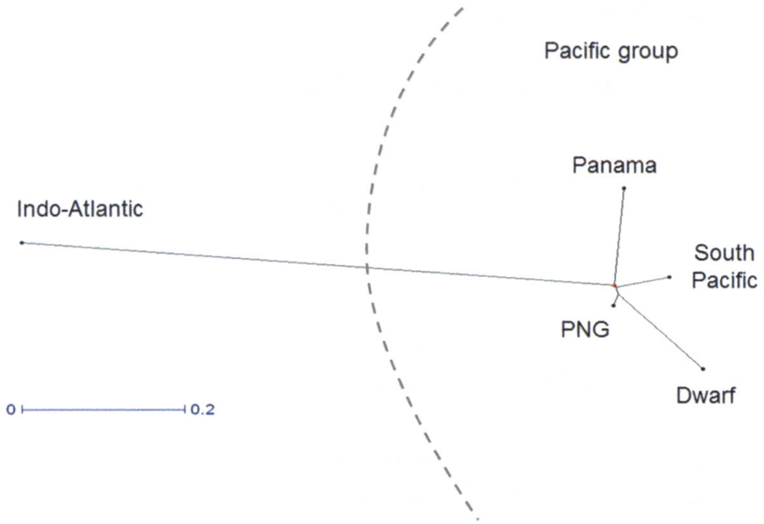

Fig. 7.1 Genetic relationship between the five populations identified by the software "Structure" (Pritchard et al. 2000). The Indo-Atlantic and Pacific groups are identified by Structure with K = 2. A deeper analysis with K = 5 reveals four weakly differentiated populations within the Pacific group. The tree representation is based on the degree of differentiation according to Jost (2009)

linguistic data. We present only the main steps (Fig. 7.2). Dwarf coconuts alone witnessed the intensity of the movements of human populations. They appear to have evolved in Southeast Asia from a local Tall population (Perera et al. 2016). They were widely disseminated, and, due to their strong tendency towards autogamy and to human selection, they generally retained their genetic structure wherever they can be found today. Coconut accompanied Austronesian migrations eastwards to deep Polynesia and even to the Pacific shores of America (Baudouin and Lebrun 2009; Baudouin et al. 2014; Patiño 1963). Austronesians also transported them westwards to Madagascar and the Comoros, and, due to active trade involving Indian and Arab merchants, the Southeast Asian coconut was brought further to East Africa and even Oman (Perera et al. 2011). There they crossed with coconuts originating from South Asia, making the region a hot spot of coconut genetic diversity. After Vasco da Gama found the route towards India around Africa, the Portuguese brought coconut into the Atlantic Ocean. Cape Verde might well have served as a hub (Harries 1977), permitting the introduction of coconut to Africa, Brazil and the Caribbean.

Finally, the "coconut craze", from 1850 to the first half of the twentieth century, resulted in the multiplication of commercial plantations. This certainly involved germplasm movement. Wherever coconut was already abundant, planters tended to use local seeds, often from a plantation of the same region. Long distance movements, however, played a significant role when the populations were low, or to restore populations after a disease (in the case of Mexico – Zizumbo, personal communication), or when specific varieties such as Dwarfs were desired. In the second

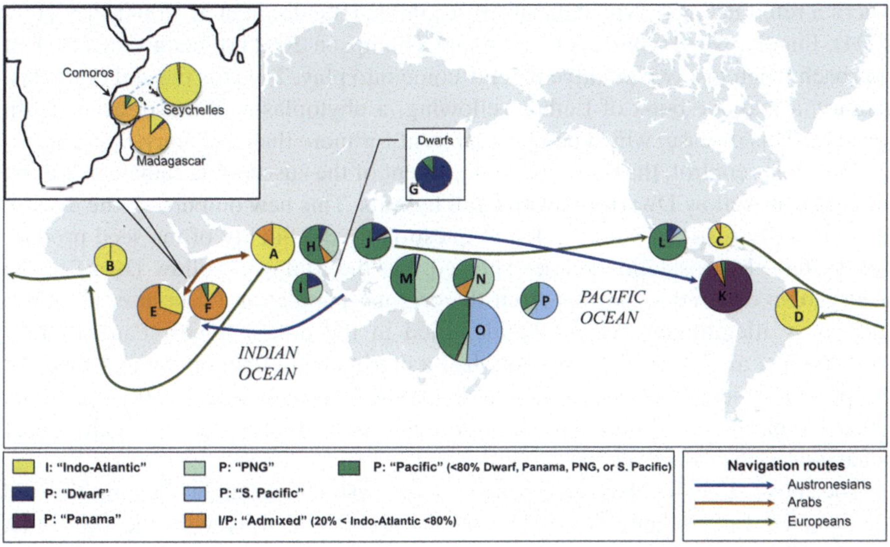

Fig. 7.2 Geographical distributions of Indo-Atlantic and Pacific coconut subpopulations. From Gunn et al. (2011). Pie-chart sizes vary according to the number of individuals studied in the region. Sectors correspond to the proportion of individuals assigned to one of seven classes based on the five populations of Fig. 7.1. Inset: subpopulation compositions for Madagascar, Comoros and Seychelles and in Dwarf coconuts, irrespective of their current locations. Note that an allelic structure like that found in Dwarfs is not infrequent in Talls from Southeast Asia

half of the twentieth century, formal breeding programmes took place, leading to the establishment of national and international collections that included imported material. Population hybrids (usually Dwarf × Tall) were also developed and widely distributed.

7.4.4 *Gene Flow and Application to Varietal Identification*

Molecular markers can be used to track the movement of germplasm. Examples were given at the end of the previous section. The simplest case is that of Dwarf coconuts, which are self-pollinating. It was found, for example, that the Kiribati Green Dwarf has the same microsatellite profile as the Sri Lanka Green Dwarf. Interestingly, Tall coconuts from Micronesia have a higher proportion of Indo-Atlantic alleles than anywhere else in the Pacific, suggesting that both types of coconuts from South Asia were transported to the region. In the cross-pollinating Talls, the task is more difficult but can be tackled using software such as Structure (Falush et al. 2003, 2007; Hubisz et al. 2009; Pritchard et al. 2000), which allows inferring population structure based on genotype information alone. Another software package GENECLASS2 illustrates a different strategy, which proved efficient where a reference genotype database is available (Baudouin et al. 2004; Piry et al. 2004). Finally, NewHybrid (Anderson and Thompson 2001) offers a very efficient approach when just two putative parents come into play. Towards the end of the last century, a massive burst of Lethal Yellowing, a phytoplasma disease, occurred in Jamaica. This disease, which had been present for more than a century, was considered as under control, thanks to the replacement of the susceptible Jamaican Tall by the Malayan Yellow Dwarf × Panama Tall hybrids. This new outburst of the disease in supposedly resistant varieties led to questioning the fidelity of the seed production system. Microsatellite studies showed that the Malayan Yellow Dwarf in the seed garden differed somewhat from the original population (Lebrun et al. 2007) and a few illegitimate trees were identified in the pollen donor (Panama Tall) (Baudouin et al. 2008). This certainly had a detrimental effect on the resistance of the produced seeds but is insufficient to explain the magnitude of the epidemic. Other factors involved may have been a mutation in the pathogen or a change of behaviour in the insect vector.

The above study involved an extensive comparison between the Panama Tall and the Jamaica Tall (Atlantic Tall). The results contrast sharply with the claim that microsatellites provide an ambiguous diagnostic between these cultivars (Mauro-Herrera et al. 2010). The latter finding might be due to genetic contamination or to an identification problem at any stage, from seed importation to sample collection and handling.

7.4.5 Markers and Agronomic Traits

Several studies were devoted to identifying loci, which could be used to identify genotypes presenting traits of economic interest. In Côte d'Ivoire, ten microsatellites of the CnCir kit make it possible to distinguish the susceptible West African Tall from the potentially resistant Sri Lanka Green Dwarf and Vanuatu Tall (Konan et al. 2007). In the 'Mohachao Narel', an Indian variety with a sweet and soft kernel and low fibre content, a genetic diversity study involving 14 microsatellite markers and 12 phenotypic traits revealed a high degree of polymorphism, indicating a good potential for genetic improvement through breeding. For the coconut mite (*Aceria guerreronis* Keifer), an acarian and a serious pest accidentally introduced into India, susceptible and resistant genotypes have been screened using 32 SSR markers and 7 RAPD primers (Shalini et al. 2007). A combination of five markers accounts for the susceptible/resistant contrast. Similar work was conducted to characterize Lethal Yellowing Disease-resistant material (Cardeña et al. 2003; Konan et al. 2007).

7.4.6 *Linkage Mapping and Quantitative Trait Loci QTL Identification*

While the above studies provide useful information for the conception of breeding programmes, they cannot be used to identify the genetic factors affecting useful traits, unlike quantitative trait loci (QTL) studies based on linkage mapping populations or on association studies. The first coconut linkage maps were constructed at the beginning of this century. A map based on 52 seedlings of a Malaysian Yellow Dwarf × Laguna Tall cross was constructed in the Philippines using a combination of ISTR, RAPD and AFLP markers (Herran et al. 2000). A total of 382 markers could be placed onto 2 maps corresponding to the segregation of alleles in each of the parents (implying that the Malayan Yellow Dwarf was heterozygous, which is not the case in most genebank accessions). In both cases 16 linkage groups were identified, but only 18 markers, belonging to 9 linkage groups, were common to both maps. Six QTLs for early germination are identified on both maps (three in each map). The percentage of total variance explained varies between 11 and 15%. Another map was produced using 67 adult palms of a Cameroon Red Dwarf × Malayan Rennell Island Tall cross (Baudouin et al. 2006; Lebrun et al. 2001), and a single map was produced, resulting from segregation among alleles from the Tall parent. A total of 274 markers (mostly SSR and AFLP) were placed on an 1,850-centimorgan (cM)-long map, with 16 linkage groups. A total of 53 QTLs were identified for 11 observed traits (fruit and bunch production and fruit composition). Due to a correlation between them, most of the QTLs involved in fruit composition traits are located at six loci with large effects. A third map was constructed based on 94 palms of an East African Tall × Rennell Island Tall cross using AFLP, SSR markers, in conjunction with a large cosmide bank (Riedel et al. 2009). Segregations of alleles from both parents were integrated to construct a 16-linkage group map based on 704 markers. A total of 46 putative QTLs involved in the cuticular wax composition of leaf surface were localized. Cuticular waxes are protective components of the leaf surface, involved in tolerance to abiotic stress. They influence water retention as well as light reflection, contributing to protect the leaf from damaging ultraviolet (UV) radiation. They also play a role in the defence against insects, bacteria and fungi. Genotyping by sequencing was also used to construct a high-density linkage map in coconut in the framework of genome sequencing (see below).

7.4.7 *Association Studies*

In a recent study, 79 local and exotic genotypes were screened using 48 mapped microsatellite markers and 10 phenotypic traits (Geethanjali et al. 2017). An association study was used to assess the extent of linkage disequilibrium and to identify a QTL close to CnCir 73, affecting various fruit traits. This confirms a finding made earlier by Baudouin et al. (2006) who had already found a QTL for the same traits at this location. Studies conducted so far have only had access to a limited set of markers; NGS (next-generation sequencing) will soon provide access to a large array of markers, with their locations on the coconut genome, opening the way to genome-wide association studies (GWAS). A necessary condition is of course to conduct large-scale phenotyping in collections as well as in situ.

7.5 Coconut Genome Expression

7.5.1 "Generalist" Transcriptomes

Gene expression studies and transcriptomics have been an active domain of research in recent years. The first large-scale study by Fan et al. (2013) used Illumina sequencing from an assemblage of spear leaves, fresh leaves and fruit flesh tissues of the Hainan Tall coconut cultivar. Assembly resulted in a 57,304 unigene set. The quality of the result was confirmed by comparison with the oil palm genome, and 68.2% was annotated based on alignment with the NCBI nr database and other databases. Gene ontologies were assigned based on the KEGG and COG databases[5]. Results were compared with model species such as *Arabidopsis thaliana*, *A. lyrata*, *O. sativa* and *Z. mays*. Gene expression varied according to the tissue sampled (Huang et al. 2014) when transcriptomic data from embryo, endosperm and leaf were observed. Of the annotated genes, 74% were tissue specific, while 6% were shared by all tissues. On examination of the small RNA-mediated epigenetic regulation pathways, the authors found that RNA-directed DNA methylation is important during coconut seed development, particularly in maturing endosperm.

In the absence of a reference genome, gene assembly is difficult, resulting in gene fragmentation (only part of the gene sequence is recovered), redundancy (one gene is represented by two different sequences) and chimerism (an assembled sequence involves fragments of different genes). The underlying idea of translational genomics is to take advantage of results obtained in a model plant to enhance the results obtained in other, related plants. In the study by Armero et al. (2017), the results obtained in the above studies were improved by aligning the original reads onto the amino-acid sequences of oil palm genes. The workflow used in that study was tested in *Arabidopsis* and used on coconut. This increased the gene sequence completion from 72–82% to 87%, often yielding complete sequences. The pipeline was able to eliminate most chimeric and redundant sequences, leaving 29,366 validated protein sequences, of which 1,246 had not been identified in the original

[5]NCBI is one of three major genomics and biological information repositories, located in the USA. The other ones are EMBL-EBI in Europe and DDBJ in Japan. NCBI proposes numerous databases such those devoted to non-redundant (nr) proteins and to Clusters of Orthologous Groups of proteins (COG). KEGG is the Kyoto Encyclopedia of Genes.

studies. It must be noted however that genes existing in coconut but not in oil palm are not identifiable using this approach.

7.5.2 Fruit Development and Characteristics

The variations of gene expression in coconut pulp were studied using suppression subtractive hybridization (SSH) libraries involving three maturation stages, 7, 9 and 11 months, following pollination. Candidate genes involved in fatty acid and carbohydrate metabolism were identified (Liang et al. 2014). Gao et al. (2014) identified and isolated a coconut fatty acid desaturase expressed in coconut endosperm. Using real-time fluorescent quantitative PCR, they found that the peak of fatty acid desaturase expression is around 8 months after pollination. Its expression increases the percentage of palmitoleic acid (16:1) and oleic acid (18:1) and reduces that of stearic acid (18:0).

The Aromatic Dwarf from Thailand has a pleasant taste due to the presence of 2-acetyl-1-pyrroline (2AP). According to Saensuk et al. (2016), an accumulation of this compound results from a point mutation in an alcohol dehydrogenase (AMADH). A similar mutation occurs in aromatic rices.

7.5.3 Resistance to Phytoplasma Diseases

Various phytoplasma diseases are devastating coconuts in several producing areas especially in Africa and in the Caribbean. Nejat et al. (2015) studied the reaction of a Dwarf coconut to the infection by Coconut Yellow Decline in Malaysia, caused by a phytoplasm of the 16SrXIV 'Candidatus Phytoplasma cynodontis' group. Overall, there were more down-regulated genes than up-regulated ones, but genes associated with defence signalling against biotic stimuli were significantly overexpressed in phytoplasma-infected leaves versus healthy coconut leaves. Other known stress-related genes were differentially expressed, and several novel defence-response candidate genes with unknown function were identified. Coconut Root Wilt Disease is also caused by phytoplasma and strongly reduces coconut production. Using the resistant Chowghat Green Dwarf, Rajesh et al. (2015) identified and studied 243 genes assigned to 6 different classes of resistance analogues (RGA). The expression of resistance gene candidates (RGCs) of the nucleotide binding site (NBS) was studied in either resistant or susceptible coconut populations. The resulting 500 to 700 base pair (bp) fragments were sequenced and submitted to a phylogenetic analysis. Their expression profile, following a salicylic acid treatment, was studied in the root, stem and leaves. Expression profiles varied in healthy and diseased plants (Puch-Hau et al. 2015, 2016). Such studies aim at understanding coconut genome interactions with these pathogens and, if possible, at defining genetic improvement strategies for disease resistance.

7.5.4 *In Vitro Propagation*

Coconut vegetative reproduction through tissue culture is notoriously difficult. The advent of transcriptomics is thought to provide a better understanding of the cellular processes coming into play during in vitro culture. Pérez-Núñez et al. (2009) identified a coconut orthologue of SERK (somatic embryogenesis receptor-like kinase) previously found (in carrot) to be specifically expressed in embryogenic tissues. Likewise, CnANT, an orthologue of the *Arabidopsis thaliana* AINTEGUMENTA-like gene, was shown to promote the transition from callus to an embryogenic structure, a key step of in vitro regeneration, in *Arabidopsis* calli (Bandupriya et al. 2014). The same team developed four expressed sequence tag (EST) libraries derived from somatic and zygotic embryos at two development stages as well as leaf tissue used as a control and identified a number of genes involved in embryo development (Bandupriya and Dunwell 2016).

7.5.5 *MicroRNAs*

Genomic analysis in coconut has not been restricted to protein genes. MicroRNAs (miRNAs) are short sequences which can inhibit genes by hybridizing themselves with their messenger RNA (mRNA) targets. Following a bioinformatic approach, 1,008 coconut ESTs were aligned onto 3777 known miRNAs through Blastn (Altschul et al. 1990) and subjected to several tests. A novel miRNA from the mir2673 family was identified as well as its putative targets (Viveka and Moossab 2016).

7.6 Coconut Genome Sequencing

7.6.1 *Cytoplasmic Genome*

The first sequence of a coconut chloroplast was obtained by Huang et al. (2013). It totalled 154,731 bp. Its guanine-cytosine content (G-C content) was 37.44% and encoded 84 protein-coding genes, 8 rRNAS, and 38 transfer RNAs (tRNAs), and it had 2 pseudogenes. In a phylogenic analysis of monocotyledons, palm chloroplastic sequences appear remarkably similar, confirming their slow evolution rate, already noted by Wilson et al. (1990). The 678,653 bp mitochondrial genome was sequenced by Aljohi et al. (2016). Its G-C content was 45.5%, and it was found to encode 72 proteins, 9 pseudogenes, 23 tRNAs, and 3 ribosomal RNAs.

7.6.2 *Nuclear Genome*

7.6.2.1 Genome Size

Genome size is a key parameter for establishing a sound strategy to assemble a genome. Gunn et al. (2011) and in Freitas Neto et al. (2016) evaluated it through flow cytometry. These papers differ by the mean values (2C = 5.97 and 5.57 pg, respectively, which correspond to 2.88 and 2.72 Gbp for a haploid genome) but agree on the existence of small but statistically significant variations in Tall coconuts. The genome size of Dwarfs is more uniform, and variations appear to result from experimental error. In addition to coconut, the former paper describes the evolution of genome size among Attaleineae.

As part of the of coconut genome sequencing efforts (see below), the coconut genome size was estimated based on K-mer frequency distribution and found to be 2.42 Gbp (Xiao et al. 2017). The estimated value depends on the method used and on protocol details such as sample quality, number of replications and the choice of control. In addition, intraspecific variations must be taken into account. Notwithstanding these differences among studies, the coconut genome was found to be large, about 1.5 times the size of the genome of oil palm (Singh et al. 2013), and requires special care to assemble its sequence, especially due to the abundance of repeated sequences.

7.6.2.2 Nuclear Genome Sequence

Efforts to map the coconut nuclear genome started around 2010. The "Joint Center for Genomics Research", representing a collaboration between KACST[6], Saudi Arabia, and CAS[7], China, announced a coconut genome sequence through a poster (Alsaihati et al. 2014), but did not provide further details. According to an IPRI[8] report cited by Perera et al. (2017), an Indonesian team also sequenced a Tall (Tenga Tall) and a Dwarf (Pati Kopyor Green Dwarf) genotype in 2015. Other initiatives were announced in 2017 by press reports (CPCRI India, no formal publication available yet) and through an Internet group (University of Los Baños, Philippines), followed by a poster (Lantican et al. 2018). The studied genotype was the Catigan Green Dwarf, but data about the Laguna Tall are also available under accession number PRJNA298457 and SRP064777. Genomic research in the Philippines is progressing, especially with a focus on insect resistance (Galvez et al. 2018).

Finally, the draft genome of coconut was published by Xiao et al. (2017). They obtained a total of 419.67 Gbp (173 times the genome size of raw reads) with the Illumina HiSeq 2,000 platform using a series of paired-end and mate-pair libraries.

---

[6] King Abdulaziz City for Science and Technology.

[7] Chinese Academy of Sciences.

[8] Indonesian Oil Palm Research Institute, Indonesia.

The genotype used was the “Hainan Tall”. The total scaffold length of 2.20 Gbp represents 90.91% of the genome. The sequencing strategy was designed especially to optimize assembly, and the typical length of the scaffolds (as expressed by the N50 statistic) was 418 Kbp (other teams obtained 114 Kb), but this was still smaller than in oil palm (1,045 Kb) due to a larger proportion of repeated sequences (73% compared to 57%), mainly long-terminal repeat retrotransposons elements (LTRs). A total of 28,039 protein-coding genes is predicted, which is somewhat less than in *P. dactylifera* and *E. guineensis*. Ab initio (i.e. based on the sequence alone) genome annotation is 74.1% complete. The study of the evolution of gene families involved in salt tolerance suggests a significant expansion of several types of antiporter genes (coding for membrane proteins, which transport actively different ions in opposite direction, thus contributing to remove one of them – e.g. Na^+ – from the cytoplasm). A second draft sequence based on the Catigan Green Dwarf was published recently (Lantican et al. 2019).

The next step is to assemble the coconut sequences into 16 pseudomolecules corresponding to the 16 chromosome pairs of coconut. Given the size of the coconut genome and the abundance of repeated sequences, this requires anchoring the scaffolds of a genome draft onto a high-density linkage map. To construct this map, a back-cross between the Malayan Yellow Dwarf (MYD) and a single MYD × West African Tall individual was produced in Côte d’Ivoire. With the recurrent parent being highly homozygous, the expected genotype at a given locus is either the homozygous, like in the MYD, or heterozygous, like in the hybrid parent. Genotyping by sequencing (GBS) consists in breaking the DNA of each individual into fragments of appropriate size using a combination of restriction enzymes and attaching an adapter and a “barcode” DNA sequence to one end of the fragments. These operations are performed separately for each individual, and the resulting sequences are bulked and sequenced together. The resulting short reads are treated by a bioinformatics pipeline, which combines them to the scaffolds of a coconut draft sequence (namely, the one produced using the Hainan Tall) and identifies polymorphism. After filtration, the resulting product consists of a table where single nucleotide polymorphic (SNP) alleles are assigned to individual palms and to a definite position on scaffolds. The next step consisted in grouping the markers by linkage groups using JOINMAP (Ooijen 2011) and in finding the most likely order of the scaffolds on each linkage group using SCAFFHUNTER (Martin et al. 2016). This procedure resulted in a map based on 8402 markers located in 16 linkage groups and genotyped in 216 individuals (Fig. 7.3).

Thanks to this map, it was possible to assign a position to a large part of the scaffolds on 16 linkage groups. The results will be published shortly. Among other things, they will enable us to trace the evolutionary history of coconut by comparing the coconut genome with those of other species, especially oil palm. In this domain, two types of events are of interest: chromosomal rearrangements involving syntenic protein blocks that are common to both species and the occurrence of insertion peaks of transposable elements, which are mainly responsible for the variations of genome size among palms. The possibility of locating genes of agronomic interest will open up new opportunities for coconut breeding.

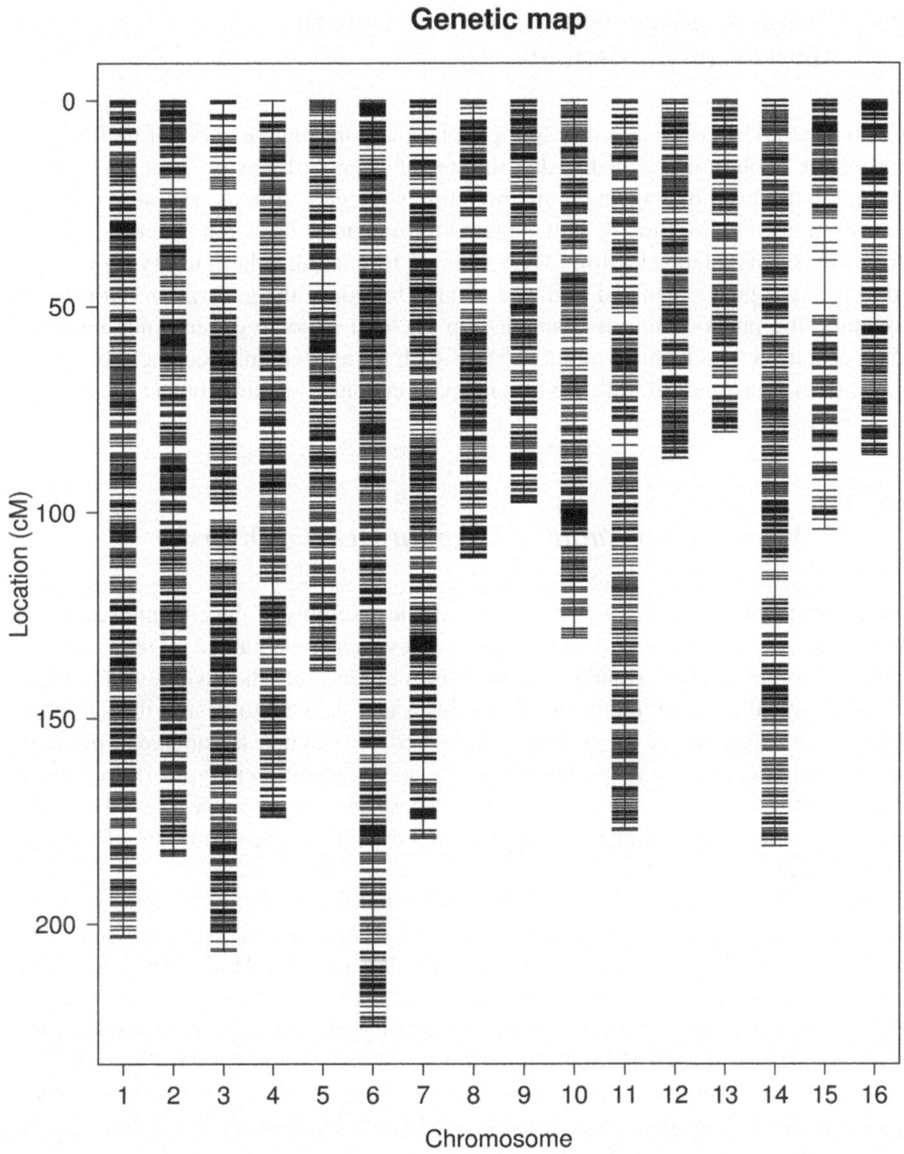

Fig. 7.3 A high-density linkage map of coconut. Each horizontal segment represents one SNP. These SNPs were aligned onto the scaffolds produced by Xiao et al. (2017), and 216 individuals from a back-cross were genotyped for 8,402 SNPs. The resulting linkage map allowed assembling 2,303 scaffold and contigs into 16 linkage groups

7.7 Towards a Renewal of Coconut Genetic Improvement Methods

Genome studies have numerous applications and contribute to a better understanding of the evolution and of the distribution of genetic diversity. This is useful to design germplasm management and breeding strategies. But, as was shown above, many of these studies directly aim at plant improvement. The main benefits expected from marker-assisted selection (MAS) are (1) to maximize heritability by directly aiming at the genes involved in a trait and (2) to reduce the generation duration by skipping the phenotyping assessment step (at least at some generations, reducing the generation time from 15 years to 5). In such a way, genomics can help by lifting some of the barriers that limit the efficiency of coconut breeding, but a certain number of steps must be taken.

7.7.1 Assessing the Value of Coconut Genetic Diversity

An important milestone will be reached with the assembly of the coconut sequence. A picture of coconut diversity at the genomic level can be obtained by resequencing many genotypes, which should be well chosen to represent the diversity of the species. The results of the microsatellite analyses can be used to design the sampling plan. Even before such a programme is achieved, we can take advantage of previous work: in this chapter we reported five sequenced coconut genotypes from various origins, and the list is probably incomplete. These sequences can be aligned onto the reference sequences and provide the first, and extremely useful, outline of the species' diversity.

Identifying genes that are responsible for economically significant traits is essential and can be done by combining two types of approaches: the candidate gene approach looks for genes which are known to play a given role in other species and the QTL (quantitative trait loci) approach which aims at identifying polymorphic markers which account for a significant part of the diversity for a given trait. Note that, even if a trait, such as fruit colour or the "aromatic" trait is in itself qualitative, the genetic contribution of a locus to this trait is quantitative (and measured by a probability), which justifies a quantitative treatment, through generalized linear models. Marker-assisted selection can be performed by selecting alleles of markers flanking the probable site of a QTL. Whenever feasible, it is more accurate to identify polymorphism within the gene that is responsible for the trait. Conversely, when polymorphism is found in candidate gene, it is still necessary to check whether it accounts for the variability of the trait.

As was shown above, several QTLs were identified using specially designed populations (back-crosses or "second filial" – F_2 – crosses). The drawback is that the whole process takes many years; this is especially true for traits than can be observed only in adult palms (production and disease resistance). An attractive alternative consists of doing genome-wide association studies (GWAS) using already-existing populations. This may require comparing genotypes growing in different locations (on farm populations or germplasm collections). On farm environmental variations may bias the results. In collections, correction for such variations can be done by referring to common germplasm. This is a good reason to systematically plant several widely used cultivars (e.g. West African Tall, Malayan Yellow Dwarf, Rennell Island Tall) as controls in collections.

Taking advantage of the advance of genome studies will require a profound evolution of the methods used for phenotyping. Standard observation methods were designed and widely used for a number of traits, such as fruit components and vegetative traits (IPGRI 1995; Santos et al. 1996), but the protocols are old, not always unambiguously described, and many useful traits, including product quality and biochemical traits, were left out. In most genetic trials, the goal was to identify – or to improve – elite population hybrids, and no attempt was made to record fruit traits or even copra yield at the individual level. Instead, fruit observations are made on a subsample and averaged for the treatment. This is unsuitable for MAS, which require individual values. Observation methods and statistical software evolve rapidly, but recording, conserving, transmitting and recovering data did not follow the advance of information technology. Dedicated software was created at the beginning of the century and maintained until recently ("CGRD Version 6.1" 2012), but is becoming outdated. Efforts should also be made to simplify and standardize data collection and utilization procedures on an individual basis. As an example, coconut would benefit from effort conducted internationally, such as the Breeding Management System (BMS) conducted by the International Breeding Platform (IBP, https://www.integratedbreeding.net/). This is, however, not straightforward, since most crops managed at IBP are annual crops and data organization would have to be adapted, requiring an active involvement of the coconut breeder community, possibly involving COGENT.

7.7.2 Crossing Strategies

One lesson that must be learned from coconut genetic diversity studies is that one third of the species' molecular diversity is accounted for by a single factor: the contrast between the Pacific and Indo-Atlantic genetic group. Genetic distances given in Gunn et al. (2011) (Table 3) imply that 90% microsatellite markers or more are heterozygous (within-population usually ranges usually from 40% to 60%). In other words, many alleles were independently fixed in each of the two groups during the period when communications between the Pacific and Atlantic Oceans were disrupted. This must also be the case within genes of agronomic interest, and this represents a tremendous genetic diversity reservoir to tap into. There is certainly room for innovative crossing plans. For example, the aim could be to introgress one gene (or a small number of genes) into a locally adapted population. This can be done by crossing this population with a donor genotype and performing repeated marker-assisted back-crosses onto the local population. The last step would consist of crossing the selected progenies among themselves to fix the desired alleles (still under marker monitoring). The success of the experiment would be tested by comparing the last generation with the unselected population. Fairly inexpensive seeds could be distributed by natural pollination in an isolated seed garden.

Other approach can be considered, for example, a "synthetic" (or "composite") breeding population combining four domestic and two exotic cultivars was developed in the Philippines (Ohler 1999), and can be a good starting point for marker-assisted breeding. Once the determinism of self-pollination in the Dwarf coconut is understood, advanced generations of Dwarf × Tall hybrids can be used and reproduced in isolated seed gardens.

Genetic engineering is a possible way of introducing specific genes or alleles into coconut varieties. This can use "conventional" transformation methods, e.g. using *Agrobacterium tumefaciens* (Andrade-Torres et al. 2011), or through gene editing using CrispR-Cas9 (Doudna and Charpentier 2014). Such an approach may offer interesting perspectives (e.g. with disease resistance), but it has to meet a triple challenge: the need to validate in the field the benefits of the edited genotype, which takes a considerable amount of time; the need for improved protocols for the regeneration of embryogenic calluses; and the risk of low public acceptance of genetically modified organisms.

Given the various uses of coconut, the range of cultivation environments and the diversity of the economic status of the growers, from the gardener to the smallholder to the large company, there cannot be a single "ideal" variety or a "best" breeding scheme. Population hybrids have provided high copra-yielding varieties and will probably continue to be used. However, they have not always been considered as adapted to the farmer's needs, and they are expensive to produce (at least until male sterile mother palms are available!). Given genetic engineering in coconut is likely to be conditional on the realization of ground-breaking innovation, we suggest that, thanks to marker-assisted selection and genomic breeding, improved varieties will be available. These can then be produced at low cost by natural pollination in cross-pollinating as well as self-pollinating genotypes. But to achieve this aim, breeders and genome scientists need to rely on publicly accessible resources on the coconut genome. We will conclude by presenting a few of these public resources, being aware (and hoping!) that this list will be outdated soon.

7.8 Conclusion

The advance of genome research in coconut in the last 20 years has been considerable. Through molecular marker studies, we have gained a much better understanding of the structure of coconut genome diversity. One of the most important findings is the existence of two well-differentiated gene pools originating from the Indian Ocean and from the Pacific Ocean, respectively. Loci involved in various useful traits such as fruit yield, fruit composition and pest and disease resistance were identified through linkage mapping or association studies. Exploration of the coconut transcriptome provides us with a comprehensive representation of the coconut enzymatic machinery and of its evolution. The variations of its expression according to the organ and to the environmental conditions were studied, providing an insight to the regulation of genome expression. In the future, this research may result in new breeding criteria.

The coconut chloroplastic and mitochondrial genomes have been sequenced, followed by the nuclear genome. The latter was recently anchored onto the 16 linkage groups of a linkage map. In addition to an insight on the evolutionary history of coconut, this chromosome-scale assembly will serve as a backbone for present and future sequencing projects. These sequences, in turn, will allow a deeper understanding of coconut genetic diversity at a genome-wide scale.

These advances offer an opportunity to boost coconut genetic improvement. Achieving this goal will require at least three conditions: an increased effort devoted to phenotyping, observing more traits at an individual level. In addition, progenies of crosses involving genetically distant populations are expected to exhibit a wide range. Early selection based on genetic markers sets can be used to speed up the production of more homogenous breeding populations enriched in desirable allele combinations. The third condition is the open availability of a wide range of genomic resources.

Appendix: Genomic Resources for Coconut – Databases, Data Sets and Useful Tools

The CGRD platform ("CGRD Version 6.1" 2012) offers a wealth of data on ex situ coconut genetic resource collections. It provides information on the microsatellite kit for coconut and data on allele frequencies in many cultivars. Much more complete is the coconut part of TropGeneDB (Hamelin et al. 2013; "Tropgene" 2013) It holds genotypic data on 1791 individuals using up to 30 microsatellite markers. The QTL section lists 63 QTLs for 15 traits, and the marker section documents 296 SSR markers, of which 72 are mapped (along with 196 AFLP and 6 RFLP markers).

PalmComparomics (Armero et al. 2017; "PalmComparomics" 2017) is devoted to comparative genomics of palms. It was constructed before the coconut genome sequence was available and uses the oil palm genome sequence (Singh et al. 2013), as its backbone. In this environment, the user can identify coconut genes matching a given sequence (blast) or, using a number of keywords (annotation terms, gene ontologies, etc.), explore coconut metabolic pathways and more. The site will be considerably enhanced once the coconut genome sequence is integrated. Of course, large amounts of data can be recovered from international databases such as NCBI ("National Center for Biotechnology Information" n.d.), DDBJ ("Bioinformation and DDBJ Center" n.d.) and EMBL-EBI ("The European Bioinformatics Institute < EMBL-EBI" n.d.) and various other specialized databases such as KEGG ("KEGG: Kyoto Encyclopedia of Genes and Genomes" n.d.) and many others.

References

Aljohi HA, Liu W, Lin Q et al (2016) Complete sequence and analysis of coconut palm (*Cocos nucifera*) mitochondrial genome. PLoS One 11(10):e0163990

Alsaihati B, Liu W, Lin Q (2014) Coconut genome de novo sequencing. In: Plant and animal genome XXII conference: plant and animal genome

Altschul S, Gish W, Miller W et al (1990) Basic local alignment search tool. J Mol Biol 215:403–410

Anderson EC, Thompson EA (2001) A model-based method for identifying species hybrids using multilocus genetic data. Genetics 160:1217–1229

Andrade-Torres A, Oropeza C, Sáenz L et al (2011) Transient genetic transformation of embryogenic callus of *Cocos nucifera*. Biologia 66(5):790

Armero A, Baudouin L, Bocs S et al (2017) Improving transcriptome de novo assembly by using a reference genome of a related species: translational genomics from oil palm to coconut. PLoS One 12(3):e0173300

Arunachalam V (2012) Genomics of cultivated palms. Elsevier, Amsterdam

Ashburner GR, Thompson WK, Halloran GM (1997) RAPD analysis of South Pacific coconut palm populations. Crop Sci 37(5–6):992–997

Baker WJ, Couvreur TL (2013) Global biogeography and diversification of palms sheds light on the evolution of tropical lineages. II. Diversification history and origin of regional assemblages. J Biogeogr 40(2):286–298

Bandupriya HDD, Dunwell JM (2016) Transcriptome analysis for discovering candidate genes involve in embryogenesis in coconut (*Cocos nucifera* L.) through 454 pyrosequencing. J Natl Sci Found Sri Lanka 43(4):319–336

Bandupriya HDD, Gibbings JG, Dunwell JM (2014) Overexpression of coconut AINTEGUMENTA-like gene, CnANT, promotes in vitro regeneration in transgenic Arabidopsis. Plant Cell Tissue Organ Cult 116(1):67–79

Batugal P (2005) Coconut genetic resources. Bioversity International, Rome

Batugal P, Bourdeix R, Baudouin L (2009) Coconut breeding. In: Breeding plantation tree crops: tropical species. Springer, New York, pp 327–373

Baudouin L, Lebrun P (2002) The development of a microsatellite kit and dedicated software for use with coconuts. Burotrop Bull 17:16–20

Baudouin L, Lebrun P (2009) Coconut (*Cocos nucifera* L.) DNA studies support the hypothesis of an ancient Austronesian migration from Southeast Asia to America. Genet Resour Crop Evol 56(2):257–262

Baudouin L, Piry S, Cornuet JM (2004) Analytical Bayesian approach for assigning individuals to populations. J Hered 95(3):217–224

Baudouin L, Lebrun P, Konan JL (2006) QTL analysis of fruit components in the progeny of a Rennell Island Tall coconut (*Cocos nucifera* L.) individual. Theor Appl Genet 112(2):258–268

Baudouin L, Lebrun P, Berger A et al (2008) The Panama Tall and the Maypan hybrid coconut in Jamaica: did genetic contamination cause a loss of resistance to Lethal Yellowing? Euphytica 161(3):353–360

Baudouin L, Philippe R, Quaicoe RN et al (2009) General overview of genetic research and experimentation on coconut varieties tolerant/resistant to lethal yellowing. OCL Ol Corps Gras Lipides 16(2):127–131

Baudouin L, Gunn BF, Olsen KM (2014) The presence of coconut in southern Panama in pre-Columbian times: clearing up the confusion. Ann Bot 113(1):1–5

Bioinformation and DDBJ Center (n.d.). Available from https://www.ddbj.nig.ac.jp/index-e.html. Accessed 1 July 2019

Bourdeix R, N'Cho YP, Sangaré A et al (1992) L'hybride de cocotier PB 121 amélioré, croisement du Nain Jaune Malais et de géniteurs Grand Ouest-Africain sélectionnés. Oléagineux 47(11):619–633

Cardena R, Oropeza C, Zizumbo Villarreal D (1998) Leaf proteins as markers useful in the genetic improvement of coconut palms. Euphytica 102:81–86

Cardeña R, Ashburner GR, Oropeza C (2003) Identification of RAPDs associated with resistance to lethal yellowing of the coconut (*Cocos nucifera* L.) palm. Sci Hortic 98(3):257–263
CGRD Version 6.1 (2012). Available from http://www.cogentnetwork.org/cgrd-version-6-0-test-version. Accessed 12 Oct 2017
De Nucé De Lamothe M, Rognon F (1973) La production de semences hybrides chez le cocotier. Exploitation des champs semenciers. Oléagineux 28(6):287–291
Devakumar K, Niral V, Jerard BA (2010) Microsatellite analysis of distinct coconut accessions from Agatti and Kavaratti Islands, Lakshadweep, India. Sci Hortic 125(3):309–315
Doudna JA, Charpentier E (2014) The new frontier of genome engineering with CRISPR-Cas9. Science 346(6213):1258096
Duran Y, Rohde W, Kullaya A et al (1997) Molecular analysis of East African Tall coconut genotypes by DNA marker technology. J Genet Breed 51:279–288
Falush D, Stephens M, Pritchard JK (2003) Inference of population structure using multilocus genotype data: linked loci and correlated allele frequencies. Genetics 164:1567–1587
Falush D, Stephens M, Pritchard JK (2007) Inference of population structure using multilocus genotype data: dominant markers and null alleles. Mol Ecol Notes 7(4):574–578
Fan H, Xiao Y, Yang Y et al (2013) RNA-Seq analysis of Cocos nucifera: transcriptome sequencing and de novo assembly for subsequent functional genomics approaches. PLoS One 8(3):e59997
Freitas Neto M, Pereira TNS, Geronimo IGC et al (2016) Coconut genome size determined by flow cytometry: tall versus dwarf types. Genet Mol Res 15(1). https://doi.org/10.4238/gmr.15017470
Galvez HF, Lantican DV, Sison MLJ (2018) Coconut genetics and genomics for host insect resistance. In PAG – plant and animal genome XXVI conference, San Diego
Gao L, Sun R, Liang Y (2014) Cloning and functional expression of a cDNA encoding stearoyl-ACP Δ9-desaturase from the endosperm of coconut (*Cocos nucifera* L.). Gene 549(1):70–76
Gaut BS, Morton BR, McCaig BC et al (1996) Substitution rate comparisons between grasses and palms: synonymous rate differences at the nuclear gene Adh parallel rate differences at the plastid gene rbcL. Proc Natl Acad Sci 93(19):10274–10279
Geethanjali S, Rukmani A, Rajakumar D et al (2017) Genetic diversity, population structure and association analysis in coconut (*Cocos nucifera* L.) germplasm using SSR markers. Plant Genet Resour 16(2):156–168
Gomez-Navarro C, Jaramillo C, Herrera F et al (2009) Palms (Arecaceae) from a Paleocene rainforest of northern Colombia. Am J Bot 96(7):1300–1312
Gunn BF (2004) The phylogeny of the Cocoeae (Arecaceae) with emphasis on *Cocos nucifera*. Ann Mo Bot Gard 91(3):505–522
Gunn BF, Baudouin L, Olsen KM (2011) Independent origins of cultivated coconut (*Cocos nucifera* L.) in the old world tropics. Plos One 6(6):e21143
Hahn WJ (2002) A phylogenetic analysis of the Arecoid Line of palms based on plastid DNA sequence data. Mol Phylogenet Evol 23(2):189–204
Hamelin C, Sempere G, Jouffe V (2013) TropGeneDB, the multi-tropical crop information system updated and extended. Nucleic Acids Res 41(D1):D1172–D1175
Harries HC (1977) The Cape Verde region (1499 to 1549); the key to coconut in the Western hemisphere? Turrialba 27(3):227–231
Harries HC, Clement CR (2014) Long-distance dispersal of the coconut palm by migration within the coral atoll ecosystem. Ann Bot 113(4):565–570
Herran A, Estioko L, Becke D et al (2000) Linkage mapping and QTL analysis in coconut (*Cocos nucifera* L.). Theor Appl Genet 101(1):292–300
Huang YY, Matzke AJM, Matzke M (2013) Complete sequence and comparative analysis of the chloroplast genome of coconut palm (*Cocos nucifera*). PLoS One 8(8):e74736
Huang YY, Lee CP, Fu JL et al (2014) De novo transcriptome sequence assembly from coconut leaves and seeds with a focus on factors involved in RNA-directed DNA methylation. G3-Genes Genom Genet 4(11):2147–2157
Hubisz MJ, Falush D, Stephens M et al (2009) Inferring weak population structure with the assistance of sample group information. Mol Ecol Resour 9(5):1322–1332

IPGRI (1995) Descriptors for coconut (*Cocos nucifera*). International Plant Genetic Resources Institute, Rome

Jay P, Bourdeix R, Potier F (1991) Polymorphism of coconut leaf polyphenols. In: Coconut breeding and management. Kerala Agricultural University, Vellanikkara, pp 60–68

Kamaral LCJ, Dassanayaka PN, Perera KLNS et al (2016) SSR markers reveal the population structure of Sri Lankan yellow dwarf coconuts (*Cocos nucifera* L.). Tree Genet Genomes 12(6):116

KEGG: Kyoto Encyclopedia of Genes and Genomes (n.d.). Available from https://www.genome.jp/kegg/. Accessed 4 July 2019

Konan KJN, Koffi KE, Konan JL et al (2007) Microsatellite gene diversity in coconut (*Cocos nucifera* L.) accessions resistant to lethal yellowing disease. Afr J Biotechnol 6(4):341–347

Konan KJN, Koffi KE, Konan K (2011) Microsatellite gene diversity within Philippines dwarf coconut palm (*Cocos nucifera* L.) resources at Port-Bouët, Côte d'Ivoire. Sci Res Essays 6(28):5986–5992

Lantican DV, Susan RS, Alma OC et al (2018) The coconut genome: providing a reference sequence towards coconut varietal improvement. In PAG – plant and animal genome XXVI conference, San Diego

Lantican DV, Strickler SR, Canama AO et al (2019) *De novo* genome sequence assembly of dwarf coconut (*Cocos nucifera* L. 'Catigan Green Dwarf') provides insights into genomic variation between coconut types and related palm species. G3-Genes Genome Genet. https://doi.org/10.1534/g3.119.400215

Lebrun P, N'Cho YP, Seguin M (1998) Genetic diversity in coconut (Cocos nucifera L.) revealed by restriction fragment length polymorphism (RFLP) markers. Euphytica 101:103–108

Lebrun P, Baudouin L, Grivet L et al (1999) Genetic diversity of coconut trees. RFLP study of the large collection of the M. Delorme station in Côte d'Ivoire

Lebrun P, Baudouin L, Bourdeix R et al (2001) Construction of a linkage map of the Rennell Island Tall coconut type (*Cocos nucifera* L.) and QTL analysis for yield characters. Genome 44:962–970

Lebrun P, Baudouin L, Myrie W et al (2007) Recent lethal yellowing outbreak: why is the Malayan Yellow Dwarf Coconut no longer resistant in Jamaica? Tree Genet Genomes 4(1):125–131

Liang Y, Yuan Y, Liu T et al (2014) Identification and computational annotation of genes differentially expressed in pulp development of *Cocos nucifera* L. by suppression subtractive hybridization. BMC Plant Biol 14:205

Liu X, Tang H, Li D et al (2011) Genetic diversity of coconut cultivars in China by microsatellite (SSR) markers. Mol Plant Breed 2. https://doi.org/10.5376/mpb.2011.02.0012

Manimekalai R, Nagarajan P (2006) Assessing genetic relationships among coconut (*Cocos nucifera* L.) accessions using inter simple sequence repeat markers. Sci Hortic 108(1):49–54

Martin G, Baurens FC, Droc G (2016) Improvement of the banana "Musa acuminata" reference sequence using NGS data and semi-automated bioinformatics methods. BMC Genomics 17:243

Martinez RT, Baudouin L, Berger A et al (2010) Characterization of the genetic diversity of the Tall coconut (*Cocos nucifera* L.) in the Dominican Republic using microsatellite (SSR) markers. Tree Genet Genomes 6(1):73–81

Masumbuko LI, Sinje S, Kullaya A (2014) Genetic diversity and structure of East African Tall Coconuts in Tanzania using RAPD markers. Open J Genet 4(02):175–181

Maurice EO, Najya M, Kimani NC et al (2016) Assessment of the genetic diversity of Kenyan coconut germplasm using simple sequence repeat (SSR) markers. Afr J Biotechnol 15(40):2215–2223

Mauro-Herrera M, Meerow AW, Perera L et al (2010) Ambiguous genetic relationships among coconut (*Cocos nucifera* L.) cultivars: the effects of outcrossing, sample source and size, and method of analysis. Genet Resour Crop Evol 57(2):203–217

Meerow AW, Wisser RJ, Brown JS et al (2003) Analysis of genetic diversity and population structure within Florida coconut (*Cocos nucifera* L.) germplasm using microsatellite DNA, with special emphasis on the Fiji Dwarf cultivar. Theor Appl Genet 106:715–726

Meerow AW, Noblick L, Borrone JW et al (2009) Phylogenetic analysis of seven WRKY genes across the palm subtribe Attaleinae (Arecaceae) identifies Syagrus as sister group of the coconut. PLoS One 4(10):e7353

Meerow AW, Noblick L, Salas-Leiva DE et al (2014) Phylogeny and historical biogeography of the cocosoid palms (Arecaceae, Arecoideae, Cocoseae) inferred from sequences of six WRKY gene family loci. Cladistics 31(5):509–534

N'Cho YP, Sangaré A, Bourdeix R et al (1993) Evaluation de quelques écotypes de cocotier par une approche biométrique. 1. Etude des populations de Grands. Oléagineux 48(3):121–132

National Center for Biotechnology Information (n.d.). Available from https://www.ncbi.nlm.nih.gov/. Accessed 1 July 2019

Nayar NM (2017) The coconut, phylogeny, origin, and spread. Academic, London

Nejat N, Cahill DM, Vadamalai G et al (2015) Transcriptomics-based analysis using RNA-Seq of the coconut (*Cocos nucifera*) leaf in response to yellow decline phytoplasma infection. Mol Gen Genomics 290(5):1899–1910

Noblick LR, Hahn WJ, Griffith MP (2013) Structural cladistic study of Cocoseae, subtribe Attaleinae (Arecaceae): evaluating taxonomic limits in Attaleinae and the neotropical genus Syagrus. Brittonia 65(2):232–261

Ohler J (1999) Coconut management. Intermediate Technology Publications Ltd. (FAO), London

Ooijen J (2011) Multipoint maximum likelihood mapping in a full-sib family of an outbreeding species. Genet Res Camb 93:343–349

PalmComparomics (2017). Available from http://palm-comparomics.southgreen.fr/. Accessed 12 Oct 2017

Parthasarathy V, Geethalakshmi P, Niral V (2004) Analysis of coconut cultivars and hybrids using isozyme polymorphism. Acta Bot Croat 63(1):69–74

Patiño VM (1963) Plantas cultivadas y animales domésticos en América equinoccial. Tomo I: frutales. Imprenta Departamental, Cali

Perera L, Russel JR, Provan J et al (1998) Evaluating genetic relationships between indigenous coconut (*Cocos nucifera* L.) accessions from Sri Lanka by means of AFLP profiling. Theor Appl Genet 96:545–550

Perera L, Russell JR, Provan J et al (1999) Identification and characterization of microsatellite loci in coconut (*Cocos nucifera* L.) and the analysis of coconut populations in Sri Lanka. Mol Ecol 8(2):335–346

Perera L, Russell JR, Provan J et al (2000) Use of microsatellite DNA markers to investigate the level of genetic diversity and population genetic structure of coconut (*Cocos nucifera* L.). Genome 43(1):15–21

Perera L, Russell JR, Provan J et al (2001) Levels and distribution of genetic diversity of coconut (*Cocos nucifera* L., var. Typica form typica) from Sri Lanka assessed by microsatellite markers. Euphytica 122(2):381–389

Perera L, Russell JR, Provan J et al (2003) Studying genetic relationships among coconut varieties/populations using microsatellite markers. Euphytica 132(1):121–128

Perera L, Baudouin L, Bourdeix R (2011) Coconut palms on the edge of the desert: genetic diversity of Cocos nucifera in Oman. CORD 27(1):9–19

Perera L, Baudouin L, Mackay I (2016) SSR markers indicate a common origin of self-pollinating dwarf coconut in South-East Asia under domestication. Sci Hortic 211:255–262

Perera L, Manimekalai R, Sudarsono S et al (2017) Biotechnology of plantation crops: coconut, pp 219–239

Pérez-Núñez MT, Souza R, Sáenz L et al (2009) Detection of a SERK-like gene in coconut and analysis of its expression during the formation of embryogenic callus and somatic embryos. Plant Cell Rep 28(1):11–19

Piry S, Alapetite A, Cornuet JM et al (2004) GENECLASS2: a software for genetic assignment and first-generation migrant detection. J Hered 95(6):536–539

Pritchard JK, Stephens M, Donnelly P (2000) Inference of population structure using multilocus genotype data. Genetics 155:945–959

Puch-Hau C, Oropeza-Salín C, Peraza-Echeverría S et al (2015) Molecular cloning and characterization of disease-resistance gene candidates of the nucleotide binding site (NBS) type from Cocos nucifera L. Physiol Mol Plant Pathol 89:87–96

Puch-Hau C, Oropeza C, Góngora-Paredes M (2016) New insights into the evolutionary history of resistance gene candidates in coconut palms and their expression profiles in palms affected by lethal yellowing disease. Genes Genomics 38(9):793–807

Rajesh MK, Arunachalam V, Nagarajan P et al (2008a) Genetic survey of 10 Indian coconut landraces by simple sequence repeats (SSRs). Sci Hortic 118(4):282–287

Rajesh MK, Nagarajan P, Jerard BA et al (2008b) Microsatellite variability of coconut accessions (*Cocos nucifera* L.) from Andaman and Nicobar Islands. Curr Sci 94(12):1627–1631

Rajesh MK, Jerard BA, Preethi P et al (2013) Development of a RAPD-derived SCAR marker associated with tall-type palm trait in coconut. Sci Hortic 150:312–316

Rajesh MK, Samsudeen K, Jerard BA et al (2014) Assessment of genetic diversity and phylogenetic relationships among coconut populations from Amini and Kadmat Islands, Lakshadweep (India). Emir J Food Agric 26(10):898–906

Rajesh MK, Rachana KE, Naganeeswaran SA et al (2015) Identification of expressed resistance gene analog sequences in coconut leaf transcriptome and their evolutionary analysis. Turk J Agric For 39:489–502

Ribeiro FE, Baudouin L, Lebrun P et al (2010) Population structures of Brazilian tall coconut (*Cocos nucifera* L.) by microsatellite markers. Genet Mol Biol 33(4):696–702

Ribeiro FE, Baudouin L, Lebrun P et al (2013) Genetic diversity in Brazilian tall coconut populations by microsatellite markers. Crop Breed Appl Biotechnol 13(4):356–362

Riedel M, Riederer M, Becker D et al (2009) Cuticular wax composition in Cocos nucifera L.: physicochemical analysis of wax components and mapping of their QTLs onto the coconut molecular linkage map. Tree Genet Genomes 5(1):53–69

Rivera R, Edwards KJ, Barker JHA (1999) Isolation and characterization of polymorphic microsatellites in *Cocos nucifera* L. Genome 42(4):668–675

Rohde W, Kullaya A, Rodriguez J (1995) Genome analysis of Cocos nucifera L. by PCR amplification of spacer sequences separating a subset of copia-lide EcoRI repetitive elements. J Genet Breed 49(2):179–186

Saensuk C, Wanchana S, Choowongkomon K et al (2016) De novo transcriptome assembly and identification of the gene conferring a "pandan-like" aroma in coconut (*Cocos nucifera* L.). Plant Sci 252:324–334

Santos GA, Batugal PA, Othman A (1996) Manual on standardized research techniques in coconut breeding. IPGRI, Rome

Shalini KV, Manjunatha S, Lebrun P (2007) Identification of molecular markers associated with mite resistance in coconut (*Cocos nucifera* L.). Genome 50(1):35–42

Singh R, Ong-Abdullah M, Low ETL et al (2013) Oil palm genome sequence reveals divergence of interfertile species in old and new worlds. Nature 500(7462):335–339

Srivastava R, Srivastava G (2014) Fossil fruit of Cocos L. (Arecaceae) from Maastrichtian-Danian sediments of central India and its phytogeographical significance. Acta Palaeobot 54(1):67–75

Teulat B, Aldam C, Trehin R (2000) An analysis of genetic diversity in coconut (*Cocos nucifera*) populations from across the geographic range using sequence-tagged microsatellites (SSRs) and AFLPs. Theor Appl Genet 100(5):764–771

The European Bioinformatics Institute < EMBL-EBI (n.d.). Available from https://www.ebi.ac.uk/. Accessed 1 July 2019

Tripathi RP, Mishra SN, Sharma MP (1999) *Cocos nucifera* like petrified fruit from the Tertiary of Amarkantak, M.P, India. Palaeobotanist 48(3):251–255

Tropgene (2013). Available from http://tropgenedb.cirad.fr/tropgene/JSP/interface.jsp?module=COCONUT. Accessed 19 May 2017

Upadhyay A, Jayadev K, Manimekalai R et al (2004) Genetic relationship and diversity in Indian coconut accessions based on RAPD markers. Sci Hortic 99(3–4):353–362

Viveka AT, Moossab F (2016) Identification of novel micro RNAs and their targets in Cocos nucifera–a bioinformatics approach. Biosci Biotechnol Res Commun 9(3):481–488

Wilson MA, Gaut B, Clegg MT (1990) Chloroplast DNA evolves slowly in the palm family (Arecaceae). Mol Biol Evol 7(4):303–314

Xiao Y, Xu P, Fan H et al (2017) The genome draft of coconut (*Cocos nucifera*). GigaScience 6(11):gix095

Yong X, Yi L, Yaodong Y et al (2013) Development of microsatellite in *Cocos nucifera* and their application in evaluating the level of genetic diversity of *Cocos nucifera*. Plant OMICS J 6(3):193–200

Chapter 8
Biotechnology Contributing to Integrated Pest Management: The Example of Two Major Coconut Pests, *Oryctes rhinoceros* and *Brontispa longissima*

Jelfina C. Alouw, Meldy L. A. Hosang, and Quang Nguyen

8.1 Introduction

Coconut (*Cocos nucifera* L.) plays an important role in the economic, social, and cultural life of people in many countries around the world. It is a major source of income for ca. 23 million farmers worldwide. Almost all parts of the palm can be used, ranging from healthy foods, beverages, medicines, cosmetics, houses, furniture, handicrafts, and bioenergy. Coconut is presently cultivated in ca. 12 million ha of land throughout the world (COGENT 2017), producing between 60 and 70 billion fruits per year. However, the present level of production has been struggling to keep up with the growing demand for both the traditional and now many new coconut-derived products. Furthermore, the rapid proliferation of significant pests and diseases, together with the use of outdated agronomic practices, and the high proportion of senile palms in the field are all contributing to this present low productivity.

Efforts to improve crop production often involve intensification and diversification. These efforts transform agricultural ecosystems which can trigger the over-abundance of certain insect pests which were not previously a problem (Kogan 1998). In Indonesia, for instance, major pests and diseases attacking coconut plants include the coconut rhinoceros beetle (CRB) (*Oryctes rhinoceros* L.), coconut leaf

J. C.Alouw (✉)
Indonesian Center for Estate Crops Research and Development, Bogor, Indonesia

M. L. A. Hosang
Indonesian Palm Crops Research Institute, Manado, Indonesia

Q. Nguyen
School of Agriculture and Food Sciences, The University of Queensland, Gatton, QLD, Australia

School of Biotechnology, International University, Vietnam National University, Ho Chi Minh City, Vietnam

S. Adkins et al. (eds.), *Coconut Biotechnology: Towards the Sustainability of the 'Tree of Life'*, https://doi.org/10.1007/978-3-030-44988-9_8

beetle [*Plesispa reichei* (Chapuis)], coconut hispid beetle (CHB) (*Brontispa longissima* Gestro), coconut treehoppers (*Sexava* spp.), and coconut bud rot disease (*Phytophthora* spp.) and nut fall disease (*Fusarium* and *Phytophthora* spp.) (Alouw and Hosang 2008a, b; Hosang and Salim 2014; Kalshoven 1981; Hosang 2013). In Indonesia, yield loss and management to control major coconut pests can cost up to US$ 86 million (Directorate General of Estate Crops 2008).

Historically, the concept of integrated pest management (IPM) emerged and developed to improve the conventional pest control concept which relied too heavily on the use of wide spectrum pesticides (Stern et al. 1959). Early IPM methods focused mainly on integration or incorporation of biological control together with chemical control. Modern approaches tend to combine a wider range of tactics for pest control. These approaches aim to keep pest populations below the economic and environmental thresholds (Norris et al. 2003) and are undertaken in consideration of their social and cultural impacts, as well as their economic sustainability to maintain good crop production. The primary aims of an IPM program are (i) to optimize agricultural productivity, (ii) to improve farmer's welfare, (iii) to manage pest populations, and (iv) to guarantee environmental protection. Pest management tools are grouped into six approaches: (1) cultural, (2) biological, (3) chemical, (4) resistant varieties, (5) physical and mechanical, and (6) control by regulation (Norris et al. 2003).

Recently, biotechnological approaches have been applied which help to identify coconut pests and to determine their origin and genetic relationships to other populations. Biotechnological interventions can also help in the monitoring of pest populations, providing pest management tools and improving the efficacy of pest control products (Gupta and Jindal 2014; Navia et al. 2005; Talukdar 2013). Moreover, recent advances in biotechnology can enable identification of novel genes related to crop resistance or even pest resistance. Up- and downregulation of the novel genes with an objective of plant protection could be undertaken through various ways, including RNA interference (RNAi) (Baum et al. 2007; Mito et al. 2011; Zhang et al. 2017). This chapter provides information on identification or diagnosis of some coconut major pests and diseases and strategies to manage them.

8.2 Coconut Rhinoceros Beetle (CRB)

8.2.1 Life Cycle of CRB

Coconut rhinoceros beetle (CRB) belongs to the order Coleoptera and the family Scarabaeidae, subfamily Dynastinae. This insect undergoes complete metamorphosis from eggs, larvae, pupae, and imago. The female imago lays eggs between 5 and 15 cm below the surface in organic matter such as sawdust, dung heaps, or compost created from rotting coconut biomass. The newly laid eggs are about 2.3 × 3.5 mm in size (Fig. 8.1a). After absorbing water, the eggs expand to 3.7 × 4.0 mm, and the

Fig. 8.1 Life cycle of coconut rhinoceros beetle. (**a**) Eggs of coconut rhinoceros beetle. (**b**) Third instar larva of coconut rhinoceros beetle. (**c**) Pupa of coconut rhinoceros beetle. (**d**) Adult stage of coconut rhinoceros beetle male (left) and female (right)

color changes from cream white to brownish yellow. The incubation period of eggs is between 8 and 12 days. After hatching, the larvae usually eat their egg shells. When CRB adults are given various types of oviposition substrates, they prefer to lay their eggs in small particles of mulch rather than in sand (Manley et al. 2018). The CRB then undergoes three larval instars (Fig. 8.1b) in about 85–200 days (Waterhouse and Norris 1989). In the adult stage, the beetle's body length is around 40 mm; after 24 h, it exits the pupa phase. Newly hatched beetles live in pupa wrappers for 17–22 days (Fig. 8.1c). Male adult can live for about 4–6 months, while female up to 5–9 months (Fig. 8.1d). Female can produce about 512 eggs (Waterhouse and Norris 1987).

8.2.2 *Symptoms and Damage of CRB Attack*

Adult beetles fly into the coconut canopy at night and enter through one of the axils at the top of the canopy (Fig. 8.2a), making a noise while entering the leaf axil. The most preferred entry point is the third, fourth, or fifth frond from the top. Because damage occurs to young leaf midribs, only few beetles are required to cause high

Fig. 8.2 Symptoms of CRB attack can be found as (**a**) damaged leaves in productive coconut palm and (**b**) damaged stem in young coconut palm

losses. In the feeding stage, five beetles ha^{-1} can cause severe damage. Beetles damage the palm by cutting the still folded developing leaves. As the leaves unfold, the damage appears as V-shaped cuts in the fronds, which is the typical symptom of CRB attack, or holes through the midrib. CRB adults could enter and eat the young tissue in the form of a vertical groove toward the growing point, leading to the death of the palm (Bedford 2013a, b; Bedford 2014).

A heavy attack causes death of the palm. At 1 year old or younger, the entry point could be at the base of the stem at ground level (Fig. 8.2b). Damage to the base of the frond or stem could result in the breaking of the frond. The beetle could damage four fronds in a week. While the tissue consumed might only be small, the amount of tissue that is damaged can be a lot (Hosang 2013). Five beetles ha^{-1} could potentially destroy half the population of newly planted palms (Balitka 1989).

The pest can attack throughout the year, being affected only by the availability of food, appropriate breeding sites, and the suitability of the environment (Gnanasegaram et al. 2015). There are several types of breeding sites for CRB which include decaying coconut or oil palm (*Elaeis guineensis* Jacq.) trunks, cow dung, sawdust, rice husks, piles of garbage, rubber tree stumps, and other organic materials. Removal of senile coconut palms could lead to CRB pest problems if the trunks are cut and left unmanaged (Balitka 1989; Kalshoven 1981; Zelazny and Alfiler 1987; Zelazny and Hosang 1991).

8.2.3 Management of CRB Linked with Pest Evolution

Whether management of CRB takes place is generally based on the level of damage on a site and its surrounding areas, the availability of CRB breeding sites and the type of breeding sites, and the climate and natural enemies of CRB in the area. Although several management tactics are available to control CRB, the recent emergence of a new type of beetle, much more aggressive and seemingly resistant to the most efficient method of biological control, poses a new problem (Marshall et al. 2016, 2017). This new form seems to have first appeared on the island of Guam in 2007 and was reported 2 years later in Papua New Guinea and in 2013 in Hawaii and the Solomon Islands.

Control tactics target either the larval stage or the adult stage. These include the use of the biological control agents *Nudivirus* and *Metarhizium*, pheromones, cultivation techniques by planting intercropping and cover crop, sanitation, and the use of naphthalene balls and powdered neem oil cake.

8.2.3.1 Biological Control

The breeding sites of CRB can be infested with entomophagous *Metarhizium anisopliae* Metsch (Sorokin) to help manage the larvae (Bedford 1986; Hosang 2013; Varma 2012). This fungus is most effective in controlling the larvae but can also infect adult beetles (Fig. 8.3). The timing and the technique of applications are important in most regions, as *M. anisopliae* needs a high humidity to grow and develop. In a typical approach, 25 g *M. anisopliae* hyphae in maize (*Zea mays* L.) grains applied per breeding site or approximately two *Metarhizium*-infected larvae to 5 m^2 of breeding site significantly decreases larval population 1 year after application (Balitka 1989; Bedford 2014).

In addition to *M. anisopliae*, there is also *Nudivirus* as a biological control agent that can be applied to control beetles (Jackson and Marshall 2017, Shyam Prasad

Fig. 8.3 *Metarhizium*-infected larvae

et al. 2008; Waterhouse and Norris 1989; Young and Longworth 1981). *Nudivirus* inoculum can be prepared by making a suspension from the midgut taken from infected beetles. The virus inoculum needs to be orally given to healthy beetles, which act as the transmitter of the virus to other adult beetles through mating or other biological interaction (Bedford 1986; Jackson and Marshall 2017; Shyam Prasad et al. 2008). The beetle itself is a very dynamic vector of the virus (Huger 2005). These biological control agents are relatively slow to give a significant result and can vary in effectiveness due to environmental factors, but they potentially can control a pest population for a long period. They are also an environmentally friendly method and can be applied along with other control methods.

Oryctes Nudivirus (OrNv), a nuclear rod-shaped virus, previously known as non-occluded baculoviruses has been widely used as an effective biocontrol agent for controlling CRB in coconut and oil palm (Huger 2005; Young and Longworth 1981) since it was firstly discovered in Malaysia in the 1960s. Swollen midgut is the typical symptom of an adult infected by OrNv. However, resistance to the virus in CRB adults was reported in 2007 from Guam, followed by some other regions and countries, such as Port Moresby, Papua New Guinea, in 2009; Oahu, Hawaii, in 2013; and Honiara, Solomon Islands, in 2015 (Marshall et al. 2017). The resistant CRB that was firstly reported from Guam was called CRB-G (Marshall et al. 2017). The origin of this CRB-G found in Guam and other places and the counteradaptation mechanism by CRB need to be further studied (Fig. 8.4).

Fig. 8.4 Different types of pheromone-based traps: (**a**) pit-fall trap with baffles (large trap surface), (**b**) pit-fall trap with PVC pipe and plastic sheet as hood (small trap surface), (**c**) pit-fall trap with PVC and plastic sheet as hood, (**d**) pit-fall trap with PVC pipe

8.2.3.2 Pheromone-Based Trap

Another alternative tactic to controlling CRB is the use of pheromone-based traps (Fig. 8.5; Alouw 2007a; Bedford 2014). Pheromones are chemical signals released by certain organisms that may stimulate a response from another individual of the same species (Alouw 2007a; Lu et al. 2009; Trona et al. 2013). Aggregation pheromone from the male adult CRB can attract both female and other male adults (Fig. 8.6; Bedford 2013a, b; Gnanasegaram et al. 2015; Hammack 1997). In addition to the role of pheromones as a control tool, they can also be used to monitor the pest populations in the field as part of the strategy to control the pest (Alouw 2007a).

Fig. 8.5 Pupae of *Brontispa longissima*

Fig. 8.6 *Brontispa longissima*: (**a**) female and male, (**b**) antennae, (**c**), adults, (**d**) tip of abdomen of female (left) and male (right)

Indonesian Palm Crops Research Institute (IPCRI) in collaboration with IPB University has evaluated previous pheromone-based traps, from which a new trap (Fig. 8.4) has been developed that can capture a significantly higher number of CRB adult beetles trapped (332–733% increase) (Hosang and Salim 2014; Hosang 2013). One trap in 2 ha of coconuts was highly effective and efficient. For comparison, when the trap was installed in an oil palm plantation, which suffered severe damage of CRB, more than a thousand male and female CRB beetles per trap were captured in 10 days (Hosang and Salim 2014).

8.2.3.3 Cover Crops

Intercrops are planted in order to cover any decaying coconut stems, thereby preventing adults from laying eggs on them (Hosang and Alouw 2005). Cover crops are planted in one part of the garden, 6 months before new coconut palms are planted. Logging of an old plantation should be done at the beginning of the rainy season. Stumps are cut as close as possible to the ground, and then the coconut stem is cut into 2 m sections so that it can be transported by two people. The cut stems are piled up in about three layers (Balitka 1989). Fertilizer can be applied to the cover crops if it is needed to facilitate quick coverage. *Centrocema pubescens* (Balitka 1989) or other local cover crops might be used.

8.3 Coconut Hispid Beetle (CHB)

8.3.1 Symptoms of Brontispa longissima *Attack and Diagnosis*

The typical early damage symptom caused by CHB (*Brontispa longissima* Gestro) (Coleoptera: Chrysomelidae) is the presence of longitudinal brown stripes on the incompletely opened frond. The leaflet then becomes brownish and looks burnt (Fig. 8.7). Leaf damage may reduce photosynthesis which in turn can lead to yield loss.

The CHB recently received international attention, because this pest, which was endemic to Indonesia, Papua New Guinea, and the Solomon Islands, has spread into Asia and the Pacific Islands (Rethinam and Singh 2005; Takano et al. 2011). Both beetles and larvae attack the leaves of seedlings and mature palms (Alouw and Hosang 2008a; Hosang and Alouw 2007). In addition to coconut, the pest can also cause serious damage to ornamental and other palm species. Severe damage can cause abnormal plant growth, lower fruit yield, and even death of the palm (Alouw and Hosang 2010; Hosang et al. 2004b). Within Indonesia the spread of the insect was initially limited to a few areas, in South Sulawesi, Lampung, South Sumatra, West Kalimantan, Irian Jaya, Bali, and Yogyakarta (Rethinam and Singh 2005). But lately the pest has spread rapidly to other coconut plantations in Indonesia.

Fig. 8.7 Symptom of coconut palm and leaflet attacked by *Brontispa longissima*. (**a**) Coconut frond attacked, (**b**) coconut seedlings attacked, (**c**) leaflet chewed by larvae

Molecular analyses using partial sequences of the mitochondrial cytochrome oxidase subunit 1 (*COI*) genes of CHB collected from several locations in Asia and Pacific revealed that CHB consists of two cryptic species now called the Asian and Pacific clades. Differences in certain life-history characteristics between the two clades suggest the need to develop different pest management strategies for each (Takano et al. 2013).

8.3.2 Toward Integrated Management of **Brontispa longissima** *(CHB)*

Potential biological control agents to control CHB are the egg parasitoid *Hispidophila (Haeckeliana) brontispae*, *Trichogrammatoidea nana*, and *Ooencyrtus podontiae*, the larvae and pupae parasitoid *Tetrastichus brontispae*, the entomopathogenic fungi *Metarhizium anisopliae* var. *anisopliae* and *Beauveria bassiana*, the bacterium *Serratia marcescens*, and predator earwig *Chelisoches morio* (Alouw and Hosang 2008a; Alouw et al. 2005; Alouw 2007b; Hosang and Alouw 2014; Hosang et al. 2004b).

8.3.2.1 Entomopathogenic Fungus

The entomopathogenic fungus *M. anisopliae* var. *anisopliae* can infect larva, pupa, and the imago of CHB. Mycelia and conidia are also a source for infection of other insect pests. This fungus can be propagated in several media including one based on corn flour or one using coconut water. The results of field research showed that spraying the suspension of the fungi 2 weeks apart each year effectively controlled CHB (Hosang et al. 2004a).

8.3.2.2 Entomopathogenic Bacterium

Larvae of CHB can be infected by the red bacterium in the field (Fig. 8.8). Molecular identification using polymerase chain reaction (PCR) amplification of 16s ribosomal RNA of the infected larvae followed by sequencing showed that *Serratia marcescens* is the causal agent of this disease (Alouw et al. 2015). This entomopathogenic bacterium was found to be attacking coconut insect pests in several locations in Indonesia (Nusa Tenggara Timur and Kaiwatu Villages, in North Sulawesi). It is thought that this bacterium could be developed into a specific biopesticide for future control of CHB (Alouw et al. 2015; Hosang and Alouw 2014).

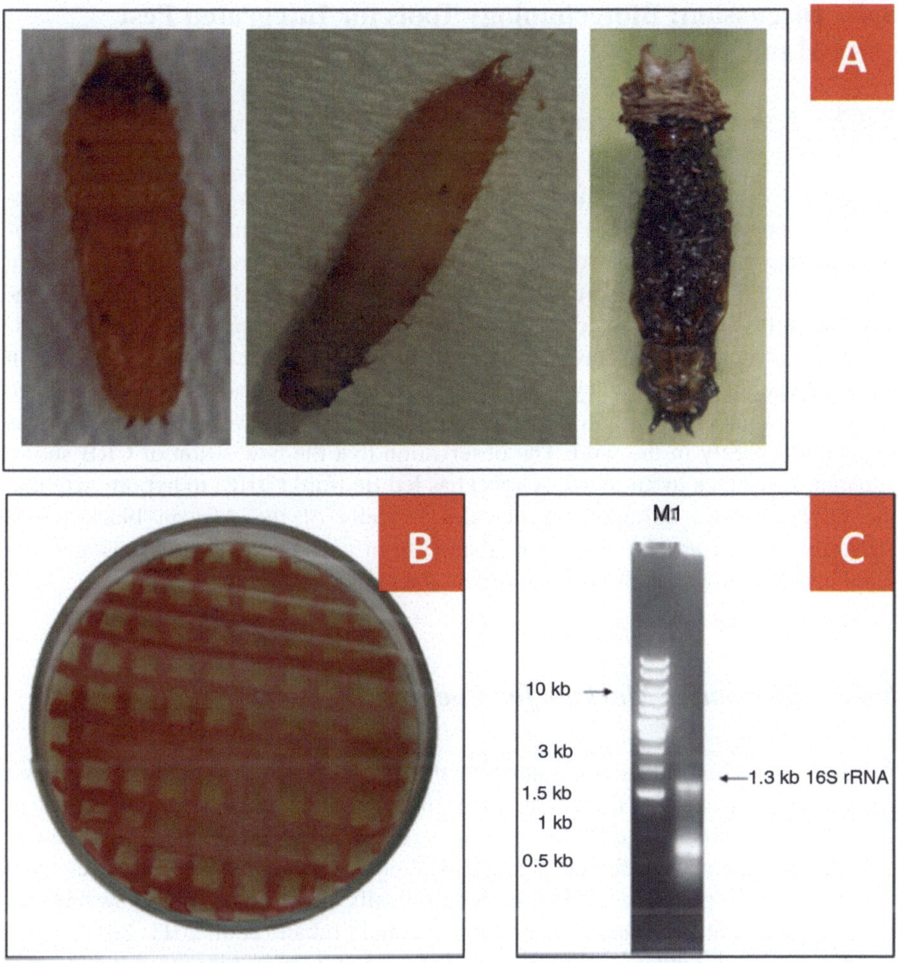

Fig. 8.8 (**a**) Symptom of *B. longissima* larva infected by red-pigmenting bacterium. (**b**) Pure culture of the red-pigmenting bacterium on Luria-Bertani agar. (**c**) Electrophoresis of PCR product of 16S rRNA red-pigmenting bacterium isolated from *B. longissima*. M marker (1 kb), 1 red-pigmenting bacterium

8.3.2.3 Predator Earwig

An earwig (*Chelisoches morio*) is a common predator that preys on all developmental stages of CHB (Alouw and Hosang 2008a; Hosang et al. 2004b; Rethinam and Singh 2005). One earwig can consume 23.5 2nd instar larvae in a single day (Alouw 2007b). The presence of *C. morio* in the coconut fields needs to be encouraged. It was reported that the coconut palm is more susceptible to CHB attack when the planting environment is not maintained well. Therefore, field management should be combined with other control components, to better suppress pest populations.

8.4 Discussion: Biotechnology Tools for Integrated Pest Management

8.4.1 *Laboratory Breeding of Insects and Their Parasites*

Biotechnology has made it possible for the laboratory breeding of many more coconut insect pests and parasites than previously possible. This has led to the development of new biological control methods that require the raising of parasites that are then released in to coconut areas. However, this breeding can have undesirable consequences, by gradually selecting individuals that are adapted to an artificial environment. A recent study in Hawaii showed behavioral differences between lab-grown and wild-caught CRB (Manley et al. 2018). Considering that capture can also induce stress, it can never be certain that captured individuals behave in the same way as those living freely in the wild. The observation that the new strain of CRB shows apparent resistance to nudivirus (OrNv) has led Bedford (2018) to hypothesize that the OrNv isolates cultured for decades in cells of the African black beetle (*Heteronychus arator* F.) may have adapted to in vitro culture conditions and suffered a weakening of their ability to infect the target pest.

8.4.2 *Molecular Markers for Coconut Pest Identification*

Accurate pest identification is fundamental to the success of pest management and entomological research (Yang et al. 2015). Molecular tools allow for the identification of species which are difficult to identify on morphological characteristics alone (Mehle and Trdan 2012; Takano et al. 2013). For example, CHB (which has evolved into two cryptic species, while still morphologically indistinguishable) now requires two different control measures to be implemented (Takano et al. 2011, 2013).

Most molecular analyses undertaken are based on various regions of the internal transcribed spacer (ITS) of the nuclear ribosomal DNA (rDNA) (Andersen et al. 2010; Armstrong and Ball 2005) or on the mitochondrial cytochrome oxidase subunit I (MtCOI) (Ali et al. 2018; Armstrong and Ball 2005; Fišer Pečnikar and Buzan 2014; Miller et al. 2016; Takano et al. 2013). The MtCOI gene was found to be a suitable marker for differentiating the two CHB morphotypes, now called the Pacific and Asian types (Takano et al. 2011). In addition, the MtCOI gene has been used in the accurate identification of the coconut spike moth (*Tirathaba* spp.) that was infesting both oil and coconut palms (Hung 2019). PCR-RFLP of the MtCOI gene following treatment with the BslI, HpyCH4III, or NlaIV restriction enzymes was found to be a suitable tool for differentiating the cryptic species of CHB (Takano et al. 2013).

Identification of various coconut insect pests, through DNA barcoding (Hebert et al. 2003), has become a widely used methodology. A popular approach includes one where a PCR amplification is followed by the sequencing of a short conserved gene region (such as MtCOI) and then comparing that sequence to a reference database (Ashfaq and Hebert 2016; Hebert et al. 2003; Miller et al. 2016). In addition, next-generation DNA barcoding is expected to speed up the identification of many more insects (i.e. > 14,000 chironomids), at a much cheaper price (US$ 0.40 specimen^{-1}) (Baloğlu et al. 2018).

8.4.3 RNA Interference (RNAi) as a Potential Tool for Coconut Pest Management

RNA interference (RNAi – a biological process in which RNA molecules inhibit gene expression or translation) is a great way to understand gene function, and it has been identified as one of the molecular tools that could be used for coconut insect pest management (Baum et al. 2007; Jaba 2018; Pitino et al. 2011; Scott et al. 2013; Yang et al. 2011; Zhao et al. 2011). RNAi has been studied in relation to controlling the red palm weevil (*Rhynchophorus ferrugineus* Olivier) (Coleoptera: Curculionidae) (Al-Ayedh et al. 2016; Malacrinò et al. 2017) and used to study its gene function (Malacrinò et al. 2017).

RNAi is a biological phenomenon naturally found in plants (Zheng et al. 2011), animals, and fungi (Rana 2007) that aids against endogenous parasitic and exogenous pathogenic nucleic acids (Fire 2007). The process, which was first discovered in the round worm nematode (*Caenorhabditis elegans* Maupas), is triggered by the presence of double-stranded RNA (dsRNA) which leads to the degradation of the targeted mRNA (Fire et al. 1998; Hannon 2002).

RNAi technology has been applied in several different research fields including plant-herbivore interactions (Mitra et al. 2008) and pest insect management (Malacrinò et al. 2017; Wang et al. 2011; Zhao et al. 2011). This new technology has good potential for coconut insect pest management. For the management of the red palm weevil, RNAi-mediated gene silencing successfully suppressed the activity of three genes (coding for α-amylase, V-ATPase, and the ecdysone receptor), which are important in different biological processes in the red palm weevil. The silencing of these three genes was achieved through injection or ingestion of the dsRNA at doses of 1,500 and 5,500 ng into the insect. Silencing an olfactory co-receptor (Orco) of *R. ferrugineus*, which functions in finding coconut as a host plant, by injecting 20 μL of 40 ng μL^{-1} ds-RferOrco in to 10-day-old pupae, reduced the expression of RferOrco and therefore the ability of the pest in finding its host (Soffan et al. 2016). Further research should now be focused on using RNAi technology to tackle a wider range of insect pests, which are hampering a sustainable production of the coconut industry.

8.5 Conclusion

The economic impact of insect pests attacking coconut remains high. Biotechnology has now become an essential component of the various systems of IPM. Molecular tools have been used for several purposes including the facilitation of the accurate identification of certain pests. In addition, RNAi technology has recently been used for the management of the red palm weevil. Recent developments in RNAi technology have devised a new tool not only for pest management. Advanced methods now used for laboratory breeding of insects and their parasites have facilitated the biological control program. Furthermore, a combination between biological control and pheromone-based trap might be a new venue to explore.

According to Bourdeix (R. Bourdeix, personal communication), what happened recently with the CRB in the Pacific Region can be compared to the history of the coconut lethal yellowing disease in Jamaica. In both cases, more than 30 years ago, researchers developed effective solutions: a viral biological control agent for the insect pest and the planting of a hybrid tolerant to the phytoplasma. As a result, and perhaps because of these successes, research on these matters has progressively been reduced. However, it should have been expected that after a few decades, and through selection, the pathogens and pests involved would become more tolerant requiring further research to continue to manage them.

The strategies developed for managing coconut pests should be built on their effectiveness, their human and environmental safety, as well as their cost-effectiveness. This requires good coordination between researchers, technicians, farmers, and other stakeholders. Public awareness and dialogue between farmers and biotechnologists should also take place. Researchers should participate, as both trainers and trainees, to address the agricultural constraints such as pest attacks in coconut production. Establishing a "Farmers Field School" would be a good way for developing strategic long-term pest management programs. Regular field evaluation of these management programs and follow-up actions to improve them are also needed. Collaboration among scientists, extension services, farmer groups, and the government could promote efficient pest management implementation. Integration of advanced pest management technologies and traditional practices will contribute to a better crop protection and provide yield stability.

References

Al-Ayedh H, Rizwan-ul-Haq M, Hussain A et al (2016) Insecticidal potency of RNAi-based catalase knockdown in *Rhynchophorus ferrugineus* (Oliver) (Coleoptera: Curculionidae). Pest Manag Sci 72(11):2118–2127

Ali H, Muhammad A, Bala NS et al (2018) Genomic evaluations of Wolbachia and mtDNA in the population of coconut hispine beetle, *Brontispa longissima* (Coleoptera: Chrysomelidae). Mol Phylogenet Evol 127:1000–1009

Alouw JC (2007a) Feromon dan Pemanfaatannya Dalam Pengendalian Hama Kumbang Kelapa *Oryctes rhinoceros* (Coleoptera : Scarabaeidae) (Pheromone and its use to control coconut beetle). Buletin Palma 32:12–21

Alouw JC (2007b) Kemampuan Memangsa Predator Celisoches morio Terhadap Hama Kelapa *Brontispa longissima* (Predatory capacity of Celisoches morio on coconut pest *Brontispa longissima*). Buletin Palma 33:1–8

Alouw JC, Hosang ML (2008a) Observasi Musuh Alami Hama *Brontispa longissima* (Gestro) di Propinsi Maluku (Observation of natural enemies of *Brontispa longissima* (Gestro) in Maluku province). Buletin Palma 35(1):34–42

Alouw JC, Hosang ML (2008b) Survei Hama Kumbang Kelapa *Brontispa longissima* (Gestro) dan Musuh Alaminya di Propinsi SUlawesi Utara (Survey on coconut Hispine beetle *Brontispa longissima* (Gestro) and its natural enemies in North Sulawesi province). Buletin Palma 34(1):9–17

Alouw JC, Lumentut N, Hosang ML (2005) Cendawan entomopatogen Metarhizium anisopliae: Ekobiologi dan penilaian mutu biakannya. Entomopathogen *Metarhizium anisopliae:* ecobiology and quality control. In: Seminar Nasional Pengendalian Hama Terpadu pada Kelapa. (National Seminar on integrated coconut pests management). Balai Peneltiian tanaman Palma, Manado

Alouw JC, Hosang MLA, Heliyanto B (2010) Hama *Brontispa longissima* (Coleoptera: Chrysomelidae): masalah dan pengendaliannya. In: Prosiding Konperensi Nasional Kelapa VII, buku 1. Balai Peneltiian tanaman Palma, Manado

Alouw JC, Novianti D, Meldy D et al (2015) Molecular identification of bacterial pathogen infecting coconut leaf beetle *Brontispa longissima* (Coleoptera:Chrysomelidae) (Identifikasi Molekular Bakteri Pathogen yang Menginfeksi Hama Daun Kelapa Brontispa longissima (Coleoptera:Chrysomelidae)). Buletin Palma 16(2):147–153

Andersen JC, Gruwell ME, Morse GE (2010) Cryptic diversity in the Aspidiotus nerii complex in Australia. Ann Entomol Soc Am 103(6):844–854

Armstrong K, Ball S (2005) DNA barcodes for biosecurity: invasive species identification. Philos Trans R Soc B 360(1462):1813–1823

Ashfaq M, Hebert PDN (2016) DNA barcodes for bio-surveillance: regulated and economically important arthropod plant pests. Genome 59(11):933–945

Balitka (1989) Pengendalian Kumbang kelapa Secara Terpadu. Direktorat Perlintan Perkebunan

Baloğlu B, Clews E, Meier R (2018) NGS barcoding reveals high resistance of a hyperdiverse chironomid (Diptera) swamp fauna against invasion from adjacent freshwater reservoirs. Front Zool 15(1):1–8

Baum JA, Bogaert T, Clinton W et al (2007) Control of coleopteran insect pests through RNA interference. Nat Biotechnol 25(11):1322–1326

Bedford GO (1986) Biological control of the rhinoceros beetle (*Oryctes rhinoceros*) in the South Pacific by baculovirus. Agric Ecosyst Environ 15(2–3):141–147

Bedford GO (2013a) Biology and management of palm dynastid beetles: recent advances. Annu Rev Entomol 58:353–372

Bedford GO (2013b) Long-term reduction in damage by rhinoceros beetle *Oryctes rhinoceros* (L.) (Coleoptera: Scarabaeidae: Dynastidae) to coconut palms at Oryctes Nudivirus release sites on Viti Levu, Fiji. Afr J Agric Res 8(49):6422–6425

Bedford GO (2014) Advances in the control of rhinoceros beetle, *Oryctes rhinoceros* in oil palm. J Oil Palm Res 26(3):183–194

Bedford GO (2018) Possibility of evolution in culture of the Oryctes nudivirus of the coconut Rhinoceros beetle *Oryctes rhinoceros* (Coleoptera: Scarabaeidae: Dynastinae). Advances in Entomology (6):27–33

COGENT (2017) A global strategy for the conservation and use of coconut genetic resources 2018–2028 Bourdeix R and Prades A, compilers). Bioversity International, Montpellier, France

Fire AZ (2007) Gene silencing by double-stranded RNA (Nobel lecture). Angew Chem Int Ed 46(37):6966–6984

Fire A, Xu S, Montgomery M et al (1998) Potent and specific genetic interference by double-stranded RNA in Caenorhabditis elegans. Nature 391(6669):806–811

Fišer Pečnikar Ž, Buzan EV (2014) 20 years since the introduction of DNA barcoding: from theory to application. J Appl Genet 55(1):43–52

Gnanasegaram M, Muhamad R, Tan SG (2015) Genetic variation studies in Oryctes rhinoceros (L.) (Coleoptera : Scarabaeidae) using single locus DNA microsatellite markers. J Entomol Zool Stud 3(2):225–237

Gupta VK, Jindal V (2014) Biotechnological approaches for insect pest management. In: Integrated pest management. Academic Press, pp 311–335

Hammack L (1997) Attractiveness of synthetic corn volatiles to feral northern and western corn rootworm beetles (Coleoptera: Chrysomelidae). Environ Entomol 26(2):311–317

Hannon G (2002) RNA interference. Nature 418(6894):244–251

Hebert PDN, Cywinska A, Ball SL et al (2003) Biological identifications through DNA barcodes. Proc R Soc Lond Biol Sci 270:313–322. https://doi.org/10.1098/rspb.2002.2218

Hosang ML (2013) Serangan Oryctes rhinoceros pada Kelapa Kopyor di Beberapa Sentra Produksi dan Potensi Metarhizium anisopliae sebagai Musuh Alami. Buletin Palma 14(1):47–53

Hosang MLA, Alouw JC (2005) Perbaikan Teknologi PHT untuk hama Oryctes. In: Prosiding Seminar Nasional PHT Kelapa. Balai Peneltiian tanaman Palma, Manado

Hosang MLA, Alouw JC (2014) Parasitoid, predator dan entomopatogen pada hama kelapa Brontispa longissima (Gestro). In: Prosiding Kongres VIII dan Sminar Nasional Perhimpunan Entomologi Indonesia. Perhimpunan Entomologi Indonesia (PEI), Bogor

Hosang MLA, Salim (2014) Penekanan populasi Oryctes rhinoceros dan Rhynchophorus ferrugineus dengan perangkap feromon. In: Prosiding Konferensi Nasional Kelapa (KNK) VIII, p 67–72

Hosang ML, Alouw JC, Novarianto H (2004a) Biological control of Brontispa longissima (Gestro) in Indonesia. In: Report of the expert consultation on coconut beetle outbreak in APPPC member countries. FAO Regional Office for Asia and the Pacific, Bangkok, pp 26–27

Hosang ML, Alouw JC, Novarianto H (2004b) Biological control of Brontispa longissima (Gestro) in Indonesia. In: Report of the expert consultation on coconut beetle outbreak in APPPC member countries, Bangkok, pp 39–52

Hosang MLA, Tumewan F, Alouw JC (2007) Efektivitas cendawan entomopatogen Metarhizium anisopliae vr. anisopliae dan Beauveria bassiana terhadap hama Brontispa longissima. In: Simposium IV Hasil Penelitian Tanaman Perkebunan. Buku 3. Pusat Penelitian dan Pengembangan Perkebunan, Bogor

Huger A (2005) The Oryctes virus: its detection, identification, and implementation in biological control of the coconut palm rhinoceros beetle, Oryctes rhinoceros (Coleoptera: Scarabaeidae). J Invertebr Pathol 89:78–84

Hung PKJ (2019) They are different: molecular approach on Tirathaba pest infesting oil palm and coconut tree. Adv Plants Agric Res 8(1):71–73. https://doi.org/10.15406/apar.2018.08.00294

Jaba J (2018) Advance towards host mediated RNA interference insect pest management. Adv Biotechnol Microbiol 5(4):1–6

Jackson TA, Marshall SDG (2017) The role of Oryctes nudivirus in management of the coconut rhinoceros beetle. Japanese Society for Insect Pathology, Tokyo

Kalshoven LG (1981) The pests of crops in Indonesia. PT. Ichtiar Baru, Van Houve

Kogan M (1998) Integrated pest management: historical perspectives and contemporary developments. Annu Rev Entomol 43(1):243–270

Lu CF, Zhang SB, Li Y et al (2009) Stereoselective synthesis of (R)-10-methyltridecan-2-one, the sex pheromone of the southern corn rootworm, using (4S)-benzylthiazolidinethione as a chiral auxiliary. Tetrahedron Asymmetry 20(19):2267–2269

Malacrinò A, Strano CP, Gatehouse AMR et al (2017) RNAi-mediated gene silencing in Rhynchophorus ferrugineus (Oliver) (Coleoptera: Curculionidae). Open Life Sci 12(1):214–222

Manley M, Melzer M, Spafford H (2018) Oviposition preferences and behavior of wild-caught and laboratory-reared coconut rhinoceros beetle, Oryctes rhinoceros (Coleoptera: Scarabaeidae), in relation to substrate particle size. Insects 9(4):141

Marshall SDG, Moore A, Vaqalo M et al (2017) A new haplotype of the coconut rhinoceros beetle, Oryctes rhinoceros, has escaped biological control by Oryctes rhinoceros nudivirus and is invading Pacific Islands. J Invertebr Pathol 149:127–134

Mehle N, Trdan S (2012) Traditional and modern methods for the identification of thrips (Thysanoptera) species. J Pest Sci 85(2):179–190

Miller SE, Hausmann A, Hallwachs W et al (2016) Advancing taxonomy and bioinventories with DNA barcodes. Philos Trans R Soc B 371(1702):20150339

Mito T, Nakamura T, Bando T et al (2011) The advent of RNA interference in entomology. Entomol Sci 14(1):1–8

Mitra S, Wuensche H, Giri A (2008) Silencing 7 herbivory-regulated proteins in Nicotiana attenuata to understand their function in plant–herbivore interactions. Funct Ecol 22:606–615

Navia D, De Moraes GJ, Roderick G et al (2005) The invasive coconut mite Aceria guerreronis (Acari: Eriophyidae): origin and invasion sources inferred from mitochondrial (16S) and nuclear (ITS) sequences. Bull Entomol Res 95(6):505–516

Norris, Robert F, Chen C et al (2003) Concepts in integrated pest management. No. 632.9 N6

Pitino M, Coleman AD, Maffei ME et al (2011) Silencing of aphid genes by dsRNA feeding from plants. PLoS One 6(10):e25709

Rana TM (2007) Illuminating the silence: understanding the structure and function of small RNAs. Nat Rev Mol Cell Biol 8(1):23–36

Rethinam P, Singh S (2005) Asian and Pacific Coconut Community current status of coconut beetle outbreak in Asia Pacific region. Asian Pac Commun 1:1–10. Retrieved from http://www.apccsec.org

Scott JG, Michel K, Bartholomay LC et al (2013) Towards the elements of successful insect RNAi. J Insect Physiol 59(12):1212–1221

Shyam Prasad G, Jayakumar V, Sharma TVRS (2008) Management of coconut rhinoceros beetle (Oryctes rhinoceros) by augmentation of Oryctes baculovirus (Kerala isolate) in little Andaman Islands. Indian J Agric Sci 78(11):962–965

Soffan A, Antony B, Abdelazim M et al (2016) Silencing the olfactory co-receptor RferOrco reduces the response to pheromones in the red palm weevil, Rhynchophorus ferrugineus. PLoS One 11(9):e0162203. https://doi.org/10.1371/journal.pone.0162203

Stern VM, Smith RF, Van den Bosch R et al (1959) The integrated control concept. Hilgardia 29(2):81–101

Takano SI, Mochizuki A, Konishi K et al (2011) Two cryptic species in Brontispa longissima (Coleoptera: Chrysomelidae): evidence from mitochondrial DNA analysis and crosses between the two nominal species. Ann Entomol Soc Am 104(2):121–131

Takano SI, Mochizuki A, Takasu K et al (2013) Rapid discrimination of two cryptic species within Brontispa longissima (Gestro) (Coleoptera: Chrysomelidae) by PCR-RFLP. J Pest Sci 86(2):151–155

Talukdar D (2013) Modern biotechnological approaches in insect research. Int Res J Sci Eng 1(3):71–78

Trona F, Anfora G, Balkenius A et al (2013) Neural coding merges sex and habitat chemosensory signals in an insect herbivore. Proc R Soc B Biol Sci 280(1760):20130267

Varma Y (2012) Oryctes rhinoceros management. J Biopest 5(1):1–6

Wang Y, Zhang H, Li H et al (2011) Second-generation sequencing supply an effective way to screen RNAi targets in large scale for potential application in pest insect control. PLoS One 6(4):e18644

Waterhouse DF, Norris KR (1987) Biological control: Pacific prospects. ACIAR, Canberra

Waterhouse DF, Norris KR (1989) Biological control: Pacific prospects. ACIAR, Canberra

Yang G, You M, Vasseur L (2011) Development of RNAi in insects and RNAi-based pest control. In: Pesticides in the modern world-pests control and pesticides exposure and toxicity assessment. IntechOpen, Shanghai

Yang HP, Ma CS, Wen H et al (2015) A tool for developing an automatic insect identification system based on wing outlines. Sci Rep 5(1):12786

Young EC, Longworth JF (1981) The epizootiology of the baculovirus of the coconut palm rhinoceros beetle (*Oryctes rhinoceros*) in Tonga. J Invertebr Pathol 38(3):362–369

Zelazny B, Alfiler A (1987) Ecological methods for adult population of Oryctes rhinoceros (Coleoptera: Scarabaeidae). Ecol Entomol 12(2):227–238

Zelazny B, Hosang ML (1991) Estimating defoliation of coconut palms by insect pests. Trop Pest Manag 37(1):63–65

Zhang J, Khan SA, Heckel DG et al (2017) Next-generation insect-resistant plants: RNAi-mediated crop protection. Trends Biotechnol 35(9):871–872

Zhao YY, Liu F, Yang G et al (2011) PsOr1, a potential target for RNA interference-based pest management. Insect Mol Biol 20(1):97–104

Zheng SJ, Zhang PJ, van Loon JJA et al (2011) Silencing defense pathways in Arabidopsis by heterologous gene sequences from *Brassica oleracea* enhances the performance of a specialist and a generalist herbivorous insect. J Chem Ecol 37:818–829

Chapter 9
Dealing with Lethal Yellowing and Related Diseases in Coconut

Carlos Oropeza-Salín, Luis Sáenz, Maria Narvaez, German Nic-Matos, Ivan Córdova, Wayne Myrie, Carlos F. Ortíz, and Eder Ramos

9.1 Introduction

The coconut is a plant species of economic and social importance that is cultivated in more than 90 countries around the world. There is a great variety of products that can be obtained from this palm. Their markets, particularly for coconut water, coconut milk, virgin coconut oil and coconut sugar, have been growing exponentially within the past decade (Prades et al. 2016). This represents a very promising future for coconut cultivation and the whole industry. Unfortunately, this growth is threatened by a reduction in fruit production. This is because most palms are now senile and declining in production. In addition, phytoplasma-associated diseases such as lethal yellowing (LY) and similar LY-type diseases (LYDs) have been devastatingly affecting coconut palms in countries of Latin America and the Caribbean (LAC), and Africa (Gurr et al. 2016; Ntushello et al. 2013). These incurable diseases have destroyed millions of coconut palms, as well as the livelihoods of the affected farmers. Thus, it is crucial to develop means and strategies to combat these diseases. This chapter reviews different aspects and approaches that are of importance to management or control.

C. Oropeza-Salín (✉) · L. Sáenz · M. Narvaez · G. Nic-Matos · I. Córdova
Centro de investigación Científica de Yucatán (CICY), Mérida, México
e-mail: cos@cicy.mx

W. Myrie
Coconut Industry Board (CIB), Kingston, Jamaica

C. F. Ortíz · E. Ramos
Colegio de Postgraduados (CP), Cárdenas, Mexico

S. Adkins et al. (eds.), *Coconut Biotechnology: Towards the Sustainability of the 'Tree of Life'*, https://doi.org/10.1007/978-3-030-44988-9_9

9.2 Palm Phytoplasma Diseases in the World

Sightings of coconut palms dying with symptoms like those of LY have been reported since the nineteenth century in Jamaica, Cuba and the Cayman Islands (Ntushello et al. 2013). The disease then spread to other Caribbean Island countries, reaching the continental land, the USA (Florida) and Mexico within the second half of the twentieth century, subsequently moving further south to Honduras (Table 9.1 and Fig. 9.1). In the Caribbean, LY has spread rapidly, entering Antigua by 2010 (Myrie et al. 2014). The outbreak of LY has resulted in the death of millions of coconut palms and corresponding damage to coconut cultivation in LAC. It is interesting that this extensive death of palms has occurred on the Atlantic Ocean side or the east of the Americas, where most coconut were of the Atlantic Tall variety (also known as Jamaican Tall or Brazilian Tall depending on the country), which has been the most extensively cultivated variety. Unfortunately, it happened to be the most susceptible variety to LY. In contrast, LY has been basically absent on the Pacific Ocean or west side of the Americas, where there is a greater diversity of coconut germplasm and the Atlantic Tall variety is not cultivated.

These differences in germplasm between Atlantic and Pacific regions are related to the origin of the introductions of coconut. In the case of Mexico, the Atlantic Tall was originally introduced only to ports on the Atlantic side of Mexico from Cape Verde in Africa, via Puerto Rico ca. 1549 (Zizumbo-Villarreal 1996). In the case of the Pacific coast, other coconut germplasm was introduced from Panama (ca. 1539), Solomon Islands (ca. 1569) and Philippines (between 1571 and 1816) (see Zizumbo-Villarreal 1996). This resulted in a single variety introduced to the east and a great diversity introduced to the west of Mexico and this was the same for the Americas.

LYDs affecting coconut have also been reported in other parts of the world (Table 9.1 and Fig. 9.1). In Africa, LYDs were first observed in Nigeria (West Africa) in the early half of the twentieth century and in Tanzania (East Africa) in the latter half of the twentieth century (Yankey et al. 2018). Other countries in Africa have been also affected. Noticeably, recent outbreaks in Mozambique have killed millions of coconuts while threatening thousands of hectares in Ivory Coast (Yankey et al. 2018) and the COGENT (Coconut Genetic Resources for Enhanced Livelihoods) International Coconut Genebank in Abidjan. Instances of LYDs have also been reported in member countries of the International Coconut Community (ICC) (see Gurr et al. 2016). Occurrences of LYDs affecting other palm species have also been reported in countries in different continents (Table 9.1).

9.3 LY Symptoms

The first visual symptom of LY infection in coconut-bearing palms is the premature drop of most of the fruit regardless of their developmental stage (Fig. 9.2a), followed by the blackening of new inflorescences (Fig. 9.2b and c). This symptom is

Table 9.1 Diversity of phytoplasmas associated with lethal yellowing-type diseases of palm and non-palm species

Species	16SrDNA group-subgroup	Country	Species-disease name	Author
Ca. *Phytoplasma palmae*	Presumably 16SrIV-A[a]	Cuba Belize Honduras Guatemala Haiti Cayman Islands	Lethal yellowing	Llauger et al. (2002) Escamilla et al. (1994) Ashburner et al. (1996) Mejía et al. (2004) No confirmed report No confirmed report
	16SrIV-A	Jamaica Florida (USA) Nevis Saint Kitts Antigua Dominican Republic	Lethal yellowing	Harrison et al. (2002a) Harrison et al. (2002a) Myrie et al. (2006) Myrie et al. (2012) Myrie et al. (2014) Feliz et al. IDIAF unpublished
	16SrIV-B	Mexico	Coconut lethal decline	Harrison et al. (2002b)
		Honduras	Coyol palm decline Coconut decline	Roca et al. (2006)
	16SrIV-D	Mexico	Carludovica palmae, lethal decline Coconut leaf yellowing *Sabal mexicana* foliar decay *Pseudophoenix sargentii* decline	Córdova et al. (2000) Harrison et al. (2002b) Vázquez-Euán et al. (2011) Vázquez-Euán et al. (2011)
		USA	Phoenix palm decline (Texas) Pigmy date palm decline (Florida) Phoenix palm decline (Louisiana)	Harrison et al. (2002c) Jeyaprakash et al. (2011) Singh (2014)
	16SrIV-E	Dominican Republic	Coconut lethal decline	Martínez et al. (2008)
	16SrIV-F	USA (Florida)	*Washingtonia robusta* decline	Harrison et al. (2008)
Ca. *Phytoplasma asteris*	16SrI-B	Colombia	Oil palm lethal wilt	Álvarez et al. (2014)
Ca. *Phytoplasma cocostanzaniae*	16SrIV-C	Kenya, Tanzania	Tanzanian coconut lethal decline	Harrison et al. (2002c)
Ca. *Phytoplasma palmicola*	16SrXXII-A	Mozambique, Nigeria	Lethal yellowing disease	Harrison et al. (2014)

[a]There is no report with subgroup identification

(continued)

Table 9.1 (continued)

Species	16SrDNA Group-Subgroup	Country	Species-Disease Name	Author
Ca. *Phytoplasma palmicola*-related strain	16SrXXII-B	Ghana Côte d'Ivoire	Cape St Paul wilt disease Côte d'Ivoire lethal yellowing	Harrison et al. (2014)
Ca. *Phytoplasma cynodontis*	16SrXIV[a]	Malaysia	Coconut yellow decline	Nejat et al. (2009)
Ca. *Phytoplasma malaysianum*	16SrXXXII-B	Malaysia	Coconut yellow decline	Nejat et al. (2012)
	16SrXXXII-C	Malaysia	Malayan oil palm disease	Nejat et al. (2012)
	16SrXI-B	India	Areca palm yellow leaf disease	Ramaswamy et al. (2013)
Ca. *Phytoplasma oryzae*	16SrXI-B	India	Kerala root wilt disease	Manimekalai et al. (2014)
	16SrXI[a]	India	Oil palm spear rot disease	Sumi et al. (2014)
Ca. *Ptytoplasma cynodontis*	16SrXIV[a]	India	Kerala root wilt disease Areca palm yellow leaf disease	Sumi et al. (2014)
Ca. *Phytoplasma asteris*	16SrI-B	India	Oil palm stunting disease	Mehdi et al. (2012)
Not identified	Not identified	Indonesia	*C. nucifera* Kalimantan wilt	Warokka et al. (2006)
Ca. *Phytoplasma oryzae*	16SrXI[a]	Sri Lanka	Weligama coconut leaf wilt disease (WCLWD)	Perera et al. (2012)
Tentative classification	16SrIV[a]	Papua New Guinea	Bogia coconut syndrome (BCS)	Kelly et al. (2011)
Ca. *Phytoplasma cynodontis*	16SrXIV[a]	Sudan	Date palm disease	Cronje et al. (2000)
Not identified	Not identified	Egypt	Date palm disease	Ammar et al. (2005)
Not identified	Not identified	Egypt	Date palm disease	Al Khazindar (2014)
Ca. *Phytoplasma palmae*	16SrIV-A	Kuwait	Date palm disease	Al-Awadhi et al. (2002)
Ca. *Phytoplasma fraxini*	16SrVI-A	Iran	Date palm disease	Zamharir and Eslahi (2019)
Ca. *Phytoplasma trifolii*	16SrVII-A	Iran	Date palm disease	Zamharir and Eslahi (2019)
Ca. *Phytoplasma australasia*	16SrII[a]	Saudi Arabia	Date palm disease	Omar et al. (2018)
Ca. *Phytoplasma asteris*	16SrI[a]	Saudi Arabia	Date palm disease	Alhudaib et al. (2007)

[a]Unclassified subgroup

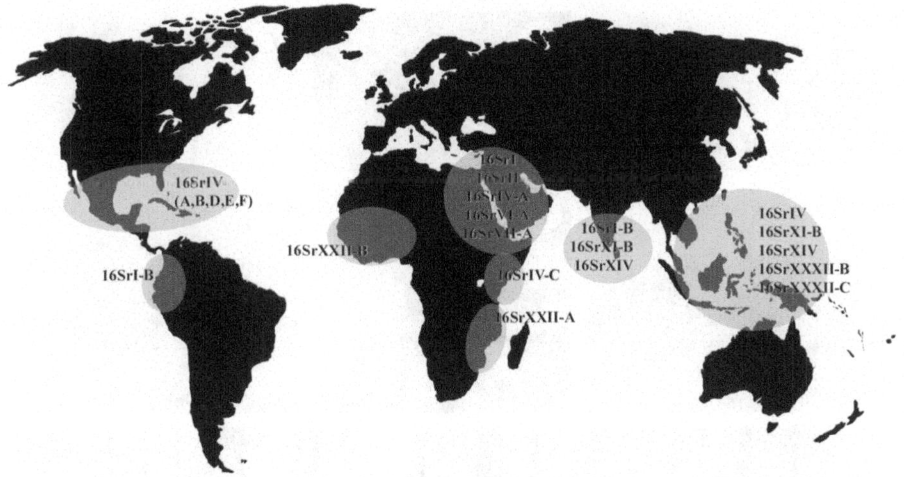

Fig. 9.1 Geographic distribution of phytoplasmas causing lethal yellowing and related diseases in the world. (This map is based on a map reported by Konan et al., COCOTECH Conference, Bali, September 2016)

most apparent when the inflorescence emerges from the spathe. The necrosis increases as the disease progresses, with younger inflorescences showing more extensive necrosis. Most of the male flowers die and no fruit set on affected inflorescences. Yellowing of the leaves usually starts after necrosis has developed in more than two inflorescences. Leaf discoloration due to LY is more rapid than normal leaf senescence. The first leaves to turn yellow are the oldest (lower) ones (Fig. 9.2d). The yellowing advances upward, affecting the younger middle leaves (Fig. 9.2e), and finally the upper ones (Fig. 9.2f). According to McCoy et al. (1983), yellow leaves are turgid and not flaccid as in the case of wilt diseases. Yellow leaves turn brown, desiccate and die. Symptom development was standardized by McCoy et al. (1983) and classified as ten different categories, from zero for healthy palms to nine for dead palms (Table 9.2). This classification system has been proven to be very useful; however, sometimes the pattern varies. For instance, inflorescence necrosis can become noticeable only after leaf yellowing has appeared as observed in Mexico (CICY, Mexico, unpublished) and Guatemala (Mejía et al. 2004).

Once foliar yellowing has reached an advanced stage, a putrid basal soft rot of the youngest leaf (spear) occurs. The spear leaf collapses, followed by an associated rot of the underlying apical meristem, invariably leaving a bare trunk standing (Fig. 9.2g). Roots also show necrosis, which becomes more extensive as the disease progresses (Islas-Flores et al. 1999). The death of the infected palm occurs within 3–6 months after the onset of visible symptoms (McCoy et al. 1983). Symptoms of LY in other palms are generally similar with some variations (McCoy et al. 1983), for example, the Manila palm (*Adonidia merrilllii* Becc.) could present necrosis of the mature leaves or necrosis of the spear leaf or spear leaf opening. However, these symptoms are apparently associated with different subgroups of LY-phytoplasmas

Fig. 9.2 Symptoms of lethal yellowing in fruit-bearing coconut palms: it starts with the premature drop of most of the fruits regardless of their developmental stage (**a**). Then there is blackening of new inflorescences (**b** and **c**). This is followed by yellowing starting with the oldest leaves (**d**), and then advances upward affecting the middle leaves (**e**) and the upper youngest leaves (**f**). Finally, the loss of the crown leaves a bare trunk standing (**g**)

Table 9.2 Rating of lethal yellowing development in coconut (as shown in McCoy et al. 1983)

Category	Stage	Symptoms
Symptomless	0	Healthy or incubating
	1	Nut fall only
Primary[a]	2	One necrotic inflorescence
	3	Two or more necrotic inflorescences
	4	Yellowing in lower leaves only
Yellowing	5	Yellowing in lower and middle leaves
	6	All leaves yellowed, spear leaf good
	7	Spear leaf dead, some green leaves left
Dying	8	Spear leaf dead, all leaves yellowed
	9	Palm dead (telephone pole)

[a]May or may not have a yellow flag leaf in the centre of crown

(Córdova-Lara et al. 2017). Symptoms of coconuts affected by LYDs in West Africa and Tanzania are similar (Yankey et al. 2018) to those described in this chapter for LY in the Americas.

9.4 Causal Agent Identification and Classification

Lethal yellowing in coconut was the first phytoplasma-associated disease found in palm species. Previously known as mycoplasma-like organisms, phytoplasmas were discovered, by electron microscopy, within the phloem vessels of diseased palms but not in healthy palms (Plavsic-Banjac et al. 1972). A cause-and-effect relationship was established when remission of symptoms was obtained in diseased palms when treated with tetracycline. However, it becomes ineffective when applying penicillin antibiotics (McCoy 1982).

As the LY-causing agent is unable to be cultured, its characterization was impossible until the advent of deoxyribonucleic acid (DNA) sequencing and relevant analytical techniques, using in silico Restriction Fragment Length Polymorphism (RFLP) analysis with the online system iPhyClassifier and data from the GenBank sequence database (Zhao et al. 2009). The LY agent has been identified and classified within the 16SrIV group, as well as other strains that are closely related to but distinguishable from the LY agent. Group 16SrIV includes the following subgroup strains: 16SrIV-A, Coconut Lethal Yellowing (LY, Florida USA) (Harrison et al. 2002a); 16SrIV-B, Yucatan Coconut Lethal Decline (LDY, México) (Harrison et al. 2002b); 16SrIV-D, *Carludovica palmata* Leaf Yellowing (CPY, Mexico) (Córdova et al. 2000) and Texas Phoenix Decline (TPD, USA) (Harrison et al. 2002c); 16SrIV-E, Coconut Lethal Decline (CLD, Dominican Republic) (Martínez et al. 2008); and 16SrIV-F, Florida *Washingtonia robusta* Lethal Disease (LD, USA) (Harrison et al. 2008). All are present in the Americas but nowhere else in the world so far, but there is another subgroup that is present in Africa but not in the Americas: 16SrIV-C, Tanzanian Coconut Lethal Decline (LDT, Tanzania) (Harrison et al. 2002c). Tables 9.1 and 9.3 presents more information on each subgroup with other cases within one country or being present in different countries in the Americas. In the case of LYDs outside the Americas, they have also been associated with phytoplasmas of other 16S rDNA groups as shown in Tables 9.1 and 9.3.

9.5 Transmission

According to the surveys conducted in LY-affected areas in Jamaica (Schuiling 1976) and in Florida (Howard and McCoy 1980), the only common species found in coconut palms in both locations was a leafhopper (*Haplaxius crudus* Van Duzee, 1907). This pest belongs to Auchenorrhyncha, sub-order of Homoptera, and is the most phytoplasma-associated vector. In addition, the apparent rate of spread of LY

Table 9.3 Palm species (other than coconut) and non-palm species hosting phytoplasmas associated with LY and related diseases

Plant species	Common name	Country	Strain	Author
Acrocomia aculeata	Coyol palm	Honduras	16SrIV-B 16SrIV-D	Roca et al. (2006) Ntushelo et al. (2013)
Adonidia merrillii	Manila palm	USA	16SrIV-A	Harrison and Oropeza (2008)
		Mexico	16SrIV-A 16SrIV-D	Córdova-Lara et al. (2017)
Aiphanes lindeniana	Ruffle palm	USA	16SrIV	Harrison and Oropeza (2008)
Alagoptera arenaria	Seashore palm			
Arenga engleri	Dwarf sugar palm			
Borassus flabellifer	Palmyra palm			
Caryota mitis	Giant fishtail palm	Puerto Rico	16SrIV-D	Rodriguez et al. (2010)
Caryota rumphiana	Round leaf palm	USA	16SrIV	Harrison and Oropeza (2008)
Caryota urens	Wine palm			
Chelyocarpus chuco	Round leaf palm			
Cocos nucifera	Coconut	–	–	See Table 9. 1
Coccotrinax readii	Nakás palm	Mexico	16SrIV-A	Narvaez et al. (2006)
Corypha taliera	Buri palm	USA	16SrIV	Harrison and Oropeza (2008)
Crysophyla warsecewiczii	Guáguara palm			
Cyphophoenix nucele	Lifou palm			
Dictyosperma album	Princess palm			
Dypsis cabadae	Cabada palm			
Dypsis decaryi	Triangle palm			
Elaeis guineensis	Oil palm	Colombia	16SrI-B;	Álvarez et al. (2014)
		Mozambique	16SrXXII-A	Bila et al. (2015)

Gaussia attenuata	Puerto Rican Gaussia palm	USA	16SrIV	Harrison and Oropeza (2008)
Howea belmoreana	Belmore sentry palm			
Howea forsteriana	Kentia palm			
Hyophorbe verschaffeltii	Spindle palm			
Latania lontaroides	Latan palm			
Livistona chinensis	Chinese fan palm			
Livistona rotundifolia	Footstool palm			
Nannorrhops ritchiana	Mazari palm			
Phoenix canariensis	Canary Island date palm	USA	16SrIV-A 16SrIV-D	Harrison et al. (2002, 2008)
			16SrIV-D	Ntushello et al. (2013)
Phoenix dactylifera	Edible date palm	USA	16SrIV-A 16SrIV-E 16SrIV-F	Harrison et al. (2008)
Phoenix reclinata	Senegal date palm	USA	16SrIV-D	Harrison et al. (2008, 2009)
Phoenix rupícola	Cliff date palm			
Phoenix silvestris	Silver date palm			
Phoenix roebelenii	Pygmy date palm			
Pritchardia pacifica	Fiji island fan palm	Mexico	16SrIV-D	Narvaez et al. (2017)
Pritchardia affinis	Kona palm	USA	16SrIV	Harrison and Oropeza (2008)
Pritchardia pacifica	Fiji island fan palm			
Pritchardia remota	Remota loulu palm			
Pritchardia thurstonii	Thurston palm			

(continued)

Table 9.3 (continued)

Plant species	Common name	Country	Strain	Author
Pseudophoenix sargentii	Florida cherry palm	USA	16SrIV-D	Harrison et al. (2008)
		Mexico		Vázquez-Euán et al. (2011)
Ravenea hildebrantii	Hildebrandt's palm	USA	16SrIV	Harrison and Oropeza (2008)
Roystonea regia	Cuban royal palm	Puerto Rico	16SrIV-D	Rodriguez et al. (2010)
		Mexico		Narvaez et al. (2016)
Syagrus schizophylla	Arikury palm	USA	16SrIV	Harrison and Oropeza (2008)
Syagrus romanzoffiana	Queen palm	USA	16SrIV-D	Harrison et al. (2008) and Ntushelo et al. (2013)
Sabal palmeto	Cabbage-palm	USA	16SrIV-D	Harrison et al. (2009)
Sabal mexicana	Mexican palmetto	Mexico	16SrIV-A 16SrIV-D	Vázquez-Euán et al. (2011)
Trachycarpus fortunei	Windmill palm	USA	16SrIV	Harrison and Oropeza (2008)
Thrinax radiata	Florida thatch palm	México	16SrIV-D	Narváez et al. (2006)
Veitchia arecina	Montgomery's Palm	USA	16SrIV	Harrison and Oropeza (2008)
Washingtonia robusta	Mexican fan palm	USA	16SrIV-D 16SrIV-F	Harrison et al. (2008) and Ntushelo et al. (2013)
Non-palm species				
Pandanus utilis	Common screwpine	USA	16SrIV	Thomas and Donselman (1979)
Carludovica palmata	Panama hat plant	Mexico	16SrIV-D	Córdova et al. (2000)
Emilia fosbergii	Florida tasselflower	Jamaica	16SrIV-A	Brown et al. (2008)
Synedrella nodiflora	Nodeweed			
Stachytarpheta jamaicensis	Blue porterweed	Jamaica	16SrIV-E (tentative)	Brown and McLaughlin (2011)
Macroptilium lathyroides	–			
Cleome rutidosperma	Fringed spider flower			
Paspalum vaginatum	Seashore paspalum	Costa de Marfil	16SrXXII-B	Arocha-Rosete et al. (2016) and Yankey et al. (2018)
Pennisetum pedicellatum	Desho grass			
Stachytarpheta indica	–			
Scoparia dulcis	Licorice weed			
Phyllanthus muellerianus	–			
Diplacrum capitatum	–			
Manihot esculenta	Cassava	Costa de Marfil	16SrXXII-B	Kra et al. (2017)

was decreased in areas where *H. crudus* populations were reduced by insecticide treatment (Howard and McCoy 1980). It was also noted that the populations of *H. crudus* in heavily affected areas were 40 times higher than in LY-free areas of Florida (Howard 1980). When coconut and other palm species, within insect-proof cages, were exposed to *H. crudus* adults captured from palms in LY-affected areas, transmission of LY to most test palms occurred within 34 months. In contrast, similar palms which were not exposed to *H. crudus* remained healthy (Howard et al. 1983). Furthermore, a Polymerase Chain Reaction (PCR)-based detection of LY-phytoplasma infection was reported in native *H. crudus* in Florida. Taken together, these studies have indicated the significance of this planthopper as a vector for LY in Florida.

Studies have also been conducted in the identification of vectors of LY phytoplasmas in other countries. In Jamaica, positive LY-phytoplasma detection has been confirmed in *H. crudus* and in another homopteran: *Cedusa* sp., captured in LY-affected sites (Brown et al. 2006). In the Yucatan peninsula of Mexico, where LY has been a devastating factor for cultivation of coconut and other palm species, wild *H. crudus* insects were captured from palm foliage and showed positive detection for 16SrIV phytoplasmas at a proportion of 2.7% (of 2726 insects analysed). Also, the detection was positive in both male and female insects, with a higher proportion found in males (Narvaez et al. 2018). In silico, RFLP and phylogenetic analyses of PCR-amplified Ribosomal DNA (rDNA) products showed that *H. crudus* insects could individually harbour one of three strains: 16SrIV-A, 16SrIV-D or 16SrIV-E, a strain diversity that has also been found in palms affected by LY-type disease syndromes in this part of Mexico (Narvaez et al. 2018). When these *H. crudus* wild insects were tested as vectors of LY phytoplasmas in insect-proof cages containing LY susceptible *Pritchardia pacifica* palms, positive transmission occurred, *P. pacifica* palms developed a LY-type syndrome and died. The phytoplasma associated strain was found as 16SrIV-A in some palms, while it was shown as 16SrIV-D in other palms (CICY, Mexico, unpublished). Similar results were obtained when using an in vitro system for testing vector transmission. After an introduction of *H. crudus* for 1 month, micropropagated coconut plantlets developed leaf yellowing. The PCR analysis of both plantlets and insects confirmed positive results and the phytoplasma strain found was the same in both plantlets and insects (CICY, Mexico, unpublished). In Tabasco, Mexico, insects of *H. crudus* captured from LY-affected palms were found to be positive with PCR detection for 16SrIV phytoplasmas. Other homopteran species, *Haplaxius skarphion*, *Haplaxius cadwuellii*, *Oeclus snowii* and *Persis foveastis*, also tested positive for 16SrIV phytoplasmas (CP, Mexico, unpublished). The results obtained in the USA, Mexico and Jamaica confirmed that *H. crudus* is a vector of 16SrIV phytoplasmas. However, there might also be other vectors involved in the transmission of these phytoplasmas.

With regard to environmental conditions favouring vector populations, a study was conducted on the wild populations of *H. crudus* growing on St Augustine grass (*Stenotaphrum secundatum* (Walt.) Kuntze), Bahia grass (*Paspalum notatum* Flüggé) and Bermuda grass (*Cynodon dactylon* (L.) Pers.). Noticeably, significantly higher numbers of adults and nymphs were collected on St Augustine grass (Reinert 1980). Another study by Howard (1990) also showed a preference of *H. crudus* for St Augustine grass over other grasses. In a coconut pathosystem in southern Mexico, para grass (*Brachiaria mutica* (Forssk.) Stapf), finger grass (*Eutachys petraea* (Sw.) Desv.), signal grass (*B. humidicola* (Forssk.) Stapf) and panic grass (*Panicum laxum* L.) were identified as the principal host species for *H. crudus* nymphs (Ramos-Hernández et al. 2018). It has been shown that grasses and St Augustine grass favour *H. crudus* populations. Therefore, management of the grasses under palm plantings is a potential method for suppressing *H. crudus* populations and thus reducing the spread of LY (Howard 1990).

The investigation of putative vectors of LYDs in Africa has led to the PCR-based detection of phytoplasmas in different insect species. In Tanzania, positive detection was obtained in the planthoppers: *Diostrombus mkurangai* and *Meenoplus* sp. (Mpunami et al. 2000), but transmission studies have not been successful. In Ghana, the presence of Cape Saint Paul Wilt Disease (CSPWD phytoplasma and transmission studies (when caged with coconut palms) were carried out with insects of four *Diostrombus* spp. and *Myndus adiopodoumeensis*. Negative results were confirmed in all PCR tests for *M. adiopodoumeensis* and exposed coconut palms. In the case of the *Diostrombus* spp., one coconut plant and one *D. mayumbensis* insect were positive to CSPWD phytoplasma. However, the coconut plant did not develop symptoms and later PCR testing was negative (Philippe et al. 2009). Although the results from this study did not support a definitive role of *D. mayumbensis* as a CSPWD phytoplasma vector, the insect should still be considered as a potential vector. In fact, further research in LY transmission needs to be undertaken with this species.

In Côte d'Ivoire, insects of *Nedotepa curta* Dmitriev (Cicadellidae: Typhlocybinae: Erythroneurini) were found to be positive not only for 16SrXXII-B phytoplasma (causing Côte d'Ivoire lethal yellowing, CILY) but also to 16SrI phytoplasma. In fact, both phytoplasmas were found as mixed infections in a group of coconut palms (Kwadjo et al. 2018). In Mozambique, early PCR screening showed positive detection of coconut lethal yellowing disease (CLYD) phytoplasmas in *Platacantha lutea* revealing this derbid as a potential insect vector of LYD phytoplasma in northern Mozambique (Dollet et al. 2011). Further research in Mozambique showed positive detection of CLYD phytoplasmas in *Diostrombus mkurangai* and that this insect could also be a potential vector of CLYD (Bila et al. 2017). These findings suggest potential vectors of LYDs in these African countries, but their capacity for transmission is yet to be confirmed.

In the case of Asian countries, research has been conducted to identify vectors of LYDs. The phytoplasmas (then known as Mycoplasma-like organisms, MLOs) causing Kerala wilt disease in India were found to be transmitted in cages by insects of *Stephanitis typica* to coconut plants. The detection was undertaken with 4′,6-diamidino-2-phenylindole (DAPI) staining, electron microscopy and serodiagnosis (Mathen et al. 1990). More recent studies reported that positive cage transmission of phytoplasmas can cause root wilt disease (16SrIX) through insects of *Proutista moesta* (Rajan 2011). These results suggest the vector role of *S. typica* and

P. moesta. In Sri Lanka, investigation of vectors associated with Weligma Coconut Leaf Wilt Disease (WCLWD) showed that eight homopteran species and a hemipteran species were positive to PCR-based detection. The DNA sequence was like WCLWD phytoplasma sequence, suggesting them as putative vector species of WCLWD (Kumara et al. 2015). In the case of Bogia Coconut Disease (BCD) in Papua New Guinea (PNG), positive Loop Mediated Isothermal Amplification (LAMP) was obtained in feeding solution and head tissue of insects from the families: Derbidae, Lophopidae, Flatidae and Ricaniidae and nested-PCR sequences obtained were identical to *Cocos nucifera* BCD phytoplasma sequences from GenBank (Lu et al. 2016). For these cases in Sri Lanka and PNG, further research is required to confirm the capacity of these insects for transmission of phytoplasmas.

9.6 Spread

Studies on gradients of LY spread in outbreaks of the disease were carried out in coconut groves in Yucatan where the disease was spreading from east to west. The results showed that as the proportion of infected palms in an outbreak increased, the distance of the disease spread from the outbreak increased and did so as an east-west symmetrical gradient (Góngora-Canul et al. 2004). Also, in Yucatan, coconut plants within a coconut grove affected by LY were studied to define the pattern of spread by following appearance of visual symptoms. The results showed that spread was randomly distributed (first 10 months) and afterward aggregates started forming, until the distribution of symptomatic palms was uniform in the whole grove (Pérez-Hernández et al. 2004). In contrast, if infected palms were searched using PCR detection, aggregate formation was already occurring at a time when symptom development was showing a random distribution (Góngora-Canul et al. 2004). In addition, studies carried out using analysis of spatial autocorrelation following visual symptom development in a coconut grove, where palms were separated 8 m from each other, showed that an infected palm could infect palms that are close to it as far as 64 m (Góngora-Canul et al. 2004).

There is also another type of spread that has been referred to as 'jump-spread' because it is associated with longer distances of up to several tens of kilometres, where a new outbreak starts with no diseased plants in between. The fact that this is occurring, supports the participation of a flying insect vector or vectors and this may be affected by the wind. Studies on long-distance spread of LY that were carried out in Yucatan showed that the gradients of LY east-west spread were asymmetrical, larger to the west than the east, coinciding with the wind direction (Pérez-Hernández et al. 2004). When patterns of spread of CSPWD were studied in coconut plantations in Ghana, it first occurred randomly in isolated palms, spreading through the entire plot in patches and then steadily to the entire plantation (Dery and Philippe 1997). And there were also jump-spreads of varying distances (Dery and Philippe 1997).

It is likely that there may be alternative paths of pathogen spread associated with human activities, particularly for long-distance dispersal that involves hundreds of kilometres. When LY first appeared in the Cancun area in Mexico and in the island of Roatan in Honduras, the disease was in both cases hundreds of kilometres away. It is suspected that LY arrived at these two sites because ornamental plants or plant parts were transported from an LY-affected region to these two tourist resorts while they were being developed; of course, this remains to be confirmed. Similarly, a particular risk could arise by taking coconut fruit from an LY-affected region or country, to an LY-free region or country. Phytoplasma DNA can be detected by PCR in embryos of fruits from LY-diseased coconut palms. In sectioned tissues from positively testing embryos, the distribution of phytoplasma DNA was shown by in situ PCR to be localized to areas corresponding to the plumule and cells enclosing it (Córdova et al. 2003). The presence of phytoplasma DNA in embryos of fruits at different stages of development from LY-diseased coconut palms was subsequently demonstrated (Oropeza et al. 2011). Research also showed that plantlets obtained by in vitro germination of embryos from seed of LY-diseased palms were infected with LY phytoplasmas as determined by PCR detection, whereas plantlets obtained from seed from LY-free palms were disease free (Oropeza et al. 2017). Therefore, an alternative path of spread is likely if seed from LY-affected areas is taken and germinated in LY-free areas.

9.7 Plant Host Species

A great number of plant species can be infected with phytoplasmas associated with LY or related diseases. Table 9.3 shows 49 palm species (besides coconut) and 14 non-palm species that have been reported as affected by phytoplasmas with group and subgroup having been identified for most of them. Most of the susceptible palm species are used as ornamentals, but some hold economic value such as oil, date and coconut palm.

A few palm species have been reported that could be infected separately by two different phytoplasmas of the same group: *A. aculeata* with 16SrIV-B or 16SrIV-D (Ntushelo et al. 2013; Roca et al. 2006), *A. merrillii* with 16SrIV-A or 16SrIV-D (Córdova-Lara et al. 2017), *S. Mexicana* with 16SrIV-A or 16SrIV-D (Vázquez-Euán et al. 2011), *P. canariensis* with 16SrIV-A or 16SrIV-D (Harrison et al. 2002, 2008, Ntushello et al. 2013) and *W. robusta* (16SrIV-D or 16SrIV-F; Harrison et al. 2008; Ntushelo et al. 2013). Also, there is a case of a species infected with phytoplasma belonging to different groups, *E. guineensis* with 16SrI-B (Álvarez et al. 2014) or 16SrXXII-A (Bila et al. 2015), each occurring in different countries. We also have the particular case of coconut with separate infections reported with as many as five phytoplasma subgroups: 16SrIV-A, 16SrIV-B, 16SrIV-C, 16SrIV-D and 16SrIV-E (Harrison et al. 1999; Harrison and Oropeza 2008; Martínez et al. 2008; Myrie et al. 2014; Ntushelo et al. 2013; Roca et al. 2006); however, subgroup 16SrIV-A is the predominant one and most damaging (Harrison and Oropeza 2008).

Fourteen non-palm species have been reported with the presence of LY-phytoplasma. Seven species can be infected with the group 16SrIV: *P. utilis* (Thomas and Donselman 1979), *E. fosbergii* and *S. nodiflora* (16SrIV-A; Brown et al. 2008), *C. palmata* (16SrIV-D; Córdova et al. 2000), *S. jamaicensis, M. lathyroides and C. rutidosperma* (16SrIV-E; Brown and McLaughlin 2011). In addition, seven species with the group 16SrXXII: *P. vaginatum, P. pedicellatum, S. indica, S. dulcis, P. muellerianus, D. capitatum* and *M. esculenta* (all with the subgroup 16SrXXII-B).

It is important to keep in mind that there are several reports producing lists of palm species of which individuals have been found susceptible to LY (Harrison and Oropeza 2008). However, most of these species have not been tested or observed for their level of susceptibility or resistance as a population.

Although there are some cases such as *E. guineensis* in Colombia (Álvarez et al. 2014), *P. dactylifera* in USA (Harrison et al. 2002c) and *P. pacifica* in Mexico (Narvaez et al. 2017), they have been shown to be very susceptible as a population, since most of the individuals were destroyed due to LY or LYDs. There are some species that are known not to be susceptible as only a few individuals die (probably less than 5%) without affecting the population size. In fact, such species, e.g. *S. Mexicana, Thrinax radiata* and *Coccothrinax readii*, could be acting as reservoirs of phytoplasmas (Narvaez et al. 2006; Vázquez-Euán et al. 2011).

9.8 Sampling of Tissue and Insects

9.8.1 Tissue Sampling

Regardless of analytical technique, detection of phytoplasmas in infected coconut plants, or other plant species, was typically carried out on tissue samples from very young leaves, unopen inflorescences or apical meristems. However, these tissues are very difficult to obtain in fully mature plants. With the advent of the very sensitive molecular methods for phytoplasma detection, the more easily obtained tissues from the lower part of the trunk (using a drill) were tested as a source of phytoplasma DNA and the results were positive (Córdova 2000). The comparative advantage of trunk shavings was confirmed by Oropeza et al. (2011). They studied the distribution of LY phytoplasmas in different parts (trunk, young leaves, inflorescences, stem apex and root apex) of infected coconut plants during symptom development, starting with plants not showing symptoms yet (stage 0, according to McCoy et al. 1983) to the stage in which they have advanced yellowing of the leaves (stage 6, according to McCoy et al. 1983). The results showed that detection by nested-PCR could be obtained in all plant parts studied, even in symptomless infected plants. However, frequency of detection was low in all parts except in trunk where it was high already at stage 0. Furthermore, the frequency of detection in all parts decreased as symptomatology progressed, except in the trunk. Detection in the trunk is still possible even when plants have lost the crown (CICY, unpublished). Therefore, trunk has become the source of choice for tissue sampling for phytoplasma detection for coconut and other plant species.

Trunk sampling is easy. One protocol according to Oropeza et al. (2010) involves the following steps: Drill a hole in the trunk, 10 cm deep with a drill bit (usually 5/8 inch in diameter) and approximately 1.5 m above the ground (Fig. 9.3a). The shavings are collected in a plastic bag avoiding the contact of these with the hands (Fig. 9.3b and c). Once the sample has been taken, the drill should be washed with 3% sodium hypochlorite (Fig. 9.3d) and then rinsed with sterile water. When finished, the hole should be blocked with a wood plug (Fig. 9.3e and f), it is advisable to apply insecticide over the sealed hole for additional protection against pests and pathogens.

The samples contain wood shavings as well as sap content where phytoplasma DNA is present in larger amount in comparison to other parts of the palm (Córdova et al. 2014). These trunk samples are adequate for standard laboratory PCR or LAMP analysis, and also for in-field detection using portable thermocycler machines coupled with dipstick technology (Zou et al. 2017).

9.8.2 *Insect Capture*

For the collection of insects, plants having easy-to-reach foliage should be chosen. In the case of *H. crudus*, it is easier to find more insects in the morning, no later than 10 a.m. However, this could depend on the insect species and local conditions. Insects are captured from the lower side or abaxial side of the leaves of palms (Fig. 9.3g). If the purpose is capturing insects that are infected with 16SrIV phytoplasmas, it is more convenient to collect insects from palms that are showing symptoms and have already been determined as positive with PCR detection. If the purpose is collecting phytoplasma-free insects, collect them from plants that are in an area that is free of LY. It is very important that the insects are captured using a system designed for trapping them. The system may include a bottomless tube inside a larger tube, in order to capture without causing any damage (Fig. 9.3h). Once captured, they can be kept in tubes within a cool container. Identification at the laboratory can be carried out under a stereoscope using key morphological characteristics (Kramer 1979; Triplehorn and Johnson 2005). Some of the insects should be sent to experts for confirmation of identity.

9.9 Detection and Diagnosis of LY Phytoplasmas

Symptoms are the first evidence for diagnosis of LY phytoplasmas. Abnormal falling of fruit, necrosis of inflorescences and yellowing of mature leaves could be indicative of the presence of LY-related phytoplasmas in palms. However, the

Fig. 9.3 Trunk sampling for phytoplasma detection: Drill a hole in the trunk at approximately 1.5 m above the ground (**a**). Collect the shavings in a plastic bag (**b**, **c**). Wash the drill with sodium hypochlorite (NaClO) at 3% (**d**). Rinse with sterile water. Block the hole with a wood plug (**e**, **f**). Insect capture: Find an insect of interest (*Haplaxius crudus* shown) on the abaxial side of a palm leave (**g**) and then capture it using a tool as the one shown (a bottomless tube inside a larger tube) with no aspiration involved to avoid damaging the insect (**h**)

disease can only be confirmed through specific diagnostic tools. In the past, transmission electron microscopy was used to confirm the presence of LY phytoplasmas (Plavsic-Banjac et al. 1972). It helped visualize them as ovoid or filamentous cells, living in the phloem of palms. However, this technique was very laborious and time-consuming and only few samples could be analysed at once. Despite these disadvantages, this technique was utilized from the 1970s to the early 1990s (Howard, 1995). The advent of molecular techniques in recent years provides better solutions for diagnosis.

The first tests of molecular diagnosis were using DNA probes marked with radioactivity (^{32}P) or fluorescent molecules that hybridize with the LY DNA. This technique allowed the detection of a great number of samples using the dot blot analysis. However, the sensitivity and specificity were limited due to low titres of phytoplasma in the samples (Harrison et al. 1994). These problems were solved when the PCR technique was applied for phytoplasma detection. It started with the development of primers that amplify rDNA from universal phytoplasma sequences and later with the development of specific primers for LY phytoplasma (Harrison et al. 2002b). This technique allowed detection of multiple samples with specificity and rapidity; however, due to the low titre of phytoplasma it was necessary to use two rounds of DNA amplification (nested-PCR) using universal and specific primers.

The establishment of quantitative PCR (qPCR) and TaqMan technology (hydrolysis probes) that enable the detection of DNA amplification in real-time using a specific probe labelled with a fluorophore helped to increase the sensitivity and specificity of phytoplasma detection, using only one round of DNA amplification. For detecting LY and LYD phytoplasmas, TaqMan probes were developed based on their 23S ribosomal gene using different sets of specific primers (Hodgetts et al. 2009). For LY phytoplasma that affect palms in the Americas, a specific qPCR assay was developed by Córdova et al. (2014), using a TaqMan probe and primers based on the 16S ribosomal gene. This assay enabled the absolute quantification of phytoplasma, showing that trunk, root meristem and immature inflorescences had the highest level of phytoplasma in coconut palm. This technique has been improved using two TaqMan probes targeting different genes (16S rRNA and GroEL) and labelled with different fluorophores that allow detection and discrimination of two strains of LY phytoplasma (16SrIV-A and -D) at the same time (Córdova et al. 2019). Other authors have developed a similar technique using SYBR Green (Asymmetrical Cyanine Dye) Fluorophore that binds to the double helix of amplified DNA using the melting point analysis of this molecule to discriminate the LY strains (Bahder et al. 2017). More recently, a new technique derived from real-time PCR has been developed and is called digital PCR. The advantage of this technique is that the detection and quantification is done by microdroplets. This provides a higher sensitivity and accuracy than the real-time PCR. Bahder et al. (2018) have developed a digital PCR protocol for LY phytoplasma detection that is more sensitive than the real-time protocol reported by Córdova et al. (2014). This technique opens up new research avenues because it would help detect very low titres of phytoplasma in plants and insects, providing a new tool for tackling this disease. An additional technique LAMP has been progressively used to detect plant pathogens in recent years. This technique consists in the amplification of six different targets of DNA that are amplified at isothermal conditions (60–65 °C) in a short time (usually 1 h or less). The addition of loop primers increases the rapidity and sensitivity of the technique. It has the potential to be simple and rapid and also requires minimal equipment and could be used in field conditions. A protocol using this technique has been reported for detection of phytoplasma of group 16SrXXII that affect palms in Africa (Tomlinson et al. 2010) and phytoplasma of group 16SrIV in insects of different coconut plantations affected with BCS in Papua New Guinea (Lu et al. 2016).

9.10 LY Resistance Screening

During the 1950s, there were great losses of coconuts in Jamaica, prompting field screening for LY resistance of coconut germplasm. Several cultivars were tested and two were selected: the Malayan Yellow Dwarf (MYD) with a very low mortality of 4% and the Panama Tall (PT) with an intermediate mortality of 44%; they were also crossed to produce the MYD × PT hybrid (Maypan), combining the LY resistance of MYD and the better agronomic characteristics of PT. Since then, the Maypan has been extensively planted in Jamaica and other countries. Unfortunately, since the 1990s, the MYD, PT and the Maypan have been dying from LY in proportions greater than expected (Broschat et al. 2002). Genetic contamination was evaluated in PT populations in Jamaica and found to be present (Baudouin et al. 2008), and similar results were obtained for MYD populations (Lebrun et al. 2008). However, it was believed to be insufficient to explain the recent LY outbreaks in these varieties and the Maypan hybrid produced from them. Hence, there are likely to be other causes. The above-mentioned authors believe that these coconut materials cannot be resistant to LY for the current situation in Jamaica.

Similarly, an approach for crossing the MYD with Pacific Tall (PT) ecotypes has been used in other countries, including Mexico where several such hybrids were produced by Instituto Nacional de Investigaciones Forestales y Agrícolas y Pecuarias (INIFAP, México). As a result, these palms have successfully survived LY for decades. The coconut germplasm on the Pacific coast of Mexico was originally introduced from East Asia and the Pacific between sixteenth and nineteenth centuries (Zizumbo-Villarreal 1996). Hence, genetic diversity and probably LY resistance could be expected. Also, coconut populations representing the genetic diversity in Mexico, particularly of the Pacific side, were screened for mortality to LY starting in 1989. The test included the Atlantic Tall (AT) and MYD as references, and 15 PT populations that were characterized and grouped into three ecotypes: MXPT1, MXPT2 and MXPT3 (Zizumbo-Villarreal et al. 2008). The MYD showed the lowest mortality and AT the highest, and MXPT1 and MXPT2 had a mortality similar to that of MYD.

Encouraged with these results, further screening by INIFAP, CP and CICY has been carried out jointly, with introduction of germplasms originally from East Asia and the Pacific. These include the first introduction of six hybrids of which two stand out: Malayan Red Dwarf × Vanuatu Tall (MRD × VTT) and Malayan Red Dwarf × Tagnanan Tall (MRD × TGT); and a second introduction with 12 varieties, currently under evaluation. The MXPT1 and MXPT2 ecotypes have been used as male parents with MYD as a female parent for hybrid production. The ecotypes and the hybrids which were used in replanting programs for nearly 20 years have survived, without being associated with any outbreak of LY so far. It is still the case even when they are cultivated in areas where different LY phytoplasma strains exist (Harrison et al. 2002b). This situation contrasts with what has happened in Jamaica with the supposedly resistant MAYPAN and parents MYD and PT.

Screening for resistance to CSPWD has also been reported in Ghana and the results showed a certain level of susceptibility for most of the varieties evaluated. West African Tall was the most susceptible, given that most of this type died. In contrast, most of the Sri Lankan Green Dwarf (SGD) palms have survived (Dery et al. 2008).

Given the information documented above, affected countries need to continue or start screening for resistance to LY or LYDs in local coconut germplasm and introduced germplasm. Given field screening for LY resistance takes a long time (up to 10 years or more), without the guarantee that within a trial disease incidence will build up to adequate levels, or natural calamities or man-made restrictions could end abruptly with the trial, we also need to develop more rapid and precise ways to assess susceptibility and resistance of coconut germplasm to phytoplasmas associated with LY or LYDs.

9.11 Translating Knowledge into Practical Use

In order to deal with LY (and related diseases), it is useful to consider what we know about the disease and other relevant aspects, particularly in relation to those species that are socially and economically important. For instance, one should look at the potential source of the pathogen, its diversity and geographic distribution, whether it needs vectors, as well as the habits of the vectors (their biology, factors for spread, the plant host species and plant-pathogen-vector interactions) and perhaps most importantly if there is resistant germplasm. Based on the aforementioned information on LY (and LYDs in general), the first action is perhaps to produce a contingency plan of general use. However, that would need to include countries or areas where these diseases are not currently present. Measures have to be taken to stop entry of infected biological materials (plant or animal). There is a contingency plan for LY in Spanish published by Organismo Internacional Regional de Sanidad Agropecuaria (OIRSA) (Oropeza et al. 2010), but it needs updating and refinement; however, it could be a convenient starting point.

There are several phytoplasmas in the Americas; however, the only one that is devastating for coconut so far is 16SrIV-A. Occurrences of coconut palms affected by other subgroup phytoplasmas (B, D and F) have also been observed in very small outbreaks in Mexico and Dominican Republic but they have not spread. It is important to monitor them, and new incursions, to define immediately if it is a phytoplasma disease. This could be done via gathering information on symptoms and context and using an in-field detection technique based on qPCR or LAMP (Córdova et al. 2014; Tomlinson et al. 2010 respectively), alternatively sampling can be carried out by a method such as the use of dipstick (Zou et al. 2017) (see tissue sampling above) and samples sent to a laboratory. Diseased plants then need to be destroyed. If we are dealing with a case in a region or country where LY or an LYD is not yet present, it would be very important to analyse surrounding symptomless palms and try to contain the outbreak (see Oropeza et al. 2010). Every step and the participants needed for these action(s) should to be very well defined as part of a contingency plan.

Importantly, special care should be taken to avoid the introduction of infected plant material or insects from a disease affected region or country to a disease-free region or country. For instance, bringing a nut, perhaps as a souvenir, from a country with LY to a country without. If that nut has an embryo infected and the nut is germinated, this could be a risk of introduction and spread of the disease. This would be worse if the material is from an LYD-affected country in one continent and moved to a country in another continent, where the disease is not present. So, it is very important that strict measures are taken by countries to avoid such occurrences.

In the cases where LY is already present in a country and spreading, measures should be taken to reduce the economic damage that can result. This could be in the form of an integrated disease management package. The Coconut Industry Board in Jamaica has implemented such a package that consists of the following: *(a) Surveillance.* This could be done while carrying out the different actions for managing the plantation, such as harvesting fruit. So, when an early symptom or symptoms are observed in a plant, it should be reported. *(b) Elimination of the symptomatic plant.* This should be done as immediately as possible after the symptomatic plant was spotted. The plant cannot be cut down and left there, since leaves that are still alive will continue being a source of phytoplasmas for vector insects feeding on them. Sampling is an option that could be convenient for later analysis. *(c) Replanting.* This should be done systematically to keep the plantation's plant density. It should be done using germplasm selected for resistance or with some level of resistance to LY. It would not be convenient to use susceptible coconuts such as the AT variety. *(d) Weeding.* Keeping the plantation free of weeds is important because it involves destruction of Gramineae plants that could host vector insects, as well as other plant species that could host the phytoplasmas as observed in Jamaica (Brown and McLaughlin 2011). *(e) Control of vectors.* Use of convenient agents for this purpose is important to reduce as much as possible the population of insects that could be vectors for the phytoplasmas. *(f) Health.* Provide the plants with good maintenance and nutritious conditions, free of other diseases and pests as much as possible. This integrated disease management package is being applied in Jamaica

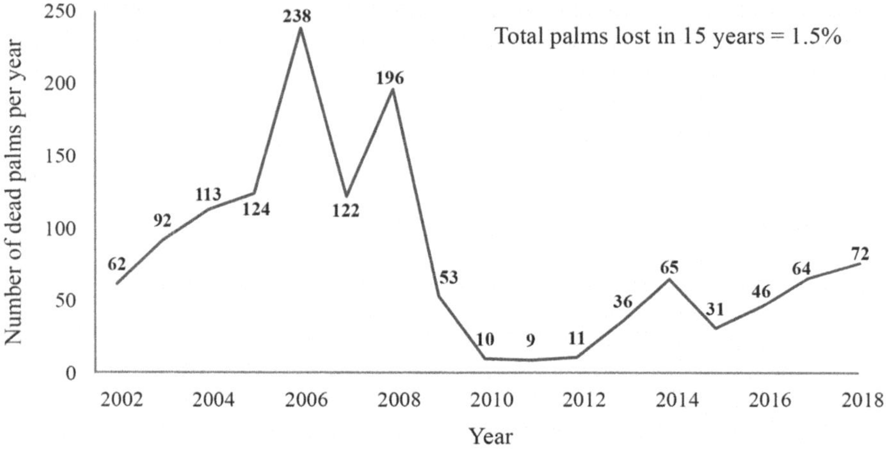

Fig. 9.4 Integrated LY management in Black's Farm, St Thomas, Jamaica, where there are nearly 80,000 plants producing fruits which are used for different products, including bottled water and virgin oil. The loss throughout 15 years does not exceed 1.5% of the 80,000 plants in the farm

on Black's Farm where about 80,000 plants have been cultivated for more than 10 years and losses due to LY have been kept below 0.5% in most years (Fig. 9.4).

However, in other countries, such as Mexico, where LY resistant ecotypes have been found and used for more than 20 years, this resistant germplasm is the main factor for dealing with LY. With the use of resistant germplasm and the addition of integrated disease management packages (described above) there would be a much lesser risk of losses due to LY.

9.12 Future Perspectives and New Venues of Research

Perhaps the most important lesson is that different institutions need to work together to develop a global strategy for identifying the resistance of coconut genotypes to the LY and LYD phytoplasmas that are devastating coconut cultivation. This could be based on studies such as that of Baudouin and Lebrun (2009) that theoretically identified the level of mortality LY could cause to different varieties, and also the phylogenetic tree reported by Lebrun and Baudouin (1999) and Baudouin and Ledbrun (2002) that uses microsatellite marker data to show grouping of the most resistant varieties known (MYD, MRD, Malayan Green Dwarf, MXPT2, etc.). These types of studies could be very helpful to select the varieties to be screened, either by traditional field testing for resistance or alternative methods, such as controlled transmission using insect-proof cages like those used by Howard (1995) or in vitro transmission (CICY, Mexico, unpublished). These last two transmission methods would require exposing the plants to be tested to feral insects captured in areas with LY or LYDs. Or even better, would be using insects reared under controlled conditions and fed with a suspension of phytoplasmas obtained by culturing them in vitro (Contaldo et al. 2016). Altogether, this integrated controlled transmission approach would allow a lot of control of the resistance screening process and more effective and precise results in shorter times. Of course, in order to achieve developing such a transmission system for screening further research is needed to identify vector insects, master rearing of vector insects and culturing in vitro of phytoplasmas causing LY and LYDs. And certainly, resistance to LY or LYDS would be a trait that should be considered in programs for coconut conservation and genetic improvement (Bourdeix 2019; COGENT 2017).

It is also important to undertake research on defence mechanisms in coconut in general. Previous research has shown that benzothiadiazole (BTH), a functional analogue of salicylic acid (SA), could activate the systemic acquired resistance (SAR) mechanism in *Arabidopsis thaliana* (Lawton et al. 1996). Also, that when plants of *A. thaliana* were treated with BTH before inoculation with X-disease phytoplasma, there was reduced infection, and also reduced survival of the phytoplasma vector *Colladonus montanus* when interacting with BHT treated plants (Bressan and Purcell 2005). Then considering that SAR could play a role in phytoplasma defence in plants, a study in coconut evaluated the occurrence of non-expresser of PR genes 1 (NPR1) homologue genes in coconut palm, two were found *CnNPR1* and *CnNPR3* and that the amount of transcripts of both were regulated positively by SA, suggesting that these homologues could be associated with the activation of SAR in coconut (Nic-Matos et al. 2017). It remains to define if this could be relevant for defence as well as the study of other mechanisms of defence in relation to LY and related diseases.

It is also essential to be able to identify resistance or susceptibility to LY and LYDs with faster methods. In a search for disease-resistance gene candidates of the nucleotide binding site (NBS) type from LY resistant or susceptible coconut ecotypes, several DNA sequences were obtained that clustered in seven different clades and their expression changed in response to SA (Puch-Hau et al. 2015). Based on these findings, two markers for susceptibility to LY have been developed and are currently being tested with promising results (CICY, Mexico, unpublished). This type of research should be extended and strengthened in order to identify coconut germplasm that is resistant to LY or LYDs rapidly and to assist other lengthier traditional methods involving vector mediated transmission of phytoplasmas.

What is mentioned in the chapter above addresses LY and related diseases, how to manage them, and the urgent need to identify, in a fast and precise fashion, coconut resistant germplasm for replanting coconut in most producing countries around the world. This is a huge task that would be very difficult to be carried out by seed propagation alone. Because of this pressure, the development of micropropagation protocols has been a priority for several years. Important advances have been obtained with the participation of several countries. In Mexico, a process has been developed that is highly efficient and has been scaled up to commercial level (Oropeza et al. 2016). Nevertheless, further research for improvements are necessary for reducing costs and managing long-distance transportation, among other things. Again, as mentioned in the above paragraphs, this requires the international collaboration of expert institutions.

Finally, we need to work in a well-organized fashion worldwide to use our resources for research in a more efficient and effective way, and this could be more easily achieved if such an effort is coordinated in collaboration with organizations such as the ICC and COGENT. It is very important to keep in mind that measures for dealing with LY or LYDs are considered within the wider scope of strengthening the coconut value chain in every producing country, but in particular aiming at improving the income and livelihoods of the people working with the coconut palm, mainly the small farmers with lower incomes.

References

Al Khazindar M (2014) Detection and molecular identification of aster yellows Phytoplasma in date palm in Egypt. Phytopathology 162:621–625

Al-Awadhi H, Hanif A, Suleman P et al (2002) Molecular and microscopical detection of phytoplasma associated with yellowing disease of date palm *Phoenix dactylifera* L. in Kuwait. Kuwait J Sci Eng 29:87–109

Alhudaib K, Arocha Y, Wilson M et al (2007) Identification and molecular characterization of a phytoplasma associated with Al-wijam disease of date palm in Saudi Arabia. Arab J Plant Prot 25:116–122

Álvarez E, Mejía JF, Contaldo N et al (2014) 'Candidatus Phytoplasma asteris' strains associated with oil palm lethal wilt in Colombia. Plant Dis 98:311–318

Ammar MI, Amer MA, Rashed MF (2005) Detection of Phytoplasma associated with yellow streak disease of date palms (*Phoenix dactylifera* L.) in Egypt. Egypt J Virol 2:74–86

Arocha-Rosete Y, Diallo HA, Konan Konan JL et al (2016) Detection and identification of the coconut lethal yellowing phytoplasma in weeds growing in coconut farms in Côte d'Ivoire. Can J Plant Pathol 38:164–173

Ashburner GR, Cordova I, Oropeza C et al (1996) First report of coconut lethal yellowing in Honduras. Plant Dis 80(8):960

Bahder BW, Helmick EE, Harrison N (2017) Detecting and differentiating phytoplasmas belonging to subgroups 16SrIV-A and 16SrIV-D associated with lethal declines of palms in Florida using qPCR and high-resolution melt analysis (HRMA). Plant Dis 101(8):1449–1454

Bahder BW, Helmick EE, Chakrabarti S et al (2018) Disease progression of a lethal decline caused by the 16SrIV-D phytoplasma in Florida palms. Plant Pathol 67:1821–1828

Baudouin L, Lebrun P (2002) The development of a microsatellite kit and dedicated software for use with coconuts. International Plant Genetic Resources Institute (IPGRI), Rome. Burotrop Bull 17:16–20

Baudouin L, Lebrun P (2009) Coconut (*Cocos nucifera* L.) DNA studies support the hypothesis of an ancient Austronesian migration from Southeast Asia to America. Genet Resour Crop Evol 56(2):257–262

Baudouin L, Lebrun P, Berger A et al (2008) The Panama Tall and the Maypan hybrid coconut in Jamaica: did genetic contamination cause a loss of resistance to Lethal Yellowing? Euphytica 161(3):353–360

Bila J, Högberg N, Mondjana A et al (2015) African fan palm (*Borassus aethiopum*) and oil palm (*Elaeis guineensis*) are alternate hosts of coconut lethal yellowing phytoplasma in Mozambique. Afr J Biotechnol 14:3359–3367

Bila J, Högberg N, Mondjana A et al (2017) First report of '*Candidatus* Phytoplasma palmicola' detection in the planthopper *Distrombus mkurangai* in Mozambique. B Insectol 70:45–48

Bourdeix R (2019) A world without coconut water? The world's trendiest nut is under threat of species collapse. https://scroll.in/article/print/822758

Bressan A, Purcell AH (2005) Effect of benzothiadiazole on transmission of X-disease phytoplasma by the vector *Colladonus montanus* to *Arabidopsis thaliana*, a new experimental host plant. Plant Dis 89:1121–1124

Broschat TK, Harrison NA, Donselman H (2002) Losses to lethal yellowing cast doubt on coconut cultivar resistance. Palms 46(4):185–189

Brown SE, McLaughlin WA (2011) Identification of lethal yellowing group (16SrIV) of phytoplasmas in the weeds *Stachytarphetaja maicensis*, *Macroptilium lathyroides* and *Cleome rutidosperma* in Jamaica. Phytopathogenic Mollicutes 1:27–34

Brown SE, Been BO, McLaughlin WA (2006) Detection and variability of the lethal yellowing group (16Sr IV) phytoplasmas in the *Cedusa* sp. (Hemiptera: Auchenorrhyncha: Derbidae) in Jamaica. Ann Appl Biol 149:53–62

Brown SE, Been BO, McLaughlin WA (2008) First report of the presence of the lethal yellowing group (16SrIV) of phytoplasmas in the weeds *Emilia fosbergii* and *Synedrella nodiflora* in Jamaica. Ann Appl Biol 57(4):770–770

COGENT (2017) A Global Strategy for the Conservation and Use of Coconut Genetic Resources, 2018–2028. (R. Bourdeix and A. Prades, compilers). Bioversity International, Montpellier, France, p 239

Contaldo N, Satta E, Zambon Y et al (2016) Development and evaluation of different complex media for phytoplasma isolation and growth. J Microbiol Meth 127:105–110

Córdova I, Oropeza C, Almeyda H et al (2000) First report of a phytoplasma-associated leaf-yellowing syndrome of palma jipi plants in southern México. Plant Dis 84(7):807–807

Córdova I, Jones P, Harrison NA, Oropeza C (2003) In situ detection of phytoplasma DNA in embryos from coconut palms with lethal yellowing disease. Mol Plant Pathol 4(2):99–108

Córdova I, Oropeza C, Puch-Hau C et al (2014) A real-time PCR assay for detection of coconut lethal yellowing phytoplasmas of group 16SrIV subgroups A, D and E found in the Americas. J Plant Pathol 96:343–352

Córdova I, Oropeza C, Harrison N et al (2019) Simultaneous detection of coconut lethal yellowing phytoplasmas (group 16SrIV) by real-time PCR assays using 16Sr-and GroEL-based TaqMan probes. J Plant Pathol 101:609–619

Córdova-Lara I, Mota NL, Puch HC et al (2017) Detection and identification of lethal yellowing phytoplasma 16SrIV-A and D associated with *Adonidia merrillii* palms in Mexico. Australas Plant Path 46:389–396

Cronje P, Dabek AJ, Jones P et al (2000) First report of a phytoplasma associated with a disease of date palms in North Africa. Plant Pathol 49(6)

Dery SK, Philippe R (1997) Preliminary study on the epidemiology of Cape St Paul wilt disease of coconut in Ghana. In: Proceedings of an International workshop on lethal yellowing-like diseases of coconut (eds) Eden-Green SJ. Ofori F, Natural Resources Institute, Chatham, United Kingdom, pp 255–260

Dery SK, Philippe R, Baudouin L et al (2008) Genetic diversity among coconut varieties for susceptibility to Cape St Paul Wilt disease. Euphytica 164:1–11

Dollet M, Macome F, Vaz A et al (2011) Phytoplasmas identical to coconut lethal yellowing phytoplasmas from Zambesia (Mozambique) found in a pentatomide bug in Cabo Delgado province. B Insectol 64:S139–S140

Escamilla JA, Oropeza C, Harrison N et al (1994) Evolución de sondas moleculares de ADN para el estudio de organismos to micoplasma causantes del amarillamiento letal. Reporte de Proyecto CONACYT, México

Góngora-Canul C, Escamilla-Bencomo J, Pérez-Hernández O et al (2004) Gradientes de diseminación del amarillamiento letal en cocotero (*Cocos nucifera* L.) en Sisal Yucatán, México. Revista Mexicana de Fitopatología 22:370–376

Gurr GM, Johnson AC, Ash GJ et al (2016) Coconut lethal yellowing diseases: a phytoplasma threat to palms of global economic and social significance. Fronti Plant Sci 7:1521

Harrison NA, Oropeza C (2008) Coconut lethal yellowing. In: Characterization, Diagnosis and Management of Phytoplasmas (eds) Harrison NA, Rao GP. Marcone C. Studium Press. Houston, USA, pp 219–248

Harrison NA, Richardson P, Jones P et al (1994) Comparative investigation of MLO's associated with Caribbean and African coconut lethal decline diseases by DNA Hybridization and PCR Assays. Plant Dis 78(5):507–511

Harrison NA, Córdova I, Richardson P et al (1999) Detection and diagnosis of lethal yellowing. In: Current Advances in Coconut Biotechnology (eds)Oropeza C, Verdeil JL, Ashburner GR, Cardeña R. Santamaria JM. Kluwer Academic Publishers, Dordrecht, The Netherlands, pp 183–196

Harrison NA, Myrie W, Jones P et al (2002a) 16S rRNA interoperon sequence heterogeneity distinguishes strain populations of palm lethal yellowing phytoplasma in the Caribbean region. Ann Appl Biol 141(2):183–193

Harrison NA, Womack M, Carpio ML (2002b) Detection and characterization of a lethal yellowing (16SrIV) group phytoplasma in Canary Island date palms affected by lethal decline in Texas. Plant Dis 86(6):676–681

Harrison NA, Narváez M, Almeyda H et al (2002c) First report of group 16SrIV phytoplasmas infecting coconut palms with leaf yellowing symptoms on the pacific coast of México. New Dis Rep 5:1–3

Harrison NA, Helmick EE, Elliott ML (2008) Lethal yellowing-type diseases of palms associated with phytoplasmas newly identified in Florida, USA. Ann Appl Biol 153:85–94

Harrison NA, Helmick EE, Elliott ML (2009) First report of a phytoplasma-associated lethal decline of *Sabal palmetto* in Florida, USA. Plant Pathol 58(4):792

Harrison NA, Davis RE, Oropeza C et al (2014) '*Candidatus* Phytoplasma palmicola', a novel taxon associated with a lethal yellowing-type disease (LYD) of coconut (*Cocos nucifera* L.) in Mozambique. Int J Syst Evol Micr 64(6):1890–1899

Hodgetts J, Boonham N, Mumford R et al (2009) Panel of 23S rRNA gene-based real-time PCR assays for improved universal and group specific detection of phytoplasmas. App Environ Microbiol 75:2945–2950

Howard FW (1980) Population densities of *Myndus crudus* Van Duzee (Homoptera: Cixiidae) in relation to coconut lethal yellowing distribution in Florida. Principes 24(4):174–178

Howard FW (1990) Evaluation of grasses for cultural control of *Myndus crudus*, a vector of lethal yellowing of palms. Entomol Exp Appl 56(2):131–137

Howard FW (1995) Lethal yellowing vector studies. I. Methods of experimental transmission. In: Lethal Yellowing Research and Practical Aspects (eds) Oropeza C, Howard FW. Ashburner GR. Kluwer Academic Publishers, Dordrecht, The Netherlands, pp 43–57

Howard FW, McCoy RE (1980) Reduction in spread in mycoplasma like organism associated lethal decline of the palm *Veitchia merrillii* by the use of insecticides. J Econ Entomol 73(2):268–270

Howard FW, Norris RC, Thomas DL (1983) Evidence of transmission of palm lethal yellowing agent by a planthopper, *Myndus crudus* (Homoptera, Cixiidae). Trop Agric 60(3):168–171

Islas-Flores I, Santamaria JM, Cordova I et al (1999) Biochemical changes in roots of coconut palms (*Cocos nucifera* L.) affected by lethal yellowing. J Plant Physiol 155(1):48–53

Jeyaprakash A, Sutton BD, Halbert SE et al (2011) First report of a 16SrIV-D phytoplasma associated with texas phoenix palm decline on pigmy date palm (Phoenix roebelenii) in florida. Plant Dis 95(11):1475–1475

Kelly PL, Reeder R, Kokoa P et al (2011) First report of a phytoplasma identified in coconut palms (*Cocos nucifera*) with lethal yellowing-like symptoms in Papua New Guinea. New Dis Rep 23:9–9

Kra KD, Toualy YMN, Kouamé AC et al (2017) First report of a phytoplasma affecting cassava orchards in Côte d'Ivoire. New Dis Rep 35:21–21

Kramer JP (1979) Taxonomic study of the planthopper genus *Myndus* in the Americas (Homoptera: Fulgoroidea: Cixiidae). T Am Entomol Soc 105:301–389

Kumara ADNT, Perera L, Meegahakumbura MK et al (2015) Identification of putative vectors of weligama coconut leaf wilt disease in Sri Lanka. In: Chakravarthy AK (ed) New horizons in insect science: towards sustainable pest management. Springer, New Delhi, pp 137–146

Kwadjo KF, Beugré NI, Dietrich CH et al (2018) Identification of *Nedotepa curta* Dmitriev as a potential vector of the Côte d'Ivoire lethal yellowing phytoplasma in coconut palms sole or in mixed infection with a '*Candidatus* Phytoplasma asteris'-related strain. Crop Prot 110:48–56

Lawton KA, Friedrich L, Hunt M et al (1996) Benzothiadiazole induces disease resistance in *Arabidopsis* by activation of the systemic acquired resistance signal transduction pathway. Plant J 10:71–82

Lebrun P, Baudouin L (1999) Studies of coconut genetic relations using microsatellite markers. 1st international workshop and laboratory course on the application of Biotechnology to Plant Breeding and Crop Protection. Organized by Max-Planck-Institut fuer Zuechtungsforschungs (MPIZ, Colonia, Germany) and Centro de Investigación Científica de Yucatán (CICY, Mérida, México). CICY, Mérida, November 21 – December 3

Lebrun P, Baudouin L, Myrie W et al (2008) Recent lethal yellowing outbreak: why is the Malayan yellow dwarf coconut no longer resistant in Jamaica? Tree Genet Genomes 4(1):125–131

Llauger R, Becker D, Cueto J et al (2002) Detection and molecular characterization of phytoplasma associated with lethal yellowing disease of coconut palms in Cuba. J Phytopathol 150(7):390–395

Lu H, You M, Wilson BAL et al (2016) Determining putative vectors of the Bogia Coconut Syndrome phytoplasma using loop-mediated isothermal amplification of single insect feeding media. Sci Rep 6:35801

Manimekalai R, Soumya VP, Nair S et al (2014) Molecular characterization identifies 16SrXI-B group Phytoplasma ('*Candidatus* Phytoplasma oryzae'-related strain) associated with root wilt disease of coconut in India. Sci Hort 165:288–294

Martínez RT, Narváez M, Fabre S et al (2008) Coconut lethal yellowing on the southern coast of the Dominican Republic is associated with a new 16SrIV group phytoplasma. Plant Pathol 57(2):366

Mathen K, Rajan P, Radhakrishnan Nair C et al (1990) Transmission of root (wilt) disease to coconut seedlings through Stephanitis typica (Distant) (Heteroptera: Tingidae). Trop Agric 67(1):69–73

McCoy RE (1982) Antibiotic treatment for control of tree diseases associated with mycoplasma-like organisms. Rev Infect Dis 4:S157–S161

McCoy RE, Howard FW, Tsai JH et al (1983) Lethal yellowing of palms, Agricultural Experiment Stations Bulletin 834. University of Florida, Gainesville

Mehdi A, Baranwal VK, Babu MK et al (2012) Sequence analysis of 16S rRNA and secA genes confirms the association of 16SrI-B subgroup phytoplasma with oil palm (*Elaeis guineensis* Jacq.) stunting disease in India. J Phytopathol 160:6–12

Mejía F, Palmieri M, Oropeza C et al (2004). First report of coconut lethal yellowing disease in Guatemala. Plant Pathol 53:80

Mpunami A, Tymon A, Jones P et al (2000) Identification of potential vectors of the coconut lethal disease phytoplasma. Plant Pathol 49(3):355–361

Myrie WA, Paulraj L, Dollet M et al (2006) First Report of coconut lethal yellowing disease in guatemala. Plant Pathol. Plant Dis 90(6):834–834

Myrie WA, Douglas L, Harrison NA et al (2012) First report of lethal yellowing disease associated with subgroup 16SrIV, a Phytoplasma on St. Kitts in the Lesser Antilles. New Dis Rep 26:25–25

Myrie W, Harrison N, Douglas L et al (2014) Identification of lethal yellowing disease of palms associated with infection by subgroup 16SrIV-A phytoplasmas in Antigua, West Indies. New Dis Rep 29:12–12

Narvaez M, Córdova I, Orellana R et al (2006) First report of a lethal yellowing Phytoplasma in *Thrinax radiata* and *Coccothrinax readii* palms in the Yucatan peninsula of Mexico. Plant Pathol 55(2):292–292

Narvaez M, Córdova-Lara I, Reyes-Martínez C et al (2016) Occurrence of 16SrIV subgroup-A phytoplasmas in *Roystonea regia* and *Acrocomia mexicana* palms with lethal yellowing-like syndromes in Yucatán, Mexico. J Phytopathol 164(11–12):1111–1115

Narvaez M, Ortíz E, Silverio C et al (2017) Changes observed in *Pritchardia pacifica* palms affected by a lethal yellowing-type disease in Mexico. Afr J Biotechnol 16(51):2331–2340

Narvaez M, Vázquez-Euán R, Harrison NA et al (2018) Presence of 16SrIV phytoplasmas of subgroups A, D and E in planthopper *Haplaxius crudus* Van Duzee insects in Yucatán, Mexico. 3Biotech 8(1):61

Nejat N, Sijam K, Abdullah SNA et al (2009) First report of a 16SrXIV, '*Candidatus* Phytoplasma cynodontis' group phytoplasma associated with coconut yellow decline in Malaysia. Plant Pathol 58(2):389–389

Nejat N, Vadamalai G, Davis RE et al (2012) '*Candidatus* Phytoplasma malaysianum', a novel taxon associated with virescence and phyllody of Madagascar periwinkle (*Catharanthus roseus*). Int J Syst Evol Microbiol 63(2):540–548

Nic-Matos JG, Narvaez M, Peraza-Echeverría E et al (2017) Molecular cloning of two novel NPR1 homologue genes in coconut palm and analysis of their expression in response to the plant defense hormone salicylic acid. G Genome 39(9):1007–1019

Ntushello K, Harrison NA, Elliott ML (2013) Palm phytoplasmas in the Caribbean basin. Palms 57(2):93–100

Omar AF, Alsohim A, Rehan MR et al (2018) 16SrII phytoplasma associated with date palm and Mexican palm fan in Saudi Arabia. J Australas Plant Pathol Soc 13(1):–39

Oropeza C, Narváez M, Echegoyén-Ramos PE et al (2010) Plan de contingencia ante un brote de amarillamiento letal del cocotero (ALC) en un país de la región del OIRSA. Organismo Internacional Regional de Sanidad Agropecuaria – OIRSA. San Salvador, El Salvador, p 149

Oropeza C, Córdova I, Chumba A et al (2011) Phytoplasma distribution in coconut palms affected by lethal yellowing disease. Ann Appl Biol 159(1):109–117

Oropeza C, Sáenz L, Chan JL et al (2016) Coconut micropropagation in Mexico using plumule and floral explants. Cord 32:21–26

Oropeza C, Córdova I, Puch-Hau C et al (2017) Detection of lethal yellowing phytoplasma in coconut plantlets obtained through in vitro germination of zygotic embryos from the seeds of infected palms. Ann Appl Biol 171(1):28–36

Perera L, Meegahakumbura MK, Wijesekara HRT et al (2012) A Phytoplasma is associated with the Weligama coconut leaf wilt disease in Sri Lanka. J Plant Pathol 94(1):205–209

Pérez-Hernández O, Góngora-Canul C, Medina-Lara MF et al (2004) Patrón espacio-temporal del amarillamiento letal en cocotero (*Cocos nucifera* L.) en Yucatán, México. Revista Mexicana de Fitopatología 22(2):231–238

Philippe R, Simon R, Deschamp S et al (2009) Study on the transmission of lethal yellowing in Ghana. OCL 16(2):102–106

Plavsic-Banjac B, Hunt P, Maramorosch K (1972) Mycoplasmalike bodies associated with lethal yellowing disease of coconut palms. Phytopathology 62(2):298–299

Prades A, Salum UN, Pioch D (2016) New era for the coconut sector. What prospects for research? OCL 23(6):D607

Puch-Hau C, Oropeza C, Peraza-Echeverria C et al (2015) Molecular cloning and characterization of disease-resistance gene candidates of the nucleotide binding site (NBS) type from *Cocos nucifera* L. Physiol Mol Plant P 89:87–96

Rajan P (2011) Transmission of coconut root (wilt) disease through plant hopper, *Proutista moesta* Westwood (Homoptera: Derbidae). Pest Manag Horticult Ecosyst 17:1–5

Ramaswamy M, Nair S, Soumya VP et al (2013) Phylogenetic analysis identifies a '*Candidatus* Phytoplasma oryzae'-related strain associated with yellow leaf disease of areca palm (*Areca catechu* L.) in India. Int J Syst Evol Micr 63(4):1376–1382

Ramos-Hernández E, Magaña-Alejandro MA, Ortiz-García CF et al (2018) The coconut pathosystem: weed hosts of nymphs of the American palm Cixiid *Haplaxius crudus* (Hemiptera: Fulgoroidea). J Nat Hist 52(5–6):255–268

Reinert JA (1980) Phenology and Density of *Haplaxius crudus* (Homoptera: Cixiidae). On Three Southern Turfgrasses. Environ Entomol 9:13–15

Roca MM, Castillo MG, Harrison NA et al (2006) First report of a 16SrIV group phytoplasma associated with declining coyol palms in Honduras. Plant Dis 90(4):526–526

Rodriguez JV, Vitoreli AM, Ramirez AL (2010) Association of a phytoplasma with dieback in palms in Puerto Rico confirmed by nested-PCR assays. Phytopathology 100(6):S110–S110

Schuiling M (1976) A survey of insect populations on *Cocos nucifera*. Principes 20(2):67

Singh R (2014) Texas Phoenix Palm Decline Confirmed in Louisiana. NPDN News 9:1–2

Sumi K, Madhupriya KS et al (2014) Molecular confirmation and interrelationship of phytoplasmas associated with diseases of palms in South India. Phytopathogenic Mollicutes 4(2):41–52

Thomas DL, Donselman HM (1979) Mycoplasma-like bodies and phloem degeneration associated with declining *Pandanus* in Florida. Plant Dis Report 63(11):911–916

Tomlinson JA, Boonham N, Dickinson M (2010) Development and evaluation of a one-hour DNA extraction and loop-mediated isothermal amplification assay for rapid detection of phytoplasmas. Plant Pathol 59(3):465–471

Triplehorn CA, Johnson NF (2005) Borror and DeLong's introduction to the study of insects, 7th edn. Thomson, Brooks/Cole, Pacific Grove, p 864

Vázquez-Euán R, Harrison NA, Narváez M et al (2011) Occurrence of a 16SrIV group phytoplasma not previously associated with palm species in Yucatan, Mexico. Plant Dis 95(3):256–262

Warokka WA, Jones P, Dickinson M (2006) Detection of phytoplasmas associated with Kalimanthan wilt disease of coconut by polymerase chain reaction. J Penelit Tanam Ind 12:154–160

Yankey EN, Bila J, Arocha Rosete Y et al (2018) Phytoplasma diseases of Palms. In: Rao GP, Bertaccini A, Fiore N, Liefting LW (eds) Phytoplasmas: plant pathogenic bacteria – I. Characterisation and epidemiology of phytoplasma – associated diseases. Springer, Singapore, pp 267–285

Zamharir MG, Eslahi MR (2019) Molecular study of two distinct phytoplasma species associated with streak yellows of date palm in Iran. J Phytopathol 167(1):19–25

Zhao Y, Wei W, Lee IM et al (2009) Construction of an interactive online phytoplasma classification tool, iPhyClassifier, and its application in analysis of the peach X-disease phytoplasma group (16SrIII). Int J Syst Evol Microbiol 59(Pt 10):2582–2593

Zizumbo-Villarreal D (1996) History of coconut in Mexico. Genet Resour Crop Evol 43(6):505–515

Zizumbo-Villarreal D, Colunga-García MP, Fernández-Barrera M et al (2008) Mortality of Mexican coconut germplasm due to lethal yellowing. Bulletin de Ressources Phytogénétiques, p 23

Zou Y, Mason MG, Wang Y et al (2017) Nucleic acid purification from plants, animals and microbes in under 30 seconds. PLoS Biol 15(11):e2003916

Chapter 10
Germplasm Reestablishment and Seedling Production: Embryo Culture

Julianne Biddle, Quang Nguyen, Zhihua H. Mu, Mike Foale, and Steve Adkins

10.1 Coconut Inflorescence, Fruit, and Embryo Morphology

Although coconut (*Cocos nucifera* L.) is the only species of the genus Cocos, it is extremely diverse and has been domesticated to improve production traits and resilience to environmental and abiotic conditions (Fig. 10.1 displays a young plant in the soil). There are no species with which interfertility (hybridization) has been reported, and there are no closely related species (Bourdeix et al. 2001). All coconut forms are diploid ($2n = 32$). Populations are referred to as varieties and are divided into two types, Tall and Dwarf. Naming nomenclature generally combines the country location and palm stature but can also indicate hybrid parentage or plant characteristics such as the color of the fruit, inflorescence form, or kernel type. Guidelines for naming can be found on the Coconut Genetic Resources for Enhanced Livelihoods (COGENT) websites: http://www.cogentnetwork.org/faq/140-naming and http://www.cogentnetwork.org/faq/139-exsitu.

Those belonging to the Tall type generally cross-pollinate, while the Dwarf group is highly self-pollinating, although the Dwarf type can be fertilized by the Tall when hand-pollinated. When the fruit morphology (husk, shell, water, and kernel) is analyzed at a set fruit development stage, coconut populations can be differentiated in the field by fruit characters; a range of deoxyribonucleic acid (DNA) profiling methods can also be used (Ashburner et al. 1997; Duran et al. 1997;

J. Biddle (✉) · Z. H. Mu · M. Foale · S. Adkins
School of Agriculture and Food Science, University of Queensland, St Lucia, QLD, Australia
e-mail: julianne.biddle@uq.edu.au

Q. Nguyen
School of Agriculture and Food Science, University of Queensland, St Lucia, QLD, Australia

Applied Biotechnology in Crop Development Research Unit, School of Biotechnology, International University, Vietnam National University-HCM, Ho Chi Minh City, Vietnam

S. Adkins et al. (eds.), *Coconut Biotechnology: Towards the Sustainability of the 'Tree of Life'*, https://doi.org/10.1007/978-3-030-44988-9_10

Fig. 10.1 A young coconut palm developed from embryo culture

Perera et al. 1998; Karp 1999; Lebrun et al. 2001; Perera et al. 1999; Rodriguez et al. 1997; Rohde et al. 1999; Teulat et al. 2000; Wadt et al. 1999).

The apical meristem of the coconut consists of a region of soft undifferentiated tissue, in the shape of a cone, positioned on the tip of the elongating axis. Surrounding the apical meristem, the inflorescence primordia form and expand, and central growth lengthens the trunk. The lateral coconut inflorescences are covered by sheathes (spathes), with the complete structures called the "spadix." These originate in the frond axil. Flowers (male and female) produce nectar, encouraging insect pollination, while also allowing for wind pollination. After pollination, the female flower develops into a fruit, growing steadily until reaching full volume at 6 months post-pollination.

A mature fruit (drupe) has a husk containing a round nut. The nutshell is about 3 mm thick, with the inside surface covered by a consistent layer of solid endosperm (kernel). When young, the husk fibers are conduits, connected to shell apertures, through which photo-assimilate is transported for kernel formation and respiration. The immature, 6 months following pollination, fruit cavity is filled with just liquid endosperm (coconut water) and free of kernel. Over time the kernel on the inside of the shell of a young fruit becomes soft and gelatinous, firming by the 10th month, and fibrous and solid by 12 months. When the kernel is fully developed in the mature fruit, the volume of liquid endosperm is reduced. The mature fruit contains approximately 50% water, 33% oil, 5% protein, and the remainder as carbohydrate, mostly fiber.

Once the fruit is mature, the vascular link to the tree halts, and abscission occurs. Once released from the mother palm, the liquid endosperm starts to evaporate and is replaced by the air over time. On an average palm, growing in ideal conditions, bunches of fruit form approximately 10 fruit of the same shape, color, and size. Fruit from immature Tall varieties are generally bronze or green, or with the recessive colors of yellow and orange, or non-recessive green and brown, common in the case of the self-pollinating Dwarf varieties.

The coconut fruit contains only one embryo (rarely two). It is contained within the solid kernel underneath the operculum (soft eye), through which the embryo emerges during germination. Embryo location can also be identified by a dimple in the kernel when the coconut is split and viewed from the inside. The coconut embryo is cylindrical, white, and ca. 8 mm long and 5 mm in diameter. As the shoot grows, it extends into the husk, and the haustorium (the single cotyledon which takes on the form of a spongy mass) grows into the nut cavity. The haustorium secretes enzymes to utilize the energy-rich liquid endosperm and kernel. About 2 months following germination, the shoot and roots are visible outside the husk, followed by the unfolding of the initial leaf lamina ca. 2 months later. For 15 months, the plant continues to use energy from the kernel. Leaves (fronds) emerge at 6 weeks and then at 22-day intervals. A Tall variety can reach peak fruit productivity from 15 years until approximately 35 years of age.

10.1.1 Embryo Culture

Plant embryos are multicellular structures, formed from an initial zygotic (in the case of a zygotic embryo) or a somatic cell (in the case of a somatic embryo), with the potential to form a new plant (Fig. 10.2 shows coconut embryos at different

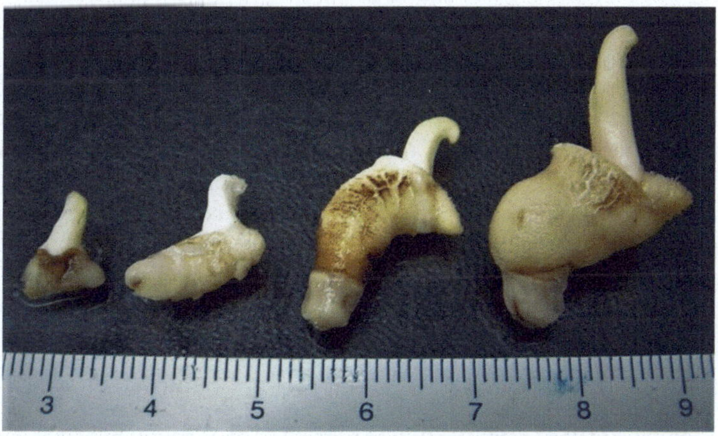

Fig. 10.2 Coconut embryos at different stages of development during embryo culture

stages of development during embryo culture). Most embryos have a bipolar organization, with primordial shoot and root structures (Sharma et al. 1996). Embryo culture (EC) or embryo rescue is a well-established form of in vitro plant culture often used to assist embryo development. This often occurs when interspecific/intergeneric crosses are made or if embryos are immature and may not survive to maturity without intervention (Hu and Zanettini 1995). During EC, embryos are generally excised and placed onto a suitable culture medium to support their growth.

Embryos develop under complex conditions in vivo (natural conditions) (Haslam and Yeung 2011b). Key considerations in determining the culture conditions to use include the maturity stage of the embryo, environmental conditions (light and temperature), and access to nutrients (medium components) (Sharma et al. 1996). Medium compositions vary but are often based on the formulation described by either by Murashige and Skoog (1962) or Gamborg et al. (1968), generally abbreviated as MS and B-5, respectively. The concentrations of some components used, such as sucrose, are varied depending on embryo age at culture initiation. Young embryos are usually grown on greater concentrations of sucrose with more complex medium components, whereas more advanced embryos can generally be grown on simple reduced sucrose medium (Haslam and Yeung 2011b). Temperature and light conditions often depend on the species, with warm climate crops needing higher temperatures than cool climate crops. Growth conditions, including the composition of the culture medium, need to be well defined. If conditions are not optimum, embryos may fail to thrive, die, or convert to undifferentiated callus (Haslam and Yeung 2011a).

As one of the foundation techniques of plant tissue culture, EC began with studies using a soil substrate in the eighteenth century (Schopfer 1943) and progressed to studies on embryo development and morphology, with steady increases in the sophistication of growth conditions (Andronescu 1919; Buckner and Kastle 1917; Dieterich 1924; Dubard and Urbain 1913; Knudson 1922; Raghavan and Torrey 1964). Modern EC utilizes simplified substrates with simpler nutrient solutions, building on a better knowledge of culture environmental conditions and improvements in the use of plant growth substances (Brown 1906; Hannig 1904). Numerous applications have been derived from the study of plant embryology; value in particular has been recognized for improving the survival of embryos formed after interspecific crosses. Embryo culture was first used for this purpose by Laibach (1929) to gain a successful cross between plants of different flax species (interspecific hybrid) *Linum austriacum* L. and *Linum perenne* L. Since then, the method has been used for various plant species to reduce generation time and overcome seed dormancy, for studies of seed development and viability, rapid breeding of species of commercial significance, and conservation purposes (Haslam and Yeung 2011a; Sharma et al. 1996).

10.1.2 Coconut Embryo Culture

Cutter and Wilson (1954) were the first to report EC of coconut; however, since then improvements in the technique have largely trailed behind other species due to challenges such as the poor and diverse response of cultured embryos, sluggish in vitro development, and weak plantlets during acclimatization (Fernando et al. 2010). Regardless of these obstacles, advances have been made in EC (Table 10.1 displays a summary of these advances). Now the technique is most commonly recognized for its application to coconut types that generally don't germinate in vivo; for the collection, transfer, and conservation of coconut genetic resources; for the regeneration of plantlets from cryopreservation and somatic embryogenesis (cloning); and, in the few cases where it has been done, genetic transformation (Nguyen et al. 2016). The technical development of coconut EC, the details regarding specific applications, and the limitations, challenges, and prospects of the technique are presented in the following sections.

10.2 Coconut Embryo Culture Progress

Several methods for coconut EC have been reported (in vitro and ex vitro); however, improvements are still required, and the success rate in using these protocols has varied among laboratories. From the earliest attempts by Cutter and Wilson (1954) to isolate and culture coconut embryos, it took a further 10 years before De Guzman and Del Rosario (1964) were able to regenerate plants successfully from coconut embryos. Since then modifications have been published in peer-reviewed journals (see Table 10.1 for specific publications), by industry organizations, such as the Food and Agriculture Organization (FAO) of the United Nations and International Plant Genetic Resources Institute (IPGRI) (Engelmann 2002; Frison et al. 1993), the Australian Centre for International Agricultural Research (ACIAR) (Ashburner et al. 1995; Samosir et al. 2008); and Bioversity International (Cueto et al. 2012), and informally in conference proceedings (Rillo 1998; Samosir and Adkins 2004; Vu 2002). It has been demonstrated that EC does not have a detrimental impact on coconut palms grown in natural conditions (Koffi et al. 2013). The following section brings together coconut EC research to date into a concise format for ease of reference (Table 10.1 lists EC publications with culture conditions).

10.2.1 Morphology and Physiology of the Zygotic Embryo (Including Phytosanitary Conditions)

In the past, the whole coconut fruit had to be transported to the research institution to initiate the conservation process. This posed a challenge relating to the size and weight of the fruit and the biosecurity issues associated with the transport of large volumes of material potentially contaminated with pests and/or diseases. The collection of, and transport of, embryos was a way of reducing biosecurity risks and transport costs associated with germplasm collection and transport, especially from hard-to-get-to locations.

Table 10.1 A summary of coconut embryo culture research, listing the varieties cultured, the medium and environmental conditions used, and the responses

Embryo description[a]	Medium and additives[b]	Environmental conditions (temperature °C and illumination[c])	Summary[d]	Publication
Unidentified	Young CW and 1.5% A	25 90 μmol m^{-2} s^{-1}	G	Cutter and Wilson (1954)
Makapuno	White, 25% CW (25%), and 1.2% A	25 D 3 weeks and then L	G and D	De Guzman and Del Rosario (1964)
Malayan Yellow D × West African T hybrid 11–12 msp	MS, MW Vit, 6% S, 0.2% AC, and 0.8% A	27 D 3 weeks and then 12:12 h L:D 55 μmol m^{-2} s^{-1}	G, D, and A	Assy-Bah et al. (1989)
Tonga and Solomon Islands	Germination: MS, MW Vit, 6% S, 0.2% AC, 0.8%, and agar Growth: Y3, 200 μM NAA, 4% S, 0.2% AC, and 0.8% A	30–31 16:8 h L:D 90 μmol m^{-2} s^{-1}	G, D, and R	Ashburner et al. (1993)
Malayan Yellow D	Modified MS (L), MW Vit, 6% S, and 0.2% AC	27 ± 1 D 8 weeks and then L 45 ± 5 μmol m^{-2} s^{-1}	G, D, and A	Triques et al. (1997)
Laguna T 10–11 msp	Germination: Y3 (L) Growth: Y3 (L), 50 μM IBA or NAA, 4.5% S, and 0.25% AC	28–30 9:15 L:D 75–90 μmol m^{-2} s^{-1}	G, D, and A	Rillo (1998)
Western Samoa and Batu Layar	Transport (up to 4 days): sterile 1 mg L^{-1} ABA Germination: Y3 (L), MW Vit, 175 mM S, and 2.5 g L^{-1} AC Subculture in Y3, MW Vit, 175 mM S, and 2.5 g L^{-1} AC	28 ± 1 Transport at low temperature D Then D 6 weeks	G	Samosir et al. (1999a, b)
Malayan Green D 12–14 msp	Modified Y3, 0.25% AC with or without 0.3% Gelrite	27 ± 2 D 1 week and then 16:8 h L:D 45–60 μmol m^{-2} s^{-1}	G, D, and A	Pech y Aké et al. (2004)
Malayan Green D 12 msp	Y3 and 4.5% S	27 ± 2 D 6–8 weeks and then 16:8 h L:D 50 μmol m^{-2} s^{-1}	G, D, and A	Fuentes et al. (2005a, b)

Malayan Green D and Malayan Yellow D 12–14 msp	Germination: modified Y3, 0.46 μM GA_3, 0.25% AC with or without 0.3% Gelrite	27 ± 2 D 5 weeks and then 16:8 h L:D 45–60 μmol m^{-2} s^{-1}	G, D, and A	Pech y Aké et al. (2007)
	Growth: modified Y3, 0.25% AC with or without 0.3% Gelrite			
Malayan Yellow D 11–12 msp	MS (L), 4% S, 0.15% AC, and 75 μM unbound lauric acid	Unknown Unknown	G	López-Villalobos et al. (2011)
Malayan Yellow D 11–12 msp	Germination: Y3 (L), Rillo Vit, 6% S, and 0.1% AC (0.1%)	27 ± 1 D 6–8 weeks and then 14:10 h L:D 90 μmol m^{-2} s^{-1} Followed by CO_2 enrichment to improve growth	G, D, and A (up to 100% success)	Samosir and Adkins (2014)
	Growth: Y3, Rillo Vit, 6% S, 0.1% AC, and 0.2% Bacto-A			
Malayan Yellow D 10–11 msp	Germination: modified Y3 (L), 0.97–7.67 μM ABA, 0.1% AC, and 6% S	28–30 14:10 h L:D 40 μmol m^{-2} s^{-1} plus red-light enrichment 660 nm Followed by CO_2 enrichment in photoautotrophic system with 14:10 h L:D 90 μmol m^{-2} s^{-1} to improve growth	G, D, and A (faster and up to 100% success)	Magdalita et al. (2015)
	Growth: modified Y3, 0.1% AC, 100 μM NAA, 0.7% A, 4.5% S and then modified Y3 (L), 100 μM NAA, 0.7% A, 0.1% AC, and 2.5% S			
	Chamber: modified Y3 (L) and vermiculite or coir dust substrate			
West African T and Malayan Orange D 10–12 msp	Y3 (L), 3% S, 0.25% AC, and 1.5 mg/L NAA	Unknown D 4 weeks and then 14:10 h L:D	G and D	Nwite et al. (2017)
Kopyor T and D 11 msp	Germination: HEC, Rillo Vit, 6% S, and 0.2% AC	27 ± 1 and then 26–29 in the mini growth chamber D 4 weeks and then 14:10 h L:D 25 ± 2 μmol m^{-2} s^{-1} Followed by a mini growth chamber 14:10 h L:D 40 μmol m^{-2} s^{-1}	G, D, and A (up to 90% success)	Sisunandar et al. (2018)
	Growth: HEC (L), Rillo Vit, and 3% S			
	Root induction/acclimatization: Mini growth chamber with aerated HEC (L), 1 μM IBA, no Vit or S, coco peat, and rice charcoal substrate (1:1 -v/v)			

[a]Embryo source, type or variety, and maturity [months since pollination (msp)]. Conditions with the best results are reported. All papers used mature embryos; if an age was specified, it is listed. Tall (T) and Dwarf (D)
[b]Sucrose (S), agar (A), activated charcoal (AC), filtered coconut water (CW), liquid medium (L), gibberellic acid (GA_3), indole-3-butyric acid (IBA), Murashige and Skoog (1962) medium (MS), Morel and Wetmore (1951) vitamins (MW Vit), naphthalene acetic acid (NAA), ascorbic acid solution (ABA), Rillo et al. (2002) vitamins (Rillo Vit), White (1943) medium (White), Eeuwens (1976) medium (Y3), and hybrid embryo culture (HEC) (Sisunandar et al. 2018)
[c]Light (L) and dark (D) growth environment and hours (h)
[d]Successful adventitious root formation only (R), embryo germination (G), plant development (D), acclimatization (A)

Fig. 10.3 Extracting the embryo-containing endosperm plug

Early forms of embryo collection in the field consisted of isolation of mature embryos and storage in water/coconut water (sterile) during movement to the research institution (Rillo and Paloma 1991). Unfortunately, there were high rates of culture contamination and death of embryos. The technique was therefore further refined, by leaving the embryos protected within a plug of endosperm (Fig. 10.3 shows the extraction of the embryo-containing endosperm plug), field surface sterilization of the plug, the addition of ascorbic acid, and storage in cool conditions (ca. 5 °C) during transport (Adkins and Samosir 2002). Generally, 10- to 14-month-old embryos are most commonly used for EC, with the greatest regeneration success in embryos harvested 12 months after pollination (Table 10.1 lists EC publications).

10.2.2 Culture Media

Coconut EC has been trialed on a range of medium types, with Y3 medium, developed by Eeuwens (1976), the most widely used. This medium contains 50% of the ammonium and nitrate nitrogen and considerably higher concentrations of microelements such as iodine, cobalt, and copper than MS, a more widely used tissue culture medium type (Murashige and Skoog 1962). The culture conditions when using Y3 better replicate the environmental conditions of coconut germination in the coastal locations where the tree is found. Like EC for other species, coconut embryos are usually grown on a medium with high sucrose, as it is needed for coconut embryo germination, and it has been demonstrated that activated charcoal (AC) reduces tissue necrosis (Table 10.1 lists EC publications including sucrose and AC concentrations). A solid medium with agar at a concentration of 1.5–0.8% has been used in some cases, but other studies use a system with liquid medium during the germination phase of EC, then subculture onto a solid medium following germination (Rillo 1998), or vermiculite soaked in nutrient solution for the growth of seedlings (Samosir and Adkins 2014). Other gelling agents have also been used, for example, gelrite (Pech y Aké et al. 2004; Pech y Aké et al. 2007). Plant growth regulators, including gibberellic acid (GA), have been commonly utilized to improve the germination of embryos (Magdalita et al. 2015), and auxins, including indole-3-butyric acid (IBA) or naphthalene acetic acid (NAA) have been used to enhance root initiation late in germination and the beginning of seedling development (Ashburner et al. 1993; Rillo 1998). In addition, a fatty acid found in the coconut kernel, lauric acid, was reported by López-Villalobos et al. (2011) to improve EC development (Table 10.1 lists EC publications).

10.2.3 Culture Environment

The optimum culture circumstances needed for germination and then embryo development in vitro reflect those that a germinating coconut fruit would experience ex vivo (in the natural environment). Optimum embryo germination and seedling growth occur in warm conditions 25–31 °C, initially in dark conditions (as would be experienced inside the fruit) for 5–8 weeks and then in illuminated conditions (45–90 µmol m^{-2} s^{-1}) following germination (Table 10.1 lists EC publications including reported temperatures).

10.2.4 Regeneration

Acclimatization conditions and soil types used for the successful regeneration of EC seedlings vary. It has been demonstrated that black-colored polyethylene bags filled with a mixture of equal weights of soil and peat moss are ideal for growing coconut plants following the tissue culture stages (Pech y Aké et al. 2004). Improvements in survival have been linked to the staged transfer of the seedlings through a sequence of environments (fogging chamber, shade, and full light) that progressively improve growth (Talavera et al. 2005). Seedling survival can be enhanced by the early formation of a photosynthetic-based metabolism (Triques et al. 1998). Growth, health, and survival can be improved using a sucrose-free photoautotrophic method with the addition of CO_2 and enrichment of 1600 μmol mol^{-1} in the light phase of the protocol (Samosir and Adkins 2014).

10.3 Applications of Embryo Culture Technique

Coconut EC is a foundation technique that has numerous applications. These include the propagation of elite varieties (Sisunandar et al. 2015; Sisunandar et al. 2018), germplasm collection and transport (Rillo 1998; Samosir and Adkins 2014), the conservation of germplasm (cryopreservation) (Sisunandar et al. 2014), and as the source of plumular tissue for the induction of somatic embryogenic callus (Pérez-Núñez et al. 2006). Embryo tissues have also been used for genetic transformation of coconut (Andrade-Torres et al. 2011; Samosir et al. 1998). Embryo culture applications are explored in more detail in the following sections.

10.3.1 Embryo Culture for High-Value Varieties

The mature coconut fruit generally has a firm endosperm; however, for some varieties, such as the Makapuno (Philippines), Kopyor (Indonesia), and Dikiri (Sri Lanka), a soft, fluffy, or jelly endosperm fills the cavity. In some countries, these varieties are revered, and the fruit and seedlings fetch a premium price. The embryos of these varieties are anatomically and visually similar to other coconut varieties, however, germination is not supported by their endosperm (Nguyen et al. 2016). Methods of EC (embryo rescue) and subsequent in vitro development have been applied successfully in commercial circumstances to produce seedlings and fruit of these varieties (Adkins 2007; Sisunandar et al. 2015). An example is the Vietnamese initiative by the Anh Dao Company for Makapuno (Fig. 10.4 shows Anh Dao staff propagating Makapuno). The company has been working on EC for more than 8 years and has had success in growing plants to full productivity (Fig. 10.5 shows mature Makapuno trees propagated by the Anh Dao Company). Makapuno palms produced by this method have the potential to bear 100% Makapuno type fruit, if pollination from other varieties is prevented (Engelmann et al. 2011).

10.3.2 Embryo Culture for Germplasm Movement and Conservation

Some plant species do not produce viable seeds, while others have seeds that do not germinate easily, or seeds are unable to be stored long term due to a high moisture content (are recalcitrant). Conservation of these types of species is challenging and

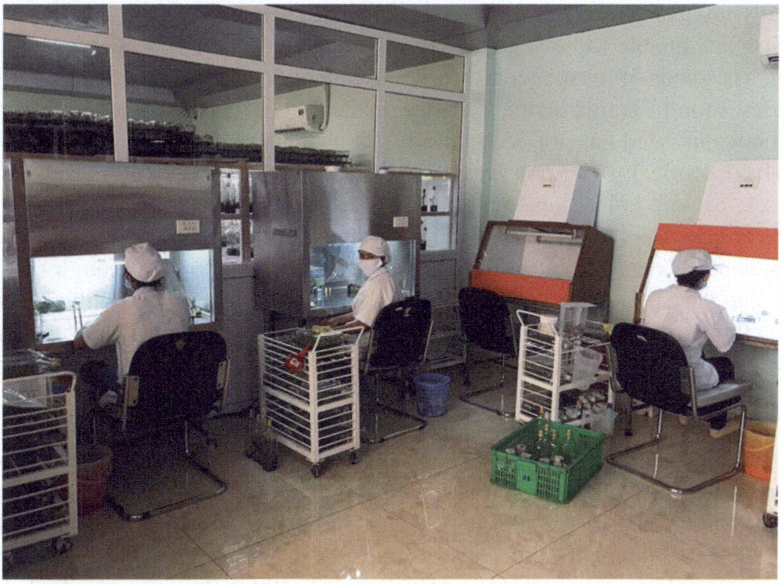

Fig. 10.4 The commercial application of embryo culture for propagation of Makapuno in Vietnam by the Anh Dao Company

Fig. 10.5 Mature Makapuno trees propagated by embryo culture in Vietnam by the Anh Dao Company

generally involves the use of seed gardens (ex situ genebank). Coconut is a recalcitrant species, unable to be vegetatively propagated, with a seed unable to tolerate drying. The seeds of most coconut varieties cannot be kept in a seed store for more than a few months before losing viability (Foale 2003). As conservation of the recalcitrant coconut seed has not been possible in a standard seed bank, coconut germplasm has historically been conserved in field collections. For long-term conservation purposes, backup collections of these field collections should also be maintained in vitro (cryopreservation), however, this has not yet been undertaken for coconut.

Coconut has one of the largest seeds in the plant kingdom. Although seed dormancy can vary between varieties, it is generally only short-lived, with most seeds germinating immediately once mature (Foale 2003). Such large and recalcitrant seeds can make germplasm collecting challenging and costly. Also, pests and diseases are more likely to be carried in the whole fruit. The collection of embryos alone for germplasm conservation and exchange has the potential to reduce phytosanitary problems and costs associated with transport (Ashburner et al. 1993; Engelmann et al. 2011). Standards have been established for coconut germplasm collection and exchange via EC, with organizations, including COGENT, the FAO, and IPGRI, recommending the use of embryos as the safest method for sharing coconut germplasm internationally (Frison et al. 1993; Pence et al. 2002; Cueto et al. 2012). These protocols provide standard techniques to collect, store, transport, prepare, germinate, and develop coconut embryos into seedlings which can be planted into the field. Despite the endorsement of these international industry bodies and the documented successes using the EC technique (Table 10.1 lists EC publications, and Fig. 10.6 shows plantlets ready for transportation), EC is not yet widely used for germplasm movement and conservation purposes, and the largest

Fig. 10.6 Embryo cultured plantlets ready for transportation overseas

feasibility trial conducted so far produced disappointing results. The study undertaken by Bioversity International to test the use of an EC protocol for the international exchange of germplasm experienced high contamination rates when germplasm was transferred from Côte d'Ivoire to three other countries (Bioversity 2012). It was proposed that protocol refinements, the availability of suitably skilled staff, and adequate laboratory equipment are crucial for the successful transfer of embryos internationally.

10.3.3 Embryo Culture for Cryopreservation

It is vital to conserve coconut genetic diversity for the future as coconut is a staple crop in many countries. Millions of families worldwide depend on it for cultural purposes and as an income source (Hocher et al. 1999). Despite the social and economic value of coconut, and the considerable research and development undertaken in many countries, including the production of high-yielding and disease-resistant varieties, farmers and conservation staff are still faced with many problems. Currently, coconut genetic diversity is only conserved in field genebank, and these banks face risks such as pests and disease, extreme weather events, and neglect arising from inadequate funding. Common pests and diseases that may impact coconut plantations and field genebank include coconut rhinoceros beetle (*Oryctes rhinoceros* Linnaeus, 1758), coconut treehopper (*Sexava* sp.), coconut hispid beetle (*Brontispa longissimi* Gestro), coconut bud rot (*Phytophthora palmivora* Butler), and lethal yellowing diseases (phytoplasmas) (Novarianto and Warokka 2006). To ensure the security of these genetic resources, these field genebank must be duplicated using an in vitro conservation technique, such as in vitro slow growth or cryopreservation of coconut tissues.

Efforts have been made to establish standard methods for the secure long-term maintenance of coconut genetic resources via cryopreservation (Sisunandar et al. 2005). Typical protocols for coconut cryopreservation begin with the pre-culture of embryos, followed by tissue dehydration and then freezing. Upon recovery, the tissues need to be regrown using EC and acclimatized. Three techniques have been trialed to dehydrate coconut tissues, namely chemical dehydration, slow physical dehydration, and fast physical dehydration. The first attempts to cryopreserve coconut embryos commenced in the 1980s, with chemical desiccation and a slow freezing method (Bajaj 1984). Since then research has focused on the use of mature embryos with a physical desiccation method (N'Nan et al. 2012; Sisunandar et al. 2014) or excised plumular tissues and a chemical dehydration method (Hornung et al. 2001).

High concentrations of sucrose, glucose, and glycerol (>10%, w/v) have been used for chemical dehydration. Excised plumular tissue from coconut embryos has successfully been encapsulated in sodium alginate (3%) after dehydration using sucrose (5%) (N'Nan et al. 2008). Drawn-out physical desiccation methods have been used for a range of drying durations of up to 48 h. Although this method has been examined for numerous coconut varieties, few plants have been regenerated using this technique. A rapid drying method (Sisunandar et al. 2010b), used directly on mature embryos, has so far been the best approach for coconut cryopreservation. Drying embryos to a 20% moisture content within 8 h maximized the embryo survival and seedling recovery, with 40% of recovered plants successfully establishing in soil. Rapid dehydration didn't cause appreciable genetic changes in the plantlets recovered (Sisunandar et al. 2010a). Once further optimized and widely used, cryopreservation has the potential to conserve coconut genetic diversity over the long term to ensure food security, and once frozen, embryos can also provide an alternative method for the transport and exchange of germplasm.

10.3.4 EC for Cloning

Embryo culture is valuable for producing tissues for Somatic Embryogenesis (SE) of coconut. It is now possible to obtain somatic embryos from cultured coconut embryos to produce clones. Coconut SE (cloning) was first documented by Eeuwens and Blake (1977) and Pannetier and Buffard-Morel (1982), using immature somatic tissues: leaves, seedlings (slices), and rachillae of inflorescence as explants (Branton and Blake 1983; Gupta et al. 1984). Later experiments have utilized both somatic and zygotic tissue types: immature inflorescences and ovaries, and embryonic tissues. Immature/mature embryos and plumular tissues from embryos, are considered easier to work with than somatic tissues. Out of these tissue types, young embryos were first thought to be most receptive, but it has been found that with longitudinal slicing, mature embryo response can be enhanced (Adkins et al. 1998; Samosir 1999), as well as the use of isolated plumules (Chan et al. 1998; Lopez-Villalobos 2002; Pérez-Núñez et al. 2006).

Coconut SE generally consists of the following steps: the induction of callus, maintenance, the development of somatic embryos, somatic embryo maturation, germination, and plantlet recovery. To produce embryogenic callus cultures, the presence of the plant growth regulator auxin in high concentrations is needed. 2,4-Dichlorophenoxyacetic acid (2,4-D) is the most commonly used, with the concentration depending on the explant tissue type, genotype, and concentration of AC in the medium. A low 2,4-D (24 μM) concentration was demonstrated to be the most successful at inducing embryogenic cultures from Sri Lanka Tall embryos (Fernando and Gamage 2000), however, an increased concentration (125 μM) was required for other varieties: Malayan Yellow Dwarf and the Buta Layar Tall (Adkins et al. 1998; Samosir 1999). It has been proposed by López-Villalobos et al. (2004) that tissues metabolize 2,4-D rapidly, resulting in triacylglycerol derivatives that can act as a stored form of 2,4-D in coconut culture with the capacity to linger and stop somatic embryo development in the absence of auxin.

Medium types Y3 (Eeuwens 1976), and to a lesser extent BM72 (Karunaratne and Periyapperuma 1989), are generally used for SE, with B5 media (Gamborg et al. 1968) or MS media (Murashige and Skoog 1962) less commonly used

(Bhallasarin et al. 1986; Branton and Blake 1983). Sucrose (3–4%), in combination with AC (0.1–0.3%), is essential for SE. Activated charcoal is needed to reduce explant and culture browning related to the release of phenols and ethylene from the cultured tissues (Samosir 1999). However, it is known that AC can inconsistently impact the effectiveness of Plant Growth Regulators and other additives in the media, resulting in unknown concentrations of these components (Pan and van Staden 1998). Variation in AC particle dimensions and effectiveness can impact the formation of embryogenic cultures (Sáenz et al. 2009). Although polyvinylpyrrolidone (PVP) does not appear to significantly reduce tissue browning (Basu et al. 1988), polyvinylpolypyrrolidone (PVPP) can have a positive impact on embryogenic culture induction (Samosir 1999). In addition, frequent sub-culturing of explants and embryogenic cultures can decrease exposure to accumulated phenols in the medium (Fernando and Gamage 2000; Pérez-Núñez et al. 2006).

Cytokinin in the medium, for example, 2-isopentyl adenine, thidiazuron (TDZ), 6-benzylaminopurine (BAP), or kinetin, is needed for the proliferation of the embryogenic cultures and maturation. It has been demonstrated that the induction of embryogenic cultures is best undertaken at 28 °C in darkness for 1 to 3 months after explanting (Adkins et al. 1998), although Chan et al. (1998) considered optimum conditions to be a 12/12-h day/night photoperiod of 45–60 μmol m^{-2} s^{-1}. The addition of polyamines, notably putrescine or spermine, seems to enhance the embryogenic response by possibly protecting explants from ethylene damage and/or by promoting the embryogenic response (Adkins et al. 1998). Inhibitors of ethylene, including aminoethoxyvinylglycine and silver thiosulfate, have been demonstrated to stimulate callus culture proliferation and somatic embryo development (Adkins et al. 1998). An absence or reduction of 2,4-D in the medium with the addition of a cytokinin initiated the transition of undifferentiated callus to embryogenic callus. In addition, the presence of benzylaminopurine (BA) (650 to 300 μM) in the culture medium can improve plant recovery (Chan et al. 1998; Pérez-Núñez et al. 2006). Growth of somatic embryos is enhanced by the addition of 5 μM ascorbic acid (ABA) (Fernando and Gamage 2000; Fernando et al. 2003; Samosir et al. 1999a, b). Osmotically active chemicals including polyethylene glycol at a concentration of 3%, when utilized with 45 μM ABA, can improve somatic embryo production, development, and germination (Samosir et al. 1998). Ancymidol (30 μM), a growth retardant, has a positive impact on somatic embryo germination frequency (Antonova 2009).

For other palm species, including oil palm (*Elaeis guineensis* Jacq.), the use of suspension cultures has improved the recovery of somatic embryos and their development into plantlets (Teixeira et al. 1995). Temporary immersion of embryogenic callus has also been used to improve plantlet production for peach palm (*Bactris gasipaes* Kunth) (Steinmacher et al. 2011), and date palm (*Phoenix dactylifera* L.) (Tisserat and Vandercook 1985). These techniques, if adapted to coconut, could advance the rate and quality of plantlet production. In addition, the acclimatization of coconut somatic embryo-derived plantlets needs optimization as rates of conversion success can be as low as 50% (Fuentes et al. 2005a, b), although Samosir and Adkins (2014) have improved the acclimatization of plantlets using a photoautotrophic culture approach and with the use of fatty acids (López-Villalobos et al. 2001, 2011), chiefly lauric acid in the culture medium.

10.3.5 Genetic Transformation

Limited research has been undertaken on the genetic transformation of coconut utilizing embryos. Embryogenic cultures have been subjected to microprojectile bombardment by Samosir et al. (1998). In that study, *Agrobacterium*-mediated transformation was attempted on various coconut explants, including zygotic tissue (embryos) and plumule-derived embryogenic cultures. It was found that a transformation approach that combined bio-ballistics to generate micro-wounds in explant tissues, followed by vacuum infiltration and culture with *A. tumefaciens* (co-culture), can be used to introduce genes of interest (Andrade-Torres et al. 2011). However, to date no genetically modified coconut palms have been produced. The complete sequence of the coconut chloroplast (Ya-Yi et al. 2013), mitochondria (Hasan Awad et al. 2016), and a draft complete genome is now available (Xiao et al. 2017). Once the genes involved in disease and/or pest resistance, resilience to environmental stresses, or key productivity traits are isolated, genetic transformation will become more useful for coconut breeding.

10.4 Limitations, Challenges, and Prospects

Coconut EC is valuable tool for the collection, transport, and storage of genetic resources and to gain viable plants from germplasm that can be difficult to germinate (Figs. 10.7, 10.8, and 10.9 show coconut plantlets at different stages of development). There are numerous publications that describe the technique (see Table 10.1). Although the method is currently being used to commercially produce plantlets of Makapuno/Kopyor types (endosperm mutants), it is not yet being used for germplasm movement or conservation despite the availability of published standards (Cueto et al. 2012; Frison et al. 1993; Samosir et al. 2008). This lack of adoption is likely due in part to the financial pressure coconut genebank and research institutions face. Short-term funding arrangements are likely to hamper long-term conservation efforts.

Although EC has now been introduced to many laboratories through numerous research projects, a large trial commissioned by the Global Crop Diversity Trust has demonstrated the shortcomings of the EC method for germplasm exchange (Bioversity 2012). To reduce contamination rates and increase embryo viability, ongoing mentoring and support from more experienced tissue culture laboratories is vital. It is crucial that operators have a strong understanding of aseptic technique. The EC technique requires regular use to maintain high success rates. There needs to be a financially secure environment for tissue culture practitioners based at gene

Fig. 10.7 A germinating coconut embryo in vitro

Fig. 10.8 Coconut in vitro plantlets at different stages of development

Fig. 10.9 Acclimatized coconut plantlets following embryo culture

banks to establish good aseptic practices and tissue culture techniques. It is recommended that although all those trained may not be working with valuable coconut germplasm in the short term, they continue to practice their technique and troubleshoot with guidance from senior tissue culture staff, in consultation with EC experts, to ensure that when they begin to work with high-value germplasm, they treat it appropriately. Mentoring is likely to help enshrine a continuous improvement culture. In addition, it should be considered how training and research techniques can be moderated for different cultural circumstances, existing equipment, and the level of in-country support available.

Germplasm conservation is likely to be important for future food security in coconut-growing regions, so further simplification of EC, perhaps in the form of a kit or with the use of pre-prepared medium, may increase the adoption of EC techniques for germplasm movement and genebank duplication.

Key research areas that require further development include coconut pathogen invasion dynamics and diagnostics. A clear understanding of the pattern of microbial pathogen infection of coconut tissues and their early detection is important, when EC is being used for germplasm exchange or conservation, as it is still uncertain whether embryos can transmit some of the microorganisms that afflict coconut. The example of phytoplasmas, where the pathogen has been detected in coconut embryos, but seed transmission has not been demonstrated (Cordova et al. 2003), and control methods currently rely on visible symptoms of the disease (Gurr et al. 2016), illustrates the research need.

It is important to note that the surface sterilization of embryos derived from a diseased palm is unlikely to fully eliminate microbial pathogens present in the embryo. Although using EC for international exchange would likely meet current phytosanitary requirements for international germplasm exchange, improvements in diagnostics are likely to be required to address increasing biosecurity and quarantine regulations for germplasm exchange and trade in many countries. Ideally, prior to collection and exchange of coconut germplasm, embryos should be tested for known pathogens.

Coconut cultivation is currently facing numerous problems. The issues include market unpredictability, aging palms, pests, disease, industry perception, and productivity decline in some areas due to reductions in the planting area. The coconut palm is susceptible to pests and disease in many parts of the world (Gurr et al. 2016; Harrison and Jones 2003; Lee 2013). Productivity is also impacted by the increasing age of many of the plantations, with most planted in the early to mid-twentieth century. Age leads to a decrease in fruit production resulting from a reduction in leaf area, as fronds that are produced are shorter and fewer. This is often seen in partnership with a decline in soil fertility and damage from extreme weather events (Samosir and Adkins 2014; Sisunandar et al. 2010a). The high value of virgin coconut oil and other new product options has provided new opportunities for farmers, and there is now a increasing need for large volumes of high-quality planting material, but it is unlikely that traditional breeding approaches alone will be able to deliver what the industry needs (Batugal et al. 2009; Thanh-Tuyen, and De Guzman 1983). Cloning holds the potential to affordably multiply elite coconut types or types resilient to pests or disease (Nguyen et al. 2015).

Biotechnology is expected to provide effective support for the selection and breeding of new coconut genotypes. However, poor regeneration of in vitro cultures remains a major hurdle, as well as the slow growth of these cultures. Clonal propagation via SE is showing great potential. Such a process, once fully developed, could be applied to the rapid production of elite coconut seedlings with superior fruit characteristics, resistance to devastating diseases and pests, or tolerance to unfavorable environments.

To overcome the poor regeneration rates, future research needs to take into consideration methods currently used for other economically important palms like oil, date, and peach palm. Like these species, a rapid clonal propagation procedure for coconut will probably require a productive embryogenic cell suspension culture system, using selected callus lines. Likewise, plantlet development and soil acclimatization of regenerants could be further enhanced by using a semi-submerging, temporary immersion system (TIS) in conjunction with photoautotrophic conditions (Nguyen et al. 2015; Samosir and Adkins 2014). Further exploration may indeed find a function for coconut water (liquid endosperm) in enhancing SE, as the seed has a valuable supply of growth factors and nutrients and has other culture applications (Sekar et al. 2013). Additionally, the identification of the genes that regulate SE is likely to be of great value.

A confirmed sequence of the whole coconut genome will be of immense value but will require intensive efforts. Even though coconut molecular genetics has lagged behind other agricultural species, constant improvements in the accessibility of high-throughput sequencing techniques are likely to accelerate the generation of genetic data for coconut (Meerow et al. 2012). Molecular selection and breeding will greatly benefit from this valuable genetic information. Although somaclonal variation is common in oil palm, it has not yet been identified in coconut. In a recent review by Sáenz-Carbonell et al. (2016), no physiological differences, e.g., germination rate, number of pollen grains, number of female flowers, etc., between somatic embryogenic and sexually produced plants have been observed. A molecular analysis based on amplified fragment length polymorphism (AFLP) markers also indicates that no significant change in genetic makeup of somatic embryo-derived coconut plants is generated (S. Fletcher, unpublished data). In addition, Sisunandar et al. (2010a) reported no genetic or epigenetic change in plantlets that had been regenerated following cryopreservation. Newer molecular tools, for example, DNA sequencing, should be used to examine for significant genetic changes that may occur during the embryogenic pathway.

There has been a shift toward coconut products with greater commercial value as they have the potential to improve the livelihood of the coconut farmers and reinvigorate the industry (Nguyen et al. 2016). Elite coconut varieties that have uniquely appealing endosperm attributes (tasty jelly-like endosperm and/or an aromatic juice) are of great interest. They can be sold for a price many times that of conventional coconuts. Embryo culture is currently a well-used tool to propagate these varieties, and with development, cloning is likely to significantly improve the process of propagation.

Emerging pests and diseases seriously endanger coconut farming and germplasm conservation worldwide. The Pacific regional coconut germplasm collection is at risk from Bogia coconut disease. The genebank does not appear to be infected, but it is at a very high risk of infection and being lost. The International Coconut Genetic Resources Network has recommended that the Madang Pacific genebank collection should be relocated within PNG and to consider safe duplication abroad in either Fiji or Samoa or both. The moving of the accessions that can be saved, the collection of new disease-free material, and their safe conservation are likely to require coconut EC and cryopreservation methods.

Despite documented successes using the EC technique (Table 10.1 lists EC publications), and the publication of standard protocols (Cueto et al. 2012; Frison et al. 1993; Pence et al. 2002), EC is not yet used for conservation purposes. Technology transfer of EC from the research laboratory to the industry, including smallholders and others in the value chain, may assist in improving the industry. In addition, further research activities to increase the applicability of coconut EC to a wider range of varieties and laboratory circumstances will also benefit the industry. Similarly, cryopreservation of embryos and plantlet regeneration have been successfully achieved with no detectable morphological, cytological, or molecular changes (Sisunandar et al. 2010a), but the technique is not yet being used for coconut conservation activities.

The adequate long-term conservation of coconut genetic diversity necessitates the use of not just EC and cryopreservation but more importantly successful international collaboration. Long-term coconut conservation will be best achieved by countries and industry bodies working together to share and provide the necessary resources, including technical competency. This has already begun with the publication of the "Feasibility Study For A Safety Back-up Cryopreservation Facility" by Acker et al. (2017) and "A Global Strategy for the Conservation and Use of Coconut Genetic Resources" edited by Bourdeix and Prades. The Global Strategy proposes that a multi-crop research team should be gathered to engage in the widespread collection and cryopreservation of coconut embryos and pollen from farmer's varieties and from isolated islands most at risk from climate change (Haeng-hoon and Engelman 2018). Consistent funding is vital to the success of these initiatives and the long-term conservation of coconut. In addition, technology transfer to genebank staff via training and rejuvenation of tissue culture facilities in coconut-growing countries is essential for successful international germplasm exchange and duplication.

10.5 Conclusion

This chapter discussed varying aspects of coconut embryo culture (embryo rescue), covering the morphology and physiology of the zygotic coconut embryo, culture medium types and growth conditions. Applications, including the use of the method for the germplasm collection, conservation (cryopreservation) and the dissemination of elite germplasm, and initiation of clonal propagation (via somatic embryogenesis) were also discussed. The technique is best known for its use for coconut types that generally don't germinate in vivo. The method is currently being used for the commercial production of elite coconut varieties such as Makapuno and Kopyor.

It has been demonstrated that embryo culture is the cornerstone coconut tissue culture technique due to its many applications (propagation, conservation, and transformation). Although the technique has been widely documented, embryo culture is being underused for germplasm movement and conservation. In addition, it has been found that most protocols have room for improvement. Key enhancements which are needed include ensuring consistently low contamination rates and high survival rates for a wide range of genotypes, in a variety of laboratories. Losses are still particularly high during acclimatization steps and when plants are transferred from tissue culture facilities into field conditions. As the expertise of the tissue culture practitioner influences the success of EC, the key to further coconut embryo culture development and widespread adoption is likely to involve the adequate training and perpetual support of genebank staff in the use of the technique.

Acknowledgments The authors greatly appreciate the support from Kokonas Indastri Koporesen (*KIK*) and the Australian Centre for International Agricultural Research (ACIAR).

References

Acker JP, Adkins S, Alves A et al (2017) Feasibility study for a safety back-up cryopreservation facility. Independent expert report, Rome, Italy, July 2017

Adkins S (2007) Coconut tissue culture for clonal propagation and safe germplasm exchange in Indonesia, Vietnam, Papua New Guinea and the Philippines. Adoption of ACIAR project outputs: studies of projects completed in 2006–07

Adkins SW, Samosir YMS (2002) Embryo culture activities at the University of Queensland. In: Engelmann F, Batugal P, Oliver L (eds) Coconut embryo in vitro culture: Part II. Merida, Mexico, pp 163–168

Adkins SW, Samosir YMS, Ernawati A et al (1998) Control of ethylene and use of polyamines can optimise the conditions for somatic embryogenesis in coconut (*Cocos nucifera* L.) and papaya (*Carica papaya* L.). In: Drew RA (ed) Proceedings of the international symposium of biotechnology in tropical and subtropical species. Brisbane, Australia, pp 459–466

Andrade-Torres A, Oropeza C, Sáenz L et al (2011) Transient genetic transformation of embryogenic callus of *Cocos nucifera*. Biologia 66:790–800

Andronescu DI (1919) Germination and further development of the embryo of *Zea mays* separated from the endosperm. Am J Bot 6:443–452

Antonova ID (2009) Somatic embryogenesis for micropropagation of coconut (*Cocos nucifera* L.). PhD thesis, The University of Queensland, Australia

Ashburner GR, Thompson WK, Burch JM (1993) Effect of alpha-naphthaleneacetic acid and sucrose levels on the development of cultured embryos of coconut. Plant Cell Tissue Org 35:157–163

Ashburner GR, Faure M, Tomlinson DR et al (1995) A guide to the zygotic embryo culture of coconut palms (*Cocos nucifera* L.). ACIAR technical report series, vol 36. Australian Centre for International Agricultural Research, Canberra, Australia

Ashburner GR, Thompson WK, Halloran GM (1997) RAPD analysis of South Pacific coconut palm populations. Crop Sci 37:992–997

Assy-Bah B, Durand-Gasselin T, Engelmann F et al (1989) The in vitro culture of coconut (*Cocos nucifera* L.) zygotic embryos. Revised and simplified method of obtaining coconut plantlets for transfer to the field. Oleagineux 44:515–523

Bajaj YPS (1984) Induction of growth in frozen embryos of coconut and ovules of citrus. Curr Sci 53(22):1215–1216

Basu A, Sethi U, Guhamukherjee S (1988) Induction of cell division in leaf cells of coconut palm by alteration of pH and its correlation with glyoxalase-I activity. J Exp Bot 39:1735–1742

Batugal P, Bourdeix R, Baudouin L (2009) Coconut breeding. In: Jain SM, Priyadarshan PM (eds) Breeding plantation tree crops: tropical species. Springer, New York, pp 327–375

Bhallasarin N, Bagga S, Sopory SK et al (1986) Induction and differentiation of callus from embryos of *Cocos nucifera* L. by IAA-conjugates. Plant Cell Rep 5:322–324

Bioversity (2012) Validation of a coconut embryo-culture protocol for the international exchange of germplasm final technical and financial report. Global Crop Diversity Trust, Maccarese (Rome)

Bourdeix R, Baudouin L, Billotte N et al (2001) Coconut. In: Harrier A, Jacquot M, Hamon S, Nicolas D (eds) Tropical plant breeding. Science Publishers, Enfield, pp 106–127

Branton RL, Blake J (1983) Development of organized structures in callus derived from explants of *Cocos nucifera* L. Ann Bot 52:673–678

Brown HT (1906) On the culture of excised embryos of barley on nutrient solutions containing nitrogen in different forms. Trans Guiness Res Lab 1:288–299

Buckner GD, Kastle JH (1917) The growth of isolated plant embryos. J Biol Chem 29:209–213

Chan JL, Saenz L, Talavera C et al (1998) Regeneration of coconut (*Cocos nucifera* L.) from plumule explants through somatic embryogenesis. Plant Cell Rep 17:515–521

Cordova I, Jones P, Harrison NA et al (2003) In situ PCR detection of phytoplasma DNA in embryos from coconut palms with lethal yellowing disease. Mol Plant Pathol 4:99–108

Cueto CA, Johnson VB, Engelmann F et al (2012) Technical guidelines for the safe movement and duplication of coconut (*Cocos nucifera* L.) germplasm using embryo culture transfer protocols. Bioversity International, Montpellier, France

Cutter VM Jr, Wilson KS (1954) Effect of coconut endosperm and other growth stimulants upon the development in vitro of embryos of *Cocos nucifera*. Bot Gaz 115:234–240

De Guzman EV, Del Rosario DA (1964) The growth and development of *Cocos nucifera* L. makapuno embryo in vitro. Philipp Agric 48:82–94

Dieterich K (1924) Uber Kutur von Embryonen ausserhalb des Samens. Flora 117:379–417

Dubard M, Urbain JA (1913) Del influence de l'albumen sur development de l'embryon. CR Acad Sci Paris 156:1086–1089

Duran Y, Rohde W, Kullaya A et al (1997) Molecular analysis of East African Tall coconut genotypes by DNA marker technology. J Genet Breed 51:279–288

Eeuwens CJ (1976) Mineral requirements for growth and callus initiation of tissue explants excised from mature coconut palms (*Cocos nucifera*) and cultured in vitro. Physiol Plant 36:23–28

Eeuwens CJ, Blake J (1977) Culture of coconut and date palm tissue with a view to vegetative propagation. Acta Hort 78:277–286

Engelmann F (2002) Chapter 10: Coconut. In: Valerie CP, Jorge AS, Victor M, Villalobos A, Engelmann F (eds) In vitro collecting techniques for germplasm conservation, vol IPGRI technical bulletin no. 7. International Plant Genetic Resources Institute, Rome, pp 68–71

Engelmann F, Malaurie B, N'Nan O (2011) In vitro culture of coconut (*Cocos nucifera* L.) zygotic embryos. In: Thorpe TA, Yeung EC (eds) Plant embryo culture: methods and protocols. Springer, Calgary, pp 3–15

Fernando SC, Gamage CKA (2000) Abscisic acid induced somatic embryogenesis in immature embryo explants of coconut (*Cocos nucifera* L.). Plant Sci 151:193–198

Fernando SC, Verdeil JL, Hocher V et al (2003) Histological analysis of plant regeneration from plumule explants of *Cocos nucifera*. Plant Cell Tissue Organ 72:281–283

Fernando SC, Vidhanaarachchi VRM, Weerakoon LK et al (2010) What makes clonal propagation of coconut difficult? Asia Pac J Mol Biol Biotechnol 18:163–165

Foale M (2003) The coconut odyssey: the bounteous possibilities of the tree of life. ACIAR monograph no. 101, Canberra

Frison EA, Putter CAJ, Diekmann M (1993) FAO/IBPGR technical guidelines for the safe movement of coconut germplasm. Food and Agricultural Organisation of the United Nations/International Board for Plant Genetic Resources, Rome

Fuentes G, Talavera C, Desjardins Y et al (2005a) High irradiance can minimize the negative effect of exogenous sucrose on photosynthetic capacity of in vitro grown coconut plantlets. Biol Plant 49:7–15

Fuentes G, Talavera C, Oropeza C et al (2005b) Exogenous sucrose can decrease in vitro photosynthesis but improve field survival and growth of coconut (*Cocos nucifera* L.) in vitro plantlets. In Vitro Cell Dev Plant 41:69–76

Gamborg OL, Miller RA, Ojima K (1968) Nutrient requirements of suspension cultures of soybean root cells. Exp Cell Res 50:151–158

Gupta PK, Kendurkar SV, Kulkarni VM et al (1984) Somatic embryogenesis and plants from zygotic embryos of coconut (*Cocos nucifera* L.) in vitro. Plant Cell Rep 3:222–225

Gurr GM, Johnson AC, Ash GJ et al (2016) Coconut lethal yellowing diseases: a phytoplasma threat to palms of global economic and social significance. Front Plant Sci 7:1521

Haeng-hoon K, Engelman F (2018) 3.3.4 Cryogenebanking – Chapter 3. Where we need to be to secure diversity and promote use. In: Bourdeix R, Prades A (eds) A global strategy for the conservation and use of coconut genetic resources 2018–2028. Bioversity International, Montpellier, France, pp 139–142

Hannig E (1904) Zur Physiologie Pflanzicher Embryonen.I. Uber die Kultur von Crucifere-Embryonen ausserhalb des Embryosacks. Bot Ztg 62:45–80

Harrison NA, Jones P (2003) Diseases of coconut. Diseases of tropical fruit crops. CABI Publishing, Cambridge, MA. https://doi.org/10.1079/9780851993904.0197

Hasan Awad A, Wanfei L, Qiang L et al (2016) Complete sequence and analysis of coconut palm (Cocos nucifera) mitochondrial genome. PLoS One 11(10):e0163990

Haslam TM, Yeung EC (2011a) Zygotic embryo culture: an overview. In: Thorpe TA, Yeung EC (eds) Plant embryo culture methods and protocols. Humana Press, Totowa, pp 3–15

Haslam TM, Yeung EC (2011b) Zygotic embryo culture: an overview. In: Thorpe TA, Yeung EC (eds) Plant embryo culture: methods and protocols. Springer, Calgary, pp 3–15

Hocher V, Verdeil JL, Rival A et al (1999) Application of in vitro techniques to the conservation and propagation of coconut palms. In: Oropeza C, Verdeil JL, Ashburner GR, Cardeña R, Santamaría JM (eds) Current advances in coconut biotechnology. Springer, Dordrecht, pp 267–288

Hornung R, Domas R, Lynch PT (2001) Cryopreservation of plumular explants of coconut (Cocos nucifera L.) to support programmes for mass clonal propagation through somatic embryogenesis. CryoLett 22:211–220

Hu C-Y, Zanettini HB (1995) Embryo culture and embryo rescue for wide cross hybrids. In: Gamborg OL, Phillips GC (eds) Plant cell, tissue and organ culture fundamental methods. Springer, Berlin/Heidelberg

Karp A (1999) The use of polymorphic microsatellites for assessing genetic diversity in coconut. In: Oropeza C, Ashburner GR, Verdeil JL, Cardena R, Santamaría JM (eds) Current advances in coconut biotechnology. Kluwer Academic Publishers, Dordrecht, pp 121–129

Karunaratne S, Periyapperuma K (1989) Culture of immature embryos of coconut, *Cocos nucifera* L.: callus proliferation and somatic embryogenesis. Plant Sci 62:247–253

Knudson L (1922) Nonsymbiotic germination of orchid seeds. Bot Gaz 73:1–25

Koffi Y, N'Nan-Alla O, Konan Konan JL et al (2013) Morphological and agronomical characteristics of coconut (*Cocos nucifera* L.) palms produced from in vitro cultured zygotic embryos. In Vitro Cell Dev Biol Plant 49(5):1–6

Laibach F (1929) Ectogenesis in plants: methods and genetic possibilities of propagating embryos otherwise dying in the seed. J Hered 20(5):201–208

Lebrun P, Baudouin L, Bourdeix R et al (2001) Construction of a linkage map of the Rennell Island Tall coconut type (*Cocos nucifera* L.) and QTL analysis for yield characters. Genome 44(6):962–970

Lee RF (2013) Cadang-cadang disease of palm and other diseases. Phytopathology 103:177–177

Lopez-Villalobos A (2002) Roles of lipids in coconut (*Cocos nucifera* L.) embryogenesis. University of London, London

López-Villalobos A, Dodds PF, Hornung R (2001) Changes in fatty acid composition during development of tissues of coconut (*Cocos nucifera* L.) embryos in the intact nut and in vitro. J Exp Bot 52:933–942

López-Villalobos A, Hornung R, Dodds PF (2004) Hydrophobic metabolites of 2,4-dichlorophenoxyacetic acid (2,4-D) in cultured coconut tissue. Phytochemistry 65:2763–2774

López-Villalobos A, Dodds PF, Hornung R (2011) Lauric acid improves the growth of zygotic coconut (*Cocos nucifera* L.) embryos in vitro. Plant Cell Tissue Organ 106:317–327

Magdalita P, Damasco O, Samosir Y et al (2015) An enhanced embryo culture methodology for coconut (*Cocos nucifera* L.). Int J Innov Res Sci 4(10):485–493

Meerow AW, Krueger RR, Singh R et al (2012) Coconut, date, and oil palm genomics. In: Genomics of tree crops. Springer, New York, pp 299–351

Morel G, Wetmore RH (1951) Fern callus tissue culture. Am J Bot 38:141–143

Murashige T, Skoog F (1962) A revised medium for rapid growth and bio assays with tobacco tissue cultures. Physiol Plant 15:473–497

N'Nan O, Hocher V, Verdeil JL et al (2008) Cryopreservation by encapsulation-dehydration of plumules of coconut (*Cocos nucifera* L.). Cryo Lett 29:339–350

N'Nan O, Borges M, Konan JLK et al (2012) A simple protocol for cryopreservation of zygotic embryos of ten accessions of coconut (*Cocos nucifera* L.). In Vitro Cell Dev Plant 48:160–166

Nguyen QT, Bandupriya HDD, López-Villalobos A et al (2015) Tissue culture and associated biotechnological interventions for the improvement of coconut (Cocos nucifera L.): a review. Planta 242(5):1059–1076

Nguyen QT, Bandupriya HDD, Foale M et al (2016) Biology, propagation and utilization of elite coconut varieties (makapuno and aromatics). Plant Physiol Biochem 109:579–589

Novarianto H, Warokka J (2006) Past, present and future coconut research in Indonesia. Paper presented at the coconut revival – new possibilities for the 'tree of life'. In: Proceedings of the international Coconut Forum held in Cairns, Australia, 22–24 November 2005. ACIAR proceedings no 125

Nwite PA, Ikhajiagbe B, Owoicho I (2017) Germination response of coconut (*Cocos nucifera* L.) zygotic embryo. J Appl Sci Environ Manag 21(6):1019–1021

Pan MJ, van Staden J (1998) The use of charcoal in in vitro culture - a review. Plant Growth Regul 26:155–163

Pannetier C, Buffard-Morel J (1982) Production of somatic embryos from leaf tissues of coconut, *Cocos nucifera* L. In: Proceedings of the 5th international Plant Tissue Culture Congress, Tokyo, Japan

Pech y Aké AE, Souza R, Maust B et al (2004) Enhanced aerobic respiration improves in vitro coconut embryo germination and culture. In Vitro Cell Dev Plant 40:90–94

Pech y Aké AE, Maust B, Orozco-Segovia A et al (2007) The effect of gibberellic acid on the in vitro germination of coconut zygotic embryos and their conversion into plantlets. In Vitro Cell Dev Plant 43:247–253

Pence VC, Sandoval JA, Villalobos VM et al (2002) In vitro collecting techniques for germplasm conservation, IPGRI technical bulletin. International Plant Genetic Resources Institute, Rome

Perera L, Russell JR, Provan J (1998) Evaluating genetic relationships between indigenous coconut (Cocos nucifera L.) accessions from Sri Lanka by means of AFLP profiling. Theor Appl Genet 96(3):545–550

Perera L, Russell JR, Provan J (1999) Identification and characterisation of microsatellite loci in coconut (Cocos nucifera L.) and the analysis of coconut populations in Sri Lanka. Mol Ecol 8:344–346

Pérez-Núñez MT, Chan JL, Sáenz L et al (2006) Improved somatic embryogenesis from *Cocos nucifera* (L.) plumule explants. In Vitro Cell Dev Plant 42:37–43

Raghavan V, Torrey JG (1964) Effects of certain growth substances on the growth and morphogenesis of immature embryos of Capsella in culture. Plant Physiol 39:691–699

Rillo EP (1998) PCA's embryo culture technique in the mass production of Makapuno coconuts. In: Batugal PA, Engelmann F (eds) Coconut embryo in vitro culture: Part I. Proceedings of the first workshop on embryo culture, Banao, Guinobatan, Albay, Philippines, 27–31 October 1997. International Plant Genetic Resources Institute (IPGRI), Rome, pp 69–78

Rillo EP, Paloma MBF (1991) Storage and transport of zygotic embryos of Cocos nucifera L. for in vitro culture. Plant Genet Resour Newslett 86:1–4

Rillo EP, Cueto CA, Medes WR et al (2002) Development of an improved embryo culture protocol for coconut in the Philippines. In: Engelmann F, Batugal P, Oliver J (eds) Coconut embryo in vitro culture: Part II. Proceedings of second international on embryo culture workshop, Mérida, Yucatán, Mexico, 14–17 March 2000. International Plant Genetic Resources Institute (IPGRI), Rome, pp 41–65

Rodriguez MJB, Estioko LP, Namia MIT et al (1997) Analysis of genetic diversity by RAPD. Philipp J Crop Sci 22:133–135

Rohde W, Becker D, Kullaya A et al (1999) Analysis of coconut germplasm biodiversity by DNA marker technologies and construction of a first genetic linkage map. In: Oropeza C, Ashburner R, Verdeil J-L, Zizumbo D (eds) Current advances in coconut biotechnology. Kluwer Academic Publishers, Dordrecht, pp 99–120

Sáenz L, Herrera-Herrera G, Uicab-Ballote F et al (2009) Influence of form of activated charcoal on embryogenic callus formation in coconut (*Cocos nucifera*). Plant Cell Tissue Organ 100:301–308

Sáenz-Carbonell L, Montero-Cortés M, Pérez-Nuñez T et al (2016) Somatic Embryogenesis in *Cocos nucifera* L. In: Loyola-Vargas VM, Ochoa-Alejo N (eds) Somatic embryogenesis: fundamental aspects and applications. Springer, Cham, pp 297–318

Samosir YMS (1999) Optimisation of somatic embryogenesis in coconut (*Cocos nucifera* L.). PhD thesis, The University of Queensland, Australia

Samosir YMS, Adkins SW (2004) Coconut embryo culture: CO_2-enrichment and root aeration for improved seedling establishment. In: Peiris TSG, Ranasinghe CS (eds) Proceedings of the international conference of the Coconut Research Institute of Sri Lanka: Part II. The Coconut Research Institute Institute of Sri Lanka, Lunuwila, Sri Lanka, pp 77–91

Samosir YMS, Adkins SW (2014) Improving acclimatization through the photoautotrophic culture of coconut (*Cocos nucifera*) seedlings: an in vitro system for the efficient exchange of germplasm. In Vitro Cell Dev Plant 50:493–501

Samosir YMS, Godwin ID, Adkins SW (1998) An improved protocol for somatic embryogenesis in coconut (*Cocos nucifera* L.). In: Drew RA (ed) Proceedings of the international symposium of biotechnology in tropical and subtropical species, vol 461, Brisbane, Australia, pp 467–475

Samosir YMS, Godwin ID, Adkins SW (1999a) The use of osmotically active agents and abscisic acid can optimise the maturation of coconut somatic embryos. In: Oropeza C (ed) Current advances in coconut biotechnology. CAB International, Wallingford, pp 341–354

Samosir YMS, Godwin ID, Adkins S (1999b) A new technique for coconut (*Cocos nucifera*) germplasm collection from remote sites: culturability of embryos following low-temperature incubation. Aust J Bot 47:69–75

Samosir Y, Mashud N, Novarianto H et al (2008) Embryo culture manual: a new embryo culture protocol for coconut germplasm conservation and elite-type seedling production. Australian Centre for International Agricultural Research, Canberra

Schopfer WH (1943) Plants and vitamins. Chronica Botanica Co, Waltham

Sekar N, Veetil SK, Neerathilingam M (2013) Tender coconut water an economical growth medium for the production of recombinant proteins in *Escherichia coli*. BMC Biotechnol 13:1472–6750

Sharma DR, Kaur R, Kumar K (1996) Embryo rescue in plants – a review. Euphytica 89:325–337

Sisunandar A, Samosir YMS, Adkins SW (2005) Towards the cryopreservation of coconut (*Cocos nucifera* L.). Paper presented at the Contributing to a Sustainable Future, Perth, Australia

Sisunandar A, Rival A, Turquay P et al (2010a) Cryopreservation of coconut (*Cocos nucifera* L.) zygotic embryos does not induce morphological, cytological or molecular changes in recovered seedlings. Planta 232:435–447

Sisunandar A, Sopade PA, Samosir YM et al (2010b) Dehydration improves cryopreservation of coconut (*Cocos nucifera* L.). Cryobiology 61:289–296

Sisunandar A, Novarianto H, Mashud N et al (2014) Embryo maturity plays an important role for the successful cryopreservation of coconut (*Cocos nucifera*). In Vitro Cell Dev Plant 50:688–695

Sisunandar A, Alkhikmah A, Husin A et al (2015) Embryo incision as a new technique for double seedling production of Indonesian elite coconut type "Kopyor". J Math Fundam Sci 47(3):252–260

Sisunandar A, Alkhikmah T, Husin A et al (2018) Ex vitro rooting using a mini growth chamber increases root induction and accelerates acclimatization of Kopyor coconut (Cocos nucifera L.) embryo culture-derived seedlings. In Vitro Cell Dev Biol Plant 54(5):508–517

Steinmacher DA, Guerra MP, Saare-Surminski K et al (2011) A temporary immersion system improves in vitro regeneration of peach palm through secondary somatic embryogenesis. Ann Bot 108(8):1463–1475

Talavera C, Contreras F, Espadas F et al (2005) Cultivating in vitro coconut palms (*Cocos nucifera*) under glasshouse conditions with natural light, improves in vitro photosynthesis nursery survival and growth. Plant Cell Tissue Organ 83:287–292

Teixeira JB, Sondahl MR, Nakamura T et al (1995) Establishment of oil palm cell suspensions and plant regeneration. Plant Cell Tissue Organ 40(2):105–111

Teulat B, Aldam C, Trehin R (2000) An analysis of genetic diversity in coconut (Cocos nucifera) populations from across the geographic range using sequence-tagged microsatellites (SSRs) and AFLPs. Theor Appl Genet 100(5):764–771

Thanh-Tuyen NT, De Guzman EV (1983) Formation of pollen embryos in cultured anthers of coconut (*Cocos nucifera* L.). Plant Sci Lett 29:81–88

Tisserat B, Vandercook CE (1985) Development of an automated plant culture system. Plant Cell Tissue Organ 5:107–117

Triques K, Rival A, Beule T et al (1997) Photosynthetic ability of in vitro grown coconut (*Cocos nucifera* L.) plantlets derived from zygotic embryos. Plant Sci 127:39–51

Triques K, Rival A, Beule T et al (1998) Changes in photosynthetic parameters during in vitro growth and subsequent acclimatization of coconut (*Cocos nucifera* L.) zygotic embryos. In: Drew RA (ed) Proceedings of the international symposium of biotechnology in tropical and subtropical species, vol 461. Acta Hort. ISHS, Leuven, pp 275–284

Vu TML (2002) Coconut embryo culture in Vietnam. Coconut embryo in vitro culture: Part II. In: Proceedings of second international on embryo culture workshop, Merida, Yucatan, Mexico, 14–17 March 2000. International Plant Genetic Resources Institute (IPGRI), Rome

Wadt LHO, Sakiyama NS, Pereira MG et al (1999) RAPD markers in the genetic diversity study of the coconut palm. In: Oropeza C, Verdeil JL, Ashburner GR, Cardena R, Santamaria JM (eds) Current advances in coconut biotechnology. Kluwer Academic Publishers, Dordrecht, pp 89–97

White PR (1943) A handbook of plant tissue culture. A handbook of plant tissue culture. The Jaques Cattell Press, Lancaster

Xiao Y, Xu P, Fan H et al (2017) The genome draft of coconut (*Cocos nucifera* L.). GigaScience 6(11):gix095

Ya-Yi H, Antonius JMM, Marjori M (2013) Complete sequence and comparative analysis of the chloroplast genome of coconut palm (*Cocos nucifera* L.). PLoS One 8(8):e74736

Chapter 11
Coconut Micropropagation for Worldwide Replanting Needs

Luis Sáenz-Carbonell, Quang Nguyen, Arturo López-Villalobos, and Carlos Oropeza-Salín

11.1 Introduction: The Growing Need for Micropropagation of Coconut

Coconut is an important oil-bearing crop contributing to the income of more than 20 million farmers and their dependents (Pham 2016; Rethinam 2006). However, leading coconut producers have been struggling to keep pace with the continuously growing demand in recent years. This is because coconut farmers commonly encounter many challenges since productivity is influenced by the senility of many of the palms, the reduction of soil fertility on which they are growing, and natural calamities (Samosir and Adkins 2014; Sisunandar et al. 2010). In addition, a reduction in the cultivated area has resulted from outbreaks of lethal pests and diseases (Gurr et al. 2016; Harrison and Jones 2003; Oropeza et al. 2011).

Even though there have been several breeding programs aiming to increase oil yield and disease resistance in many coconut-producing countries, these conventional methods involving multiple generations of inbreeding and hybridization are unlikely to be a solution for the present needs (Batugal et al. 2009; Samosir and Adkins 2004). In this context, micropropagation via SE holds a greater hope of producing many palms, of a desired genotype in a relatively short period of time.

L. Sáenz-Carbonell (✉) · C. Oropeza-Salín
Biotechnology Unit, Centro de Investigación Científica de Yucatán, Mérida, Yucatán, Mexico
e-mail: vyca@cicy.mx

Q. Nguyen
School of Agriculture and Food Sciences, The University of Queensland, Gatton, QLD, Australia

School of Biotechnology, International University, Vietnam National University, Ho Chi Minh City, Vietnam

A. López-Villalobos
Department of Biology, Purdue University, West Lafayette, IN, USA

S. Adkins et al. (eds.), *Coconut Biotechnology: Towards the Sustainability of the 'Tree of Life'*, https://doi.org/10.1007/978-3-030-44988-9_11

Among a range of tissue culture techniques, SE is commonly considered to be the only way to meet the increasing demand as it can yield multiple clonal palms from one single explant (Nguyen et al. 2015). In fact, this technique has developed significantly in the past two decades, and there have also been some attempts to scale up the existing system (Sáenz-Carbonell et al. 2016). Plumular tissue can routinely be used to attain plantlets through SE; however, the recent focus has been on somatic tissues such as immature inflorescences to enhance the production of true-to-type clones (Antonova 2009; Sandoval-Cancino et al. 2016). Furthermore, SE holds promise for the mass propagation of outstanding genotypes (Nguyen et al. 2016). A better understanding of SE in coconut at the molecular level will contribute substantially to the efficiency of the technique (e.g. conversion of embryogenic structures into mature somatic embryos) (Sáenz-Carbonell et al. 2016). In this chapter, we aim to deliver an inclusive summary of the advances in coconut micropropagation and discuss the possibilities of optimizing and scaling up the existing scheme. Further recommendations on how collaborative research programs can help convey the technology are also discussed.

11.2 Attempts to Develop Micropropagation Protocols Using Different Types of Explants

Although SE has been achieved in many plant species, it has been extremely challenging in others, and the coconut is one such species. Somatic embryogenesis in coconut was first attempted over 40 years ago at Wye College in England (Eeuwens and Blake 1977) and subsequently by the Office de la Recherche Scientifique et Technique Outre-Mer (ORSTOM) in France (Pannetier and Buffard-Morel 1982). In early years, different types of somatic tissues such as young leaves and stem slices from seedlings and from rachillae of inflorescences were chosen as explants to form callus (Branton and Blake 1983; Gupta et al. 1984).

More recently, researchers have concentrated their work on either zygotic tissues (e.g., embryo slices/plumules) or somatic tissues (e.g., immature inflorescences) to obtain SE in coconut. Early studies suggested that immature embryos were more responsive in producing callus, when compared to mature embryos (Karunaratne and Periyapperuma 1989). The responsiveness of the mature embryos when longitudinally sliced was significantly enhanced (Adkins et al. 1998; Samosir 1999). Over time, it was discovered that the plumular tissue isolated from the mature embryo was the optimal explant type to form embryogenic callus (Chan et al. 1998; Pérez-Núñez et al. 2006). However, this type of explant may not be ideal to obtain true-to-type clones as some coconut varieties outcross. More recent attention has shifted back to the use of somatic tissue explants such as those of immature inflorescences (Antonova 2009; Sandoval-Cancino et al. 2016). Even though callus formation from these types of explants can be difficult to obtain, true-to-type clones are produced.

11.3 Development of Somatic Embryogenesis from Plumules

The protocols produced before the 1990s were non-reproducible with very low rates of SE callus formation and with little evidence of plantlet formation (Blake 1990). Therefore, the *Centro de Investigación Científica de Yucatán* (CICY) in conjunction with Wye College, ORSTOM, and the French Agricultural Research Centre for International Development (*CIRAD*) decided to try different parts of the coconut zygotic embryo as explants. One of these was plumule (shoot meristem plus leaf primordia) which proved to be a very responsive tissue for the formation of embryogenic callus with high efficiency (22%) and was capable of forming somatic embryos and plantlets (Chan et al. 1998). Acclimatization of the plantlets was successful, and field trials of tissue-cultured plants were established in southeast Mexico. This SE process was studied in detail and reported by Sáenz et al. (2006). Further improvement of this protocol was reported by Pérez-Núñez et al. (2006) through secondary somatic embryogenesis and consecutive proliferation of embryogenic callus.

11.4 Further Developments in Coconut Somatic Embryogenesis

Addition of abscisic acid (ABA) and osmotic agents such as polyethylene glycol (PEG), mannitol, and sorbitol increased the formation and maturation of somatic embryos produced by zygotic embryo slices (Samosir 1999). For plumule explants, addition of 22(S), 23(S)-homobrassinolide, a brassinosteroid compound, to pretreatment medium for 3 days was shown to increase the formation of SE callus from 60 to 90% and the formation of somatic embryos from 3.8 to 10.8 per callus (Azpeitia et al. 2003). Gibberellic acid (GA_3) was also tested on plumule explants and shown to improve SE callus formation and the percentage germination of embryos (Montero-Cortés et al. 2011; Perera et al. 2009).

The effect of polyamines, such as putrescine and spermidine, was tested by Rajesh et al. (2014) on plumule explants from Malayan Yellow Dwarf and Chowghat Green Dwarf. These polyamines increased the rate of SE callus formation from 37 to 50% for Malayan Yellow Dwarf explants and from 55 to 63% for Chowghat Green Dwarf. The authors also reported the formation of shoots and then plantlets with these treatments.

Different sources of activated charcoal (AC), an additive used in media formulation to avoid browning of explants, were evaluated by Sáenz et al. (2010). Of the eight different AC sources tested, the best results were obtained with AC from SIGMA (plant cell and tissue culture acid washed), DARCO, and the United States Pharmacopeia; they increased the embryogenic callus formation in 60% of the explants but with different optimal 2,4-D concentrations. The particle profile was also determined for the eight AC sources, showing that the smallest particles

(<38 μm) were the most abundant and this particle size fraction from the SIGMA charcoal added to the culture medium produced 70% of embryogenic callus formation compared with 40% using the whole SIGMA AC.

11.5 Understanding Coconut Somatic Embryogenesis

11.5.1 Morpho-histological Development

Plant morphogenesis is controlled by a series of conditions such as the plant growth regulator content of the culture media, temperature, and photoperiod that create a microenvironment which ensures the completion of successive morphogenetic events required to achieve SE (Fig. 11.1). The first event, callus formation, is attained with optimized concentrations of 2,4-dichlorophenoxy acetic acid (2,4-D; 0.10–0.65 mM) and AC (0.10–0.25%), promoting in explants cell differentiation

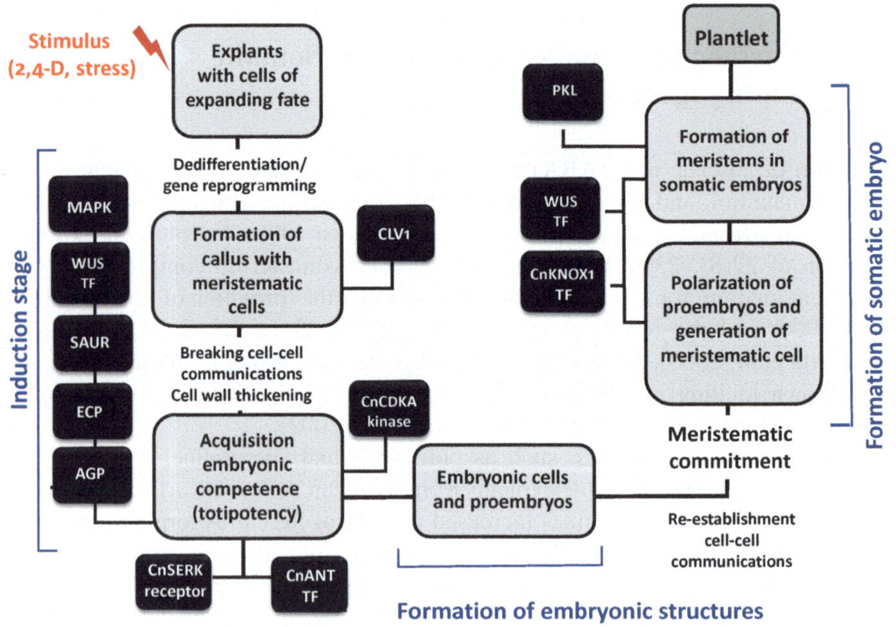

Fig. 11.1 Coconut somatic embryogenesis is regulated genetically and physiologically at various stages. Genes that have been isolated in coconut are shown in violet rectangles. *Aintegumenta* (*ANT*), *Apetala2/Ethylene-Responsive Factor* (*AP2*), *Arabinogalactan Protein* (*AGP*), *Class I Knotted-Like Homeobox* (*KNOX1*), *Cyclin-Dependent Kinases A* (*CDKA*), *Embryogenic Cell Protein* (*ECP*), *Late Embryogenesis-Abundant Protein* (*LEA*), *Mitogen-Activated Protein Kinase* (*MAPK*), *Pickle* (*PKL*), *Saur Family Protein* (*SAUR*), *Somatic Embryogenesis Receptor Kinase* (*SERK*), *Wuschel* (*WUS*)

and gene reprograming in dark conditions (Chan et al. 1998; Perera et al. 2007; Rajesh et al. 2016). Only specific cells within the explant respond to auxin to initiate SE callus formation. Cell response is mainly dependent on the cell's capacity to become totipotent (Rajesh et al. 2016). Apparently, in most coconut explants (including immature male inflorescence tissues, young leaves, and ovaries), callus forms from cells of the vascular tissue, most notably from procambial cells (Chan et al. 1998; Perera et al. 2007; Sandoval-Cancino et al. 2016). The initial callus formed contains meristematic cells that subsequently divide to form translucent structures that enclose a distinct subepidermal layer of meristematic cells (Perera et al. 2007; Sandoval-Cancino et al. 2016). At this stage, embryogenic competence is acquired by cells to produce later somatic embryos by two pathways, the uni- and multicellular pathways, which may or may not coexist in the callus mass. In the unicellular pathway, single cells located above the meristematic layer undertake ultrastructural changes to become embryogenic competent. These changes include plasmodesmata closure, Golgi apparatus-mediated vesicle deposition, and cell wall thickening that eventually leads to the breaking of cell-to-cell communications (Perera et al. 2007). In the multicellular pathway, the mechanism that cells follow to acquire embryogenic competence is not clear, but it starts with the fragmentation of the meristematic layer into discrete meristematic nodules that later are converted into embryonic structures (Sáenz et al. 2006). After cell proliferation, proembryos are formed following the reestablishment of intercellular signaling. Proembryos then undergo polarization, when storage reserves, like starch, are deposited at the proximal end where root and haustorial tissue would be generated. At the distal end, meristematic cells are multiplied to give rise to the shoot pole (Chan et al. 1998; Perera et al. 2007; Verdeil et al. 1994). Further cell differentiation is induced by omitting or decreasing the 2,4-D concentration, resulting in the formation of a tripolar somatic embryos with defined shoot and root apical meristems as well as a rudimentary haustorium (Perera et al. 2007). In most published protocols, the somatic embryo-bearing callus needs to be cultured in a medium containing ABA or the cytokinin benzyl aminopurine (BAP) to ensure the completion of the formation of the three morphogenetic poles (Chan et al. 1998; Sandoval-Cancino et al. 2016).

11.5.2 Biochemical and Physiological Control

The auxin 2,4-D is considered essential in promoting coconut SE, probably due to its reported activation of genes required for cell dedifferentiation and gene reprograming necessary to acquire embryogenic competence (Fig. 11.2). Many experiments in coconut in vitro culture have determined that the uptake of 2,4-D by cells is an active process, for which energy is mainly derived from sucrose supplied in the culture media. Simultaneously, a positive feedback effect of 2,4-D is exerted on sucrose utilization by increasing the uptake of the derived glucose by callus cells (Dussert et al. 1995; López-Villalobos et al. 2004; Magnaval et al. 1995). Because of this influx, 2,4-D and glucose promote acquisition of embryogenic competence

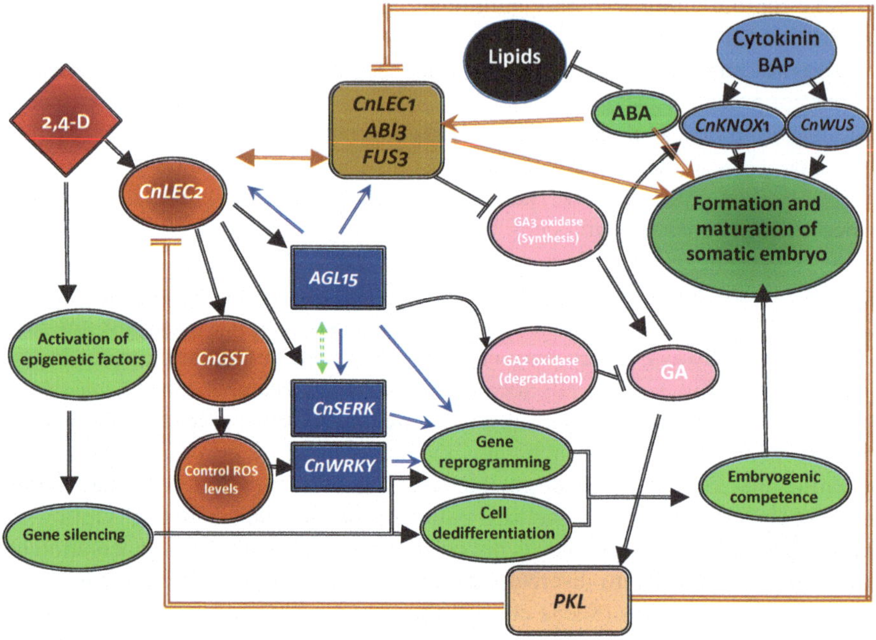

Fig. 11.2 Hormonal control of gene expression during coconut somatic embryogenesis. Genes isolated in coconut have the prefix Cn. Abscisic acid (ABA), *Abscisic Acid Insensitive 3* (*ABI3*), *Agamous-Like 15* (*AGL15*), 6-benzylaminopurine (BAP), 2,4-dichlorophenoxyacetic acid (2,4-D), gibberellins (*GAs*), *Leafy Cotyledon 1* (*LEC1*), and *Leafy Cotyledon 2* (*LEC2*)

and formation of proembryos by regulating gene expression (Fig. 11.2) and providing the energy and carbon demanded to synthesize complex carbohydrates for cell wall thickening, cell division, and polarization (Dussert et al. 1995; Magnaval et al. 1995; Rajesh et al. 2016). In the later stages of SE, BAP and ABA are major regulators of somatic embryos formation (Chan et al. 1998; Perera et al. 2007), although gibberellins also have a positive role (Montero-Cortés et al. 2010b). ABA may also repress lipid breakdown to induce maturation of somatic embryos, as reported in other plant species. However, the role of this plant growth regulator in regulating lipid activity in coconut cells is yet to be determined. Such research would represent a promising avenue to enhance coconut SE, as evidenced by reported benefits of lipid supplementation in coconut zygotic embryogenesis (López-Villalobos 2002).

11.5.3 Molecular Control

Key regulatory genes of coconut SE have been studied by the traditional gene-by-gene approach and, more recently, functional genomics, specifically by transcriptome analysis (Nguyen et al. 2015; Rajesh et al. 2016). Using these tools, the elucidation of genes involved in the regulation of each morphogenetic event of coconut SE is progressing rapidly (Fig. 11.1). The initial stage of callus formation appears to be controlled by *Clavata1* (*CLV1*), as indicated by its high level of expression in coconut plumule explants at the beginning of in vitro culture (Rajesh et al. 2016). The gene *CLV1* encodes a receptor kinase involved in a negative loop mechanism responsible for the repression of *Wuschel* (*WUS*), a transcription factor that affects the acquisition of embryonic competence in coconut (Bhavyashree et al. 2015). Therefore, *CLV1* may have an import role in controlling the timing of the acquisition of SE competence by downregulating *WUS*. In addition, the genes *Cyclin-Dependent Kinases A* (*CDKA*), *Somatic Embryogenesis Receptor Kinase* (*SERK*), *Mitogen-Activated Protein Kinase* (*MAPK*), *Apetala2/Ethylene-Responsive Factor* (*AP2*), *Saur Family Protein* (*SAUR*), *Embryogenic Cell Protein* (*ECP*), *Arabinogalactan Protein* (*AGP*), *Late Embryogenesis-Abundant Protein* (*LEA*), and *Aintegumenta* (*ANT*) are also highly expressed during embryogenic callus formation and therefore may assist *WUS* function (Bhavyashree et al. 2015; Montero-Cortés et al. 2010a; Pérez-Núñez et al. 2009; Rajesh et al. 2014, 2016). Some of these genes, for instance, *CDKA*, *AGP*, and *LEA*, regulate cell division and deposition of cell wall materials and storage proteins, which are all processes required to achieve embryonic cell competence (López-Villalobos et al. 2016; Perera et al. 2007; Pérez-Núñez et al. 2009; Montero-Cortés et al. 2010a). The final stages of coconut SE, formation, maturation, and germination of somatic embryos, are affected by the genes *Class 1 Knotted-Like Homeobox* (*KNOX1*), *Germin-Like Protein* (*GLP*), *Glutathione S-Transferase* (*GST*), *Pickle* (*PKL*), and *Wrky Transcription Factor* (*WRKY*), in addition to *WUS* that also continues to function in these later stages (Montero-Cortés et al. 2010b; Rajesh et al. 2016). Among these genes, *KNOX1* seems to be a major regulator as it controls the shoot apical meristem formation of the somatic embryo in response to the action of cytokinins and gibberellins (Montero-Cortés et al. 2010b). Besides, *KNOX1* and *WUS* may complement each other in shoot meristem ontogenesis.

11.5.4 Plantlet Formation

Once somatic embryogenic callus has been formed, germination of the embryos and the start of shoot growth can be promoted by transferring the callus to a medium with 6 μM 2,4-D, 300 μM BAP, and 2.8 μM GA_3 for 3 months. Callus with shoots can then be transferred for 2 further months to a medium with 6 μM 2,4-D and 300 μM BAP for further growth; then shoots are separated from the callus and transferred to fresh medium with the same concentrations of plant growth regulators for another 2 months. Then shoots are transferred for 2 months to a medium without plant growth regulators and subcultured a further two times (each for 2 months) to the same medium, to achieve the formation of fully developed plantlets (Oropeza et al. 2018; Sáenz et al. 2018). Full media formulation is described in Oropeza et al. (2018) and Sáenz et al. (2018). When the root system is well developed, plantlets are ready for successful acclimatization to field conditions.

11.6 Field Trials

Several field trials have been established with farmers. The first trial included plants produced by micropropagation (1 ha) and plants obtained from seed germination, both of Yucatan Green Dwarf (YGD), and was established in San Crisanto, Yucatan (Mexico), in a lethal yellowing disease (LYD)-affected area about 12 years ago. Growth and development in the fruit-bearing stage was observed to be normal without differences between the two; plants started bearing fruit before 3 years after planting, and micropropagated plants started about 6 months earlier than plants from the seed (CICY, unpublished). Monitoring of the trial was carried out only for 5 years after planting, but up to now no single plant has died due to LYD infection, although plants dying of LYD in the area are frequently found. In addition, analysis of the viability of the pollen and the histology of the female flowers of the micropropagated plants were carried out without any abnormality detected (Chuc-Armendariz et al. 2006). A second trial, complementary to the first one, was established in Ticul, Yucatan, to test the progeny from the YGD micropropagated plants of the first trial; seed was obtained by control pollination, germinated, and seedlings planted (2.5 ha), and again growth and development was normal, and plants are starting to bear fruit. A third trial was established also in Ticul (2 ha) with Mexican Pacific Tall by Mexican Pacific Tall (MPT × MPT) plants produced by micropropagation; unfortunately it was neglected and abandoned by the owner.

Two other trials have been established outside Yucatan. One was established 7 years ago in Cuyutlan, Colima (Mexico), on the Mexican Pacific Coast (about 1.5 ha) with micropropagated MPT × MPT plants and YGD (seed produced by controlled pollination from micropropagated plants from the first trial. Growth and development has been typical of these two types of coconuts. They have been grown under very limiting conditions (no fertilization and no watering) as local farmers do. As a result, fruit bearing has taken longer; Dwarf plants started bearing after 5 years and MPT × MPT plants 1 year later. Finally, a trial was established in Bacalar, Quintana Roo, with MPT × MPT plants (1 ha) grown without fertilization and watering. Plants are starting to bear fruit. So, according to the trials mentioned here, micropropagated coconut plants are doing well even under limiting growth conditions, and so far only minor losses or no losses due to LYD have been observed in these trials. The next tests that will be established will be under ideal conditions to determine the full potential of the plants.

11.7 Scaling Up of Plumule-Based Process

Primary SE obtained by using plumule explants has been scaled up by introducing a secondary SE step and using an approach to multiply embryogenic calli (Pérez-Nuñez et al. 2006). Secondary embryogenesis was obtained by using somatic embryos as explant and inducing callus formation on the embryogenic structures. The morpho-histological development of primary embryogenesis compared with secondary embryogenesis and callus multiplication was observed to be similar. The protocol reported by Pérez-Nuñez et al. (2006) consisted of a primary embryogenesis step, followed by three cycles of secondary embryogenesis and three cycles of callus multiplication. The final callus produced somatic embryos, which could germinate and produce plantlets. By comparing the yields of somatic embryos in the new protocol to the old one, a 50,000-fold increase was obtained (Pérez-Núñez et al. 2006). A process based on this protocol has now been scaled up to a semicommercial scale in the "Biofactory," a facility in Mexico that is associated with CICY.

11.8 Current Development of Protocols Based on Floral Tissue Explants

The studies carried out in the 1970s and 1980s gave inconsistent results for the formation of embryogenic callus and plantlet regeneration from inflorescence explants (Blake 1990). However, in 1994 Verdeil et al. reported a reproducible protocol using immature inflorescences but with a low regeneration efficiency. More recently, Perera et al. (2007) reported that immature female flowers could be an alternative responsive explant. They reported embryogenic callus formation (41%) and SE formation using this type of explant from stage −4 immature inflorescences. Anther explants from different inflorescence stages and temperature pretreatments have also been reported to be responsive to SE (Perera et al. 2008). Only the anthers from stage −3 immature inflorescences and with a 38 °C pretreatment produced callus (22%) but with low efficiency of regeneration (7%). They also tested different combinations of plant growth regulators, with the best results achieved with naphthalene acetic acid (NAA) in combination with 100 μM 2,4-D for the formation of SE callus and 2-isopentyl adenine to produce somatic embryos (Perera et al. 2009).

More recently, Sandoval-Cancino et al. (2016) reported regeneration of plantlets by SE using rachillae explants from immature inflorescences from different inflorescence maturity stages. This protocol was based on the strategy used with plumule explants using different cycles of secondary embryogenesis and callus multiplication. Formation of SE callus was reported using three cycles of multiplication of callus (from 40 to 70%) using a coconut hybrid (Malayan Yellow Dwarf × Mexican Pacific Tall). Similar results were found using another hybrid (Malayan Red Dwarf × Tagnanan Tall). These results suggest that the rachillae explant protocol could be scaled up.

11.9 Prospects for the Use of Micropropagation for Satisfying the Replanting Needs Worldwide

An efficient micropropagation protocol would be very valuable for aiding the renewal of coconut plantations. There are about 12 million ha of coconut palms worldwide, of which most are dated. The International Coconut Community (ICC, previously the Asian and Pacific Coconut Community) considers that at least half of these palms need to be replaced within the next 20 years to satisfy the growing demand of markets around the world. The ICC considers that such a task would not be possible by conventional propagation alone. Therefore, a micropropagation protocol such as the one developed by CICY (using plumule or rachillae explants) operating in several "biofactories" could be an alternative for the massive propagation of selected coconut palms (at least 1 billion) that are desperately needed.

11.10 What Do We Need to Micropropagate?

There would be a need to select the appropriate genotypes and the individuals with the desirable agronomic characteristics (such as high productivity, early flower production, resistance to diseases, etc.) for micropropagation. These genotypes will also need to be selected in the regions where they will be planted. For instance, hybrids such as the Malayan Red Dwarf × Tagnanan Tall (MATAG), Brazilian Green Dwarf (BGD), and high-value coconut varieties such as aromatics and makapuno would be a profitably option (Nguyen et al. 2016). Once selected, it is important to have appropriate infrastructure for large-scale micropropagation, such as laboratories for in vitro culture, greenhouses, nurseries, etc. It is also very important that facilities be specifically designed for this task, such as the "biofactories" mentioned above. The personnel must be well trained and supervised.

11.11 Prospects for an International Collaborative Research Program

Although coconut micropropagation has already been developed to a semicommercial level in one country, it is desirable for researchers in different regions to apply and improve the existing system. It is essential to establish a research program that includes (i) steps to increase the efficiency of the protocol, (ii) approaches to monitor clones' quality, (iii) methods to improve in vitro germplasm exchange, (iv) selection methods to improve productivity and disease resistance, (v) methods to cryopreserve embryogenic tissues, and (vi) field testing to evaluate the quality and genetic stability of tissue-cultured plants. This expanded program will require the collaboration of the most advanced institutions on coconut micropropagation research and of other institutions with advanced research in complementary fields of science and technology. Such an effort involving different countries would need guidance and coordination from international organizations such as ICC and COGENT. This could be part of a virtual international institute of coconut research with a wider scope and at least two coordinating centers. Such an initiative could be proposed at the next ICC's COCOTECH meeting.

11.12 Conclusion

Coconut markets have been growing rapidly in the past decade and will continue to soar, particularly with products (and derivatives) such as coconut water, virgin oil, milk, sugar, as well as the more traditional products such as coconut crude oil, desiccated coconut, and shell charcoal. Unfortunately, coconut cultivation is threatened with coconut plantations facing challenges by several factors, namely, old age of palms, emerging pests and diseases, and the damaging effects of climate change. Therefore, efforts need to be undertaken worldwide to replant aging plantations with coconuts selected for their high productivity and resistance/tolerance to pests and diseases. This requires an innovative propagation system. Fortunately, over the last two decades, laboratories in different countries have progressed and established protocols for the clonal propagation of coconut using different types of explants. These methods include not only plumules taken from zygotic embryos but also tissues from immature inflorescences to be able to clone true-to-type palms. Hence, it is suggested that "biofactories" should be developed in different countries. More importantly, further research should be undertaken, through a worldwide collaboration, to keep improving protocols and making them more efficient and affordable. This should include further insights into the basic knowledge of SE and plantlet development and the conservation of embryogenic lines. In addition, we need to strengthen our research on coconut breeding for traits related not only to productivity and managing pests/disease but also tolerances to drought and heat to cope with changing climate conditions. Also, we need to continue working on factors that help us to improve protocols for the safe international movement of plants, those produced in vitro in particular.

Finally, we need to work in a well-organized fashion worldwide to use our coconut resources in a more efficient and effective way. This could be more easily achieved if such an effort is coordinated or in collaboration with organizations such ICC and COGENT. It is very important to keep in mind that these considerations for micropropagation are within the wider scope of strengthening the coconut value chain in every producing country, by ultimately aiming at improving the income and livelihoods for the people working with the coconut palm, mainly those small farmers with lower incomes.

References

Adkins SW, Samosir YMS, Ernawati A et al (1998) Control of ethylene and use of polyamines can optimise the conditions for somatic embryogenesis in coconut (*Cocos nucifera* L.) and papaya (*Carica papaya* L.). In: Drew RA (ed) Proceedings of the international symposium of biotechnology in tropical and subtropical species. Brisbane, Australia, pp 459–466

Antonova ID (2009) Somatic embryogenesis for micropropagation of coconut (*Cocos nucifera* L.). PhD Thesis, The University of Queensland, Australia

Azpeitia A, Chan JL, Sáenz L et al (2003) Effect of 22(S), 23(S)-homobrassinolide on somatic embryogenesis in plumule explants of *Cocos nucifera* (L.) cultured in vitro. J Hortic Sci Biotechnol 78:591–596

Batugal P, Bourdeix R, Baudouin L (2009) Coconut breeding. In: Jain SM, Priyadarshan PM (eds) Breeding plantation tree crops: tropical species. Springer, New York, pp 327–375. https://doi.org/10.1007/978-0-387-71201-7_10

Bhavyashree U, Jayaraj KL, Rachana K et al (2015) Maintenance of embryogenic potential of calli derived from shoot meristem of West Coast Tall cv. of coconut (*Cocos nucifera* L.). J Plant Crop 43:105–116

Blake J (1990) Coconut (*Cocos nucifera* L.): micropropagation. In: Bajaj YPS (ed) Legumes and oilseed crops I, Biotechnology in agriculture and foresty, v 10. Springer, Berlin, pp 538–554

Branton RL, Blake J (1983) Development of organized structures in callus derived from explants of *Cocos nucifera* L. Ann Bot 52:673–678

Chan JL, Saenz L, Talavera C et al (1998) Regeneration of coconut (*Cocos nucifera* L.) from plumule explants through somatic embryogenesis. Plant Cell Rep 17:515–521

Chuc-Armendariz BH, Oropeza C, Chan JL et al (2006) Pollen fertility and female flower anatomy of micropropagated coconut palms. Rev Fitotec Mex 29(4):373–378

Dussert S, Verdeil JL, Rivel A et al (1995) Nutrient uptake and growth of in vitro coconut (*Cocos nucifera* L.) calluses. Plant Sci 106:185–193

Eeuwens CJ, Blake J (1977) Culture of coconut and date palm tissue with a view to vegetative propagation. Acta Hortic 78:277–286

Gupta PK, Kendurkar SV, Kulkarni VM et al (1984) Somatic embryogenesis and plants from zygotic embryos of coconut (*Cocos nucifera* L.) in vitro. Plant Cell Rep 3:222–225

Gurr GM, Johnson AC, Ash GJ et al (2016) Coconut lethal yellowing diseases: a phytoplasma threat to palms of global economic and social significance. Front Plant Sci 7:1521. https://doi.org/10.3389/fpls.2016.01521

Harrison NA, Jones P (2003) Diseases of coconut. Diseases of tropical fruit crops. CABI Publishing, 44 Brattle Street, 4th Floor, Cambridge, MA, 02138, USA. https://doi.org/10.1079/9780851993904.0197

Karunaratne S, Periyapperuma K (1989) Culture of immature embryos of coconut, *Cocos nucifera* L.: callus proliferation and somatic embryogenesis. Plant Sci 62:247–253

López-Villalobos A (2002) Roles of lipids in coconut (*Cocos nucifera* L.) embryogenesis. University of London, London

López-Villalobos A, Hornung R, Dodds PF (2004) Hydrophobic metabolites of 2,4-dichlorophenoxyacetic acid (2,4-D) in cultured coconut tissue. Phytochemistry 65:2763–2774

López-Villalobos A, López-Quiróz A, Yeung E (2016) Asymmetric cell division in the zygote of flowering plants: the continuing polarized event of embryo sac development. In: Molecular cell biology of the growth and differentiation of plant cells, pp 257–283

Magnaval C, Noirot M, Verdeil JL et al (1995) Free amino acid composition of coconut (*Cocos nucifera*) calli under somatic embryogenesis induction condition. J Plant Physiol 146:155–161

Montero-Cortés M, Rodríguez-Paredes F, Burgeff C et al (2010a) Characterisation of a Cyclin-dependent kinase (CDKA) gene expressed during somatic embryogenesis of coconut palm. Plant Cell Tissue Organ Cult 102:251–258

Montero-Cortés M, Sáenz L, Córdova I et al (2010b) GA_3 stimulate the formation and germination of somatic embryos and the expression of a KNOTTED-like homeobox gene of *Cocos nucifera* (L.). Plant Cell Rep 29:1049–1059

Montero-Cortés M, Chan JL, Cordova I et al (2011) Addition of benzyladenine to coconut explant cultured in vitro improves the formation of somatic embryos and their germination. Agrociencia 45(6):663–673

Nguyen QT, Bandupriya HDD, López-Villalobos A et al (2015) Tissue culture and associated biotechnological interventions for the improvement of coconut (*Cocos nucifera* L.): a review. Planta 242(5):1059–1076

Nguyen QT, Bandupriya HDD, Foale M et al (2016) Biology, propagation and utilization of elite coconut varieties (makapuno and aromatics). Plant Physiol Biochem 109:579–589

Oropeza C, Cordova I, Chumba A et al (2011) Phytoplasma distribution in coconut palms affected by lethal yellowing disease. Ann Appl Biol 159(1):109–117

Oropeza C, Sandoval-Cancino G, Sáenz L et al (2018) Coconut (*Cocos nucifera* L.) somatic embryogenesis using immature inflorescence explants. In: Jain SM, Gupta P (eds) Step wise protocols for somatic embryogenesis of important woody plants, Forestry sciences, 85. Springer, Cham, pp 103–111

Pannetier C, Buffard-Morel J (1982) Production of somatic embryos from leaf tissues of coconut, *Cocos nucifera* L. In: Proceedings of the 5th International Plant Tissue Culture Congress, Tokyo, Japan

Perera PIP, Hocher V, Verdeil JL et al (2007) Unfertilized ovary: a novel explant for coconut (*Cocos nucifera* L.) somatic embryogenesis. Plant Cell Rep 26:21–28

Perera PIP, Perera L, Hocher V et al (2008). Use of SSR markers to determine the anther-derived homozygous lines in coconut. Plant Cell Rep 27:1697–1703

Perera PIP, Yakandawala DMD, Hocher V (2009) Effect of growth regulators on microspore embryogenesis in coconut anthers. Plant Cell Tissue Organ Cult 96:171–180

Pérez-Núñez MT, Chan JL, Sáenz L et al (2006) Improved somatic embryogenesis from *Cocos nucifera* (L.) plumule explants. In Vitro Cell Dev Biol Plant 42:37–43

Pérez-Núñez M, Souza R, Sáenz L et al (2009) Detection of a *SERK*-like gene in coconut and analysis of its expression during the formation of embryogenic callus and somatic embryos. Plant Cell Rep 28:11–19

Pham LJ (2016) Coconut (*Cocos nucifera*). In: Hayes DG, Hildebrand DF, Weselake RJ (eds) Industrial oil crops. AOCS Press, New York, USA. pp 231–242

Rajesh MK, Radha E, Sajini KK et al (2014) Polyamine-induced somatic embryogenesis and plantlet regeneration in vitro from plumular explants of dwarf cultivars of coconut (*Cocos nucifera*). Indian J Agric Sci 84(4):527–530

Rajesh MK, Fayas TP, Naganeeswaran S et al (2016) De novo assembly and characterization of global transcriptome of coconut palm (*Cocos nucifera* L.) embryogenic calli using Illumina paired-end sequencing. Protoplasma 253(3):913–928

Rethinam P (2006) Asian and Pacific coconut community activities, achievements and future outlook. ACIAR Proceedings Series 125, pp 15–21

Sáenz L, Azpeitia A, Chuc-Armendariz B et al (2006) Morphological and histological changes during somatic embryo formation from coconut plumule explant. In Vitro Cell Dev Biol-Plant 42:19–25

Sáenz L, Herrera-Herrera G, Uicab-Ballote F et al (2010) Influence of form of activated charcoal on embryogenic callus formation in coconut (*Cocos nucifera*). Plant Cell Tissue Org Cult 100:301–308

Sáenz L, Chan JL, Narvaez M et al (2018) Protocol for the micropropagation of coconut from plumule explants. In: Loyola-Vargas VM, Ochoa-Alejo N (eds) Plant cell culture protocols, Methods in molecular biology, vol 1815. Springer Science+Business Media, LLC, part of Springer Nature, New York, pp 161–170

Sáenz-Carbonell L, Montero-Cortés M, Pérez-Nuñez T et al (2016) Somatic Embryogenesis in *Cocos nucifera* L. In: Loyola-Vargas VM, Ochoa-Alejo N (eds) Somatic embryogenesis: fundamental aspects and applications. Springer International Publishing, Cham, pp 297–318

Samosir YMS (1999) Optimisation of somatic embryogenesis in coconut (*Cocos nucifera* L.). PhD thesis, The University of Queensland, Australia

Samosir YMS, Adkins SW (2004) Embryo transplantation and ex vitro germination for germplasm exchange and the production of high value, endosperm mutant coconuts. In: Peiris TSG, Ranasinghe CS (eds) Proceedings of the international conference of the Coconut Research Institute of Sri Lanka: part II. The Coconut Research Institute of Sri Lanka, Lunuwila, pp 92–102

Samosir YMS, Adkins SW (2014) Improving acclimatization through the photoautotrophic culture of coconut (*Cocos nucifera*) seedlings: an in vitro system for the efficient exchange of germplasm. In Vitro Cell Dev Biol Plant 50:493–501

Sandoval-Cancino G, Sáenz L, Chan JL et al (2016) Improved formation of embryogenic callus from coconut immature inflorescence explants. In Vitro Cell Dev Biol-Plant 52(4):367–378

Sisunandar, Rival A, Turquay P et al (2010) Cryopreservation of coconut (*Cocos nucifera* L.) zygotic embryos does not induce morphological, cytological or molecular changes in recovered seedlings. Planta 232:435–447

Verdeil JL, Huet C, Grosdemange F et al (1994) Plant regeneration from cultured immature inflorescences of coconut (*Cocos nucifera L.*): evidence for somatic embryogenesis. Plant Cell Rep 13:218–221

Chapter 12
Towards Innovative Coconut Breeding Programs

H. D. Dharshani Bandupriya, Chandrika Perera, Messias G. Pereira, and Roland Bourdeix

12.1 Introduction

Since 2010, under the leadership of Brazil and Thailand, and thanks to the rise of the coconut water market, coconut cultivation has undergone an important qualitative transformation. Some plantations of Dwarf or hybrid varieties are now managed using intensive methods that rely on the use of large amounts of fertilizers (mineral or organic) and water management (using irrigation or channel systems). Many planters still use heterogeneous and unproductive varieties, but a change in behavior and the varietal requirements are emerging. The International Coconut Genetic Resources Network (COGENT), who represent 39 coconut-producing countries, recently published a global strategy and stated that "most growers are no longer interested in cultivating mixed populations of early and late maturing palms that produce unpredictable and varying yields, with variably sized nuts, containing varying kernel and water contents" (Perera 2018). Most farmers now would prefer to plant recognized elite cultivars whose yield will be high, uniform, and predictable if grown under suitable conditions. This chapter discusses how new biotechnologies can help coconut breeders to fulfill this wish of coconut farmers.

H. D. D. Bandupriya (✉)
Department of Plant Sciences, Faculty of Science, University of Colombo, Colombo, Sri Lanka
e-mail: dbandupriya@pts.cmb.ac.lk

C. Perera
Faculty of Agriculture, University of Peradeniya, Peradeniya, Sri Lanka

M. G. Pereira
Universidade Estadual do Norte Fluminense, Campos dos Goytacazes, Rio de Janeiro, Brazil

R. Bourdeix
CIRAD – UMR AGAP, Agricultural Research for Development, Montpellier, France

AGAP, University Montpellier, CIRAD, INRA, Montpellier SupAgro, Montpellier, France

S. Adkins et al. (eds.), *Coconut Biotechnology: Towards the Sustainability of the 'Tree of Life'*, https://doi.org/10.1007/978-3-030-44988-9_12

The coconut palm is a perennial crop with a long juvenile phase. Most of the Tall-type coconut palms are allogamous and show varying levels of heterozygosity due to their cross-pollinating breeding behavior. Most Dwarf-type coconut palms are autogamous and self-pollinating due to the occurrence of male and female stages at the same time, giving rise to homozygous populations. Within these two major types of varieties, several other groups do exist which are referred to as either forms (Liyanage 1958) or variants (Bourdeix et al. 2005). According to recent studies by Bourdeix and Kumar (2018), coconut varieties are now classified in five main categories:

1. Tall types generally grow at a rate of more than 50 cm annually from at least 6 to 15 years old. They flower at 5–10 years. Their productive life span is 60–70 years. These cross-pollinating and quite heterogeneous varieties represent more than 90% of all coconut palms.
2. Preferentially self-pollinating Dwarf types. They generally grow at a rate of 15–30 cm year^{-1}. Their economic life span is 25–40 years, and they start flowering 12–30 months after field planting. They are often called Dwarfs, Fragile Dwarfs, or Malayan-type Dwarfs. The Malayan Dwarfs are the prominent varieties of this category. Apart from short height, these varieties have a combination of common traits: self-breeding; reduced size of trunk, leaves, and fruits; early bearing; and fast production of bunches. Because of these last two favorable traits, they are frequently used in coconut breeding programs.
3. Originating from the Pacific region, Compact Dwarf types are much rarer. They are mostly cross-pollinating, but not always. Breeders and scientists sometimes call them Niu Leka-type Dwarfs. The Niu Leka Dwarf from Fiji is the first and most widely known cultivar of this type, but it seems that there are plenty of distinct varieties. The words *Leka* and *Leha* are registered in many traditional Polynesian languages.
4. Semi-Tall types gather a few forms which are intermediate between Talls and Dwarfs. The autogamous King Coconut cultivar found in Sri Lanka is one of the most famous varieties of this type. Some of these varieties may have recently been created from the progenies of Dwarf × Tall hybrids.
5. "Super Dwarf" is an emerging category. It seems that Pacific farmers created these new varieties by selecting within the progenies of crosses between the Malayan-type Dwarfs and the Compact Dwarfs. The palms seem to gather dwarfism genes from both Compact and Malayan types. Some of them have very slow vertical growth and fruits of excellent quality. A few progenies were also planted in Jamaica.

12.2 Constraints Linked to the Biology of the Palm

Long selection cycles of about 15 years make coconut breeding a slow process. It is said that "a coconut breeder often analyses the trials established by his/her predecessor and establishes trials for his/her successor" (Ruas 2018). The process becomes even more tedious as, due to low-density planting, coconut needs huge ground areas for field trials. Because of their long duration, the field trials can be threatened by unfavorable weather conditions and other natural calamities. In addition, these field experiments may be prematurely terminated due to revolutions, turnover of personnel, or lack of funds providing no returns for all the investment and time spent on breeding programs: "consequently, coconut research not only needs high investments but also greater functional stability. For obtaining convincing results in the field of coconut breeding, a research station should be operational for at least 20 years" (Ruas 2018).

Researchers frequently work in the labs to study crops with shorter production cycles and to deliver quicker publishable outcomes. Funding agencies and ministries regularly favor financing ventures that can be accomplished in 3–5 years, rather than subsidizing coconut experiments that need 15 years to give full outcomes.

Another technical constraint associated with coconut breeding is the cost and the low yield of controlled pollinations. Unavailability of a consistent and commercially viable mass propagation technology for the multiplication of individual palms with desired traits is a major constraint hampering the benefits of coconut breeding. Most traditional varieties produce fewer than 100 nuts annually, which makes seed nut production generally costly. In the case of controlled hand-pollination with bagging, the annual yield is no more than 30 fruits per palm. Only two-thirds of seed nuts finally give palms planted in the fields. Screening of available coconut populations to explore male sterility or self-incompatibility for incorporating them in the hybrid programs could be an option to avoid huge investments in hand-pollination in seed gardens. However, male sterility has not been so far recorded in coconut.

12.3 Objectives of Coconut Breeding

High yield, with respect to yield components, was the primary objective of coconut breeding over the near century-long history of coconut improvement. Coconut has a multitude of uses, but until the recent development of the coconut water market, the weight of either the kernel or copra (dried kernel) was considered the main economically significant yield parameter. Fruit and nut components and number of nuts are the other main factors considered for coconut yield. However, a negative correlation has been observed between the number and the size of the fruits. As a result, possibilities for increasing the number of inflorescences produced per year (which ultimately determines the number of bunches) have been predicted as a measure to overcome this problem. Another additional possibility is to keep a correct balance between number of nuts and the size of the nut. In addition to kernel and copra yield, coconuts for other uses such as beverage, coir, etc. have become important criteria for assessment in coconut breeding experiments.

Another prime goal of coconut breeding is to reduce the long juvenile (vegetative) stage associated with the coconut palm. This is important to make coconut an appealing commercial venture for coconut growers. Early flowering and fruit bearing characters of the Dwarf types offer the best opportunity to incorporate these characters into Tall types to produce more precious hybrids with early fruit bearing characters. Enhancing biotic stress tolerance (pest and disease) through breeding programs is vital for managing pest and disease control.

The most destructive groups of coconut pathogens reported to date are the intracellular phytoplasmas and viroids that cause incurable lethal or debilitating diseases. It is recorded that the phytoplasma group is the causal organism for diseases like lethal yellowing in Africa, Kalimantan wilt in Indonesia, Kerala root wilt in India, and Weligama coconut leaf wilt disease in Sri Lanka. In their breeding programs, the nations plagued by phytoplasma diseases include tolerance breeding for these diseases as a primary objective (Nair et al. 2010). Another coconut breeding goal is the short stature of the palms. Most of the traditional coconut varieties are Tall varieties, and the scarcity of labor for coconut picking has now become a significant issue in many coconut-growing countries. Moreover, short- or Dwarf-type cultivars are more favored as home garden palms. In certain cases, controlling or managing pests and diseases is facilitated by having short stature palms. For example, managing a new strain of the coconut rhinoceros beetle, which threatened the coconut cultivation in the Solomon and Guam in the Pacific region, was much easier on Dwarf palms rather than on Tall palms.

With global climate change issues, breeding coconuts for moisture stress and prolonged droughts has become a breeding objective of several countries including India, Sri Lanka, and Tanzania. Irrigation also helps to expand coconut cultivation into areas which have not been considered ameliorative for coconut growing and productivity in the past.

12.4 Conventional Coconut Breeding

12.4.1 Intravarietal Selection

Mass selection within Tall varieties has been designated as the most basic technique of coconut breeding (Liyanage 1954). Most coconut plantations in the globe were obtained from mass selection. Per palm copra yield, weight and the shape of the nuts, numbers of fruits produced per palm, and per nut copra content were used as the most common selection criteria. Successive choices of fruit characteristics were selected from the resulting coconut populations. The scheme of selecting the best trees in the best plots began to be implemented by the coconut breeders around the mid-twentieth century. Depending on the reproductive system used, three techniques of mass selection have been defined: mass selection using open pollination, selfing, or intercrossing (Bourdeix 1988).

Due to its simplicity, mass selection through open pollination strategies is the method practiced quite often. During this process, seed nuts are collected from the palms having desirable characteristics, and resulting progenies form the basis of an improved population. The highest yield improvement recorded in this method was 14% per cycle with strict selection criteria limiting the parental palm population.

Even though Tall coconut varieties are said to be cross-pollinated, there is always a chance for selfing to occur, depending on the inflorescence production rate. The higher the inflorescence production, the higher the rate of selfing. Depending on the palm vigor and the environmental conditions (specially the climatic conditions), the inflorescence emission rate will differ in individual palms. The practice of selecting the best trees in the best plots would, however, increase the tendency to self-pollination. An inbreeding depression resulting in a lower productivity could occur as a result of this.

In addition to mass selection, many other traditional techniques of breeding are based on progeny tests. In coconut, such methods were applied mainly in Indonesia, Sri Lanka, and Vanuatu (Batugal et al. 2009).

12.4.2 Test of Hybrids Between Traditional Varieties

Hybridization is a favorite option available for coconut breeders. The very first coconut hybrid was recorded in 1926. Malayan Red Dwarf was crossed with the Niu Leka Dwarf in Fiji to produce the hybrid (Marechal 1926). Dwarf × Tall hybridization was first done in India (Patel 1938). The word "hybrid" is used in its widest sense in coconut breeding. It is defined as "a cross between two structures belonging to two different coconut varieties. The term structure here means a variety, a population of palm selected from a particular variety, a family, or an individual" (Bourdeix et al. 2018a).

Three main categories of coconut hybrids can be considered as Dwarf × Tall (D × T), Tall × Tall (T × T), and Dwarf × Dwarf (D × D) hybrids. The international codes of the parental varieties often serve to name the hybrids. As an example, a popular coconut hybrid, MRD × TAGT (also called Matag or PCA15-2), is produced by crossing the Malayan Red Dwarf (MRD) with Tagnanan Tall (TAGT). The MRD is a self-pollinating variety which is considered to have a single homozygous genotype, whereas TAGT is a cross-pollinating variety; thus, palms show differences in their genotypes despite showing similar traits.

Coconut hybrids brought advantages with respect to higher copra yield and shortening of the vegetative phase of coconut. Superior performance of hybrids under a range of climate and soil conditions has been reported over many traditional Tall varieties (Nuce de Lamothe and Rognon 1986). Self-pollination is prevented during the process of hybrid seed production, ensuring that farmers effectively benefit from the intended genetic progress (Fig. 12.1).

Currently, many breeding programs devoted to coconut hybrids are limited to testing new crosses between traditional coconut varieties. Breeders collect new traditional varieties and introduce them into their collection; then they create and test new hybrids, generally by favoring Dwarf ×Tall types and more recently Dwarf × Dwarf types. Individual palms chosen in parental varieties are selected on their

Fig. 12.1 1 and 2. PB121 (Malayan Yellow Dwarf × West African Tall). This Dwarf × Tall hybrid is the most frequently planted worldwide. 3. Hybrid Cameroon Red Dwarf × Laccadive Micro Tall.

phenotype and not on the value of their individual progenies. This type of methodology will not continue to maintain substantial genetic progress. Geraldo Santos, a famous former coconut breeder, already depicted such situation in the 2000s: "in the Philippines, we tested more than 100 hybrids between Tall and Dwarf types, now this is OK, we do not want to test more!" (R. Bourdeix, personal communication). If breeders continue to use the same method again and again, the results will be limited. Yields will cap.

12.4.3 Improvement of the Best Hybrids

Hybrids which are considered the "best" between traditional varieties have been improved using the individual combining ability testing method, which takes advantage of the genetic variability within Tall varieties (Gascon and De Nuce de Lamothe 1976). This can be explained using the following example: PB113 is hybrid developed by crossing the Cameroon Red Dwarf (CRD) and a selected population of Rennell Island Tall (RIT). Their excellent performance has been further enhanced. Forty-five RIT parent palms have been individually crossed with CRD. The progenies obtained are, therefore, half-sib families each from one RIT as male and several CRD as female. In the absence of a secure cloning method that does not destroy the sampled palms, the individually tested male parents have been conserved and multiplied by selfing. The best self-families, each constituted of about 100, are conserved as pollen donors for seed nut production. The results of these experiments show that selecting the best families gives 15–30% genetic progress on yields (Bourdeix et al. 1989). It is to be noted that these experiments were mainly planted in Côte d'Ivoire before 1990. Vanuatu also planted a few experiments but has not yet released their results to farmers.

The best half-sib families located in these experiments are among the progenies where the best individuals could be chosen. This choice would benefit from one more breeding generation. In Côte d'Ivoire, this selection was not done, and experiments are now endangered by the probable expropriation of the "Marc Delorme Research Centre" in Côte d'Ivoire.

The Marc Delorme Research Centre allowed the release of tens of millions of improved hybrid seed nuts, in the country and at the international level. Unfortunately, this African program was not continued due to a political crisis that occurred in the

Fig. 12.1 (continued) 4. MATAG (Malayan Red Dwarf × Tagnanan Tall). This hybrid is frequently planted in Southeast Asia. 5. and 6. Two Dwarf × Tall hybrids in Vanuatu: Brazil Green Dwarf × Rennell Island Tall and Vanuatu Red Dwarf × Vanuatu Tall. 7. MAREN (Malayan Red Dwarf × Rennell Island Tall). This hybrid is the most frequently planted in the Pacific Islands. 8. CAMREN (Cameroon Red Dwarf × West African Tall). 9. Hybrid Brazil Green Dwarf × Rangiroa Tall used in French Polynesia. 10 and 11. Hybrid West African Tall × Rennell Island Tall. Although high yielding, such Tall × Tall hybrids, are rarely planted by farmers because they are less precocious than Dwarf ×Tall hybrids and more costly to produce. 12–14. Hybrid between Malayan Yellow and Red Dwarfs. Such Dwarf × Dwarf hybrid types will probably be extensively used in the future

2000s. The best parent palms should have been crossed, and a new phase of selection should have been launched. It would have produced in 2015 hybrids producing another 15–20% more than the improved hybrids currently available. Not having achieved this program amounts to millions of dollars of loss for the coconut sector worldwide.

Some private companies, notably in Malaysia and India, say they improved some coconut hybrids, but the detailed methods they used remain unknown. The scientific community has no idea of the methods used and the volume that these private experiments represent. The "improved hybrids" produced by those private companies have never been compared with other existing hybrids in experiments supervised by independent institutions. Everybody can claim "I have the best improved hybrid" when there is no available certification process. The vagueness and lack of regulation in the coconut seed sector disconcert stakeholders and undermine the effectiveness of the coconut value chain.

12.4.4 Complex Hybrids and Synthetic Varieties

A coconut hybrid which is crossed either with another hybrid or a variety is referred to as "complex hybrids." Three-way hybrids, (D × T) hybrid crossed with a Tall variety and (T × T) hybrid crossed with a Dwarf variety, have been formed in Thailand and Côte d'Ivoire in the 1980s even though the results were not published (Petchpiroon and Thirakul 1994).

In Côte d'Ivoire, crosses (D × T) × (D × T) from best parent palms of four distinct varieties have been planted specifically for preparing clonal selection. Although carefully observed during more than 12 years, these experiments were never used to produce any clones; they are now threatened by the probable expropriation of the Marc Delorme Coconut Research Centre.

In Cote d'Ivoire again, progenies of hybrids between Dwarf varieties have been planted to create new Dwarf varieties tolerant to lethal yellowing disease and for having better fruit quality than the Sri Lanka Green Dwarf. The third generation started to be planted in 2019 (E. Issali, personal communication) and is now under observation (Fig. 12.2).

The Philippines tried to produce Makapuno hybrids which show short stature by crossing Makapuno Tall with normal Dwarf varieties (Nunez and de Paz 2004). The process was continued with success up to the third generation. More recently,

Fig. 12.2 (continued) from the Pilipog parent. 3, 4, and 10. Natural selfing of the hybrid between the Malayan Yellow Dwarf and the Tahiti Red Dwarf. 5, 6, and 8. Natural selfing of the hybrid between the Pilipog Green Dwarf and the Sri Lanka Green Dwarf. 7. Natural selfing of the hybrid between the Cameroon Red Dwarf and the Sri Lanka Green Dwarf, seen by researchers at the CNRA Marc Delorme Research Centre (Côte d'Ivoire) and the former COGENT coordinator. 9. Natural selfing of the hybrid between the Malayan Yellow Dwarf and the Sri Lanka Green Dwarf and researchers at the CNRA Marc Delorme Research Centre (Côte d'Ivoire)

Fig. 12.2 1. Natural selfing of the hybrid between the Malayan Yellow Dwarf and the Pilipog Green Dwarf. 2. Same progeny. Inside the young fruits, some palms have a pink color inherited

researchers found already existing autogamous Dwarf types with Makapuno and Kopyor characteristics in Indonesia, Vietnam, and the Philippines.

Breeding a coconut synthetic variety was initiated in the Philippines (Santos 1990; Santos and Rivera 1994). A total of six Tall varieties identified both from the Philippines and Africa were crossed to create 15 F1 hybrids. Those F1 hybrids were planted in the same field with the intention of producing a mixture of four-way hybrids (through open pollination) to distribute among farmers as the first coconut synthetic variety.

12.4.5 Recurrent Reciprocal Selection

Many coconut improvement programs are based on short-term views without defining a sustainable strategy to maintain regular genetic progress. New coconut breeding guidelines were suggested in the 1990s by the French Agricultural Research Centre for International Development (CIRAD) based on reciprocal recurrent selection (RRS) (Bourdeix et al. 1990, 1991a, b). In RRS populations, reproductive isolation is maintained and mutually enhanced (Comstock et al. 1949). The RRS technique is useful for overcoming some of the plant's reproductive biology limitations and other genetic factors. The proposed RRS methods require the evaluation of many half sib-families. Including hybridization tests, the selfing and the intercrossing of the parent palms, and a full breeding cycle of Dwarf × Tall hybrids which would require planting at least 250 ha of field experiments over a period of about 25 years. This area could be reduced if the parent palms are cloned instead of being selfed. However, there are currently only a few research programs allocating such amount of resources to coconut breeding, and they are currently selecting other methodological alternatives.

12.4.6 Effectiveness of Current Coconut Breeding Programs

Many national breeding programs presently have a very limited size. An effective and consistent program of coconut genetic improvement should plan the annual planting of at least 10 hectares of experiments per year, or about 1600 trees. The ideal would be to reach 20 ha year^{-1}, or 3000 trees. The STANTECH Manual has provided recommendations on experimental designs for coconut hybrid tests. It stated two conditions in a randomized complete block design for yield trial: either using a plot size of 24 palms or 16 palms. The number of replicates for each condition is four or six, respectively, so 96 palms per progeny (Santos et al. 1996). For the improvement of the best hybrids, this number was often reduced to about 60 palms per progeny tested (Bourdeix et al. 1989, 1990).

Planting 3000 trees makes it possible to test a maximum of 50 progenies per year; among these progenies, the best three to five could be chosen and used for the continuation of the improvement program. Coconut is a very versatile plant, with production varying a lot according to soil and climate; multilocational experiments should be preferably conducted before releasing varieties to farmers.

Such field genetic experiments, if well managed, can be very profitable plantations. Although, it seems that many scientists involved in coconut breeding do not know or have forgotten how to appropriately plant and maintain coconut plantations. Many spend most of their time in laboratories and have limited experience of field planting. Yields and survival rates at research stations are often lower than good farmers achieve. For instance, in an international experiment conducted by the COGENT network, seven countries (Benin, Brazil, Côte d'Ivoire, Jamaica, Mexico, Mozambique, and Tanzania) have planted different hybrids (Batugal et al. 2005). Most of those countries obtained relatively low average yields and survival rates (Bourdeix, personal observation). In Brazil, hybrid farms often reach 150 coconuts per palm per year at adult age, while it seems that the hybrids in this international experiment did not reach 100 coconuts per year on average for all planted palms. Unfortunately, the results of this international experiment have never been published in detail. It may not be too late to make this collaborative assessment.

In the years 1980–2010, only a few countries (Ivory Coast, the Philippines, and Vanuatu) achieved substantial annual planting areas in their breeding programs. Since 2010, Ivory Coast and Vanuatu have strongly reduced their programs. Only a few countries are now reaching the plantation size of 10 ha or more of breeding experiments per year. The Philippines is planting about 50 ha year^{-1}of coconut breeding experiments mainly focused on developing genomic-assisted breeding: Dwarf varieties, D × T hybrids, a synthetic variety, and specific Tall-type breeding lines are evaluated using a multilocational design (R. Rivera, personal communication). India is planting about 15 ha year^{-1} of coconut breeding experiments scattered in many research centers, including the Central Plantation Crops Research Institute (CPCRI) and its regional stations, Central Island Agricultural Research Institute (CIARI) Port Blair, 24 All India Coordinated Research Project on Palms (AICRP) centers under different State Agricultural Universities, and a few private firms. The experiments are mainly multilocational and/or on-farm tests of D × T and D × D hybrids and some breeding lines of Tall and Dwarf varieties (A. Jerard, personal communication). Indonesia plants about 5 ha of coconut breeding experiments per year (H. Novarianto, personal communication). China is planting approximately 10 ha. Brazil carries outbreeding experiments in large industrial plantations (S. Ramos, personal communication), but the average area was not estimated.

Coconut breeding programs conducted in many countries seem not to be planned in terms of optimal efficiency. A breeding generation (cycle) should be achieved every 13–15 years: 3–8 years are considered as "juvenile period," while 9–12 years are considered as the "adult period" for coconut breeding trials (Bourdeix 1999). This quite long characterization period could probably be reduced if relevant dedicated studies are undertaken. One and a half year is needed for controlled pollination and another 1 year for seed nut raising in the nursery. The successive steps of such a breeding cycle are:

1. Raise seed nuts in the nursery.
2. Plant progenies in breeding experiments.
3. Characterize the progenies at juvenile and adult age.
4. Use this characterization data to choose the best palms (within the progenies or within the parents from these progenies according to the type of experiment).
5. Cross the best palms or harvest open-pollinated seed nuts from them.
6. Transfer these improved seed nuts in the nursery to start a new cycle.

For instance, in Sri Lanka, the improvement of the CRIC60 variety (Sri Lanka Tall) started around 1930 (Liyanage et al. 1988). As of 2019, only three breeding cycles have been fully achieved (L. Perera, personal communication), when this number could have reached five to six. There is a need to review all the coconut breeding programs worldwide using the same systematic approach.

Around the world and over more than 80 years, breeders have created many hybrid combinations between traditional coconut varieties. The last systematic review estimated that about 400 hybrids were created during the twentieth century (Bourdeix 1999). Among those 400, breeders have selected a tenth that produce about twice than that of the traditional varieties. Most of the hybrids presently used were discovered before 2000. A few breeders have started to improve some of these hybrids by testing the aptitude of individual palms to hybrid combination. They have obtained half-sib families producing 20% more than the former hybrids between traditional varieties.

At the same time, Brazilian agronomists have sought to optimize production for a Dwarf variety, the Brazilian Green Dwarf, using irrigation and a high mineral fertilization. This Dwarf was created by gardeners, in the Philippines and later in Brazil, not by scientists or coconut breeders. However, by defining optimal crop conditions specifically adapted to this variety, Brazilian agronomists have managed to obtain higher fruit yields than coconut breeders achieved with their improved hybrids. This has disrupted coconut breeders, challenging their work. At present, no improved hybrid has been selected under the agronomic conditions developed by Brazilians. The improved hybrids produced in Côte d'Ivoire and Vanuatu have never even been grown using these Brazilian methods; the yields they could achieve under these conditions remain unknown.

12.5 Integration of Biotechnologies in Coconut Breeding

For the last several decades, conventional coconut breeding has been progressed mostly via mass selection and hybridization. To overcome the limitations and to produce vastly better cultivars, several research groups have proposed alternative and novel strategies based on biotechnologies. Some of these are already in practice at a limited scale.

12.5.1 Molecular Approaches to Enhance Coconut Breeding

As discussed, coconut breeding is a very long process. Breeding depends on the amount of genetic variability, referred to as "the amount of genetic variability among individuals of a variety or population of a species" (Brown 1983). Genetic diversity is first assessed by using morphological characteristics. In the case of coconut, those are strongly affected by environmental effects, stage of development of the plant, and the type of plant material. The evaluation of genetic diversity with molecular markers provides an additional diversity assessment.

In the specific case of molecular breeding of coconut, genetic research has been ongoing since the early 1990s. The first step was to identify suitable markers for the use of genetic diversity research and then to identify a group of markers that could be used for the characterization of cultivars. Molecular approaches when used for breeding could help better understand and make use of the genes controlling the different forms of dwarfism, in order to reduce the size and increase the early production of fruit. Molecular approaches could also be used to find tolerance to diseases such as lethal yellowing or discern the quality of fruits and coconut products, such as the composition of the oil. It is very likely that such fruit characteristics are highly heritable and that a corresponding quantitative trait locus (QTL) can be detected.

Initially, coconut molecular markers were developed for the purpose of evaluating genetic variability; these included random amplified polymorphic deoxyribonucleic acid (DNA) (RAPD) (Ashburner et al. 1997; Duran et al. 1997; Everard 1996); restriction fragment length polymorphism (RFLP) (Lebrun et al. 1998); amplified fragment length polymorphism (AFLP) (Perera et al. 1998; Teulat et al. 2000); and inverse sequence-tagged repeat (ISTR) (Duran et al. 1997). Subsequently, microsatellite markers (SSRs) specific to coconut were developed. With the purpose of standardizing techniques from different laboratories to compare results, making more efficient the detection of genetic diversity and identification of varieties, different sets of microsatellite markers were developed: CnZ series in the Philippines (Rivera et al. 1999), CAC series developed in Scotland (Perera et al. 1999, 2000, 2003), and CnCir series developed in France (Baudouin and Lebrun 2002), all of which primers are available in GenBank database. Out of these marker systems, SSRs have become the most widely used because of their codominant nature (Dasanayake et al. 2003; Meerow et al. 2003; Perera et al. 2000; Teulat et al. 2000). In addition, a study using SNPs (single nucleotide polymorphism) markers using new-generation sequencing data to assess the genetic diversity in Brazilian Dwarf coconut has been reported (Santos 2016).

12.5.2 Genetic/QTL Maps in Coconut

Gene/QTL maps provide the marker trait associations needed to proceed with the marker-assisted selection and breeding of crops. As discussed earlier, traditional coconut breeding relies on identifying palms with desired characters which will be then evaluated by looking at the phenotypic traits in their next generation. These characters are observable at the latter part of the palm developmental process due to the long juvenile process; thus they are costly to measure. Coconut breeding research has been focused on detecting markers associated with traits with the intention of avoiding problems associated with the long juvenile phase. Efforts for the development of genetic maps of coconut have been made by several groups (Bandaranayake and Kearsey 2005; Baudouin et al. 2006; Herran et al. 2000; Lebrun et al. 2001).

Initial work done on the construction of a linkage map for coconut was reported just before 2000 (Rohde et al. 1999), using a cross between a Laguna Tall (LAGT) genotype from the Philippines and a Malayan Yellow Dwarf (MYD). Three hundred and eighty-two markers were used to generate 16 linkage groups for the two parents. Estimated genome length for the LAGT map was 2226 cM and for the MYD map was 1266 cM. Six QTLs correlating with traits including early flowering and yield were identified, and this study was the first attempt at marker-assisted selection in coconut (Herran et al. 2000).

The second linkage map was based on a cross between Cameroon Red Dwarf (CRD) genotype and Rennell Island Tall (RIT) (Lebrun et al. 2001). In this study, a total of 227 markers were placed into 16 linkage groups. Analysis of QTLs for yield characteristics recognized nine loci including QTLs for the number of bunches and the number of nuts, traits which are important in breeding programs (Lebrun et al. 2001). The same map was then modified by adding another 53 SSR markers to identify QTLs linked with fruit traits and 52 putative QTLs for 11 different traits. Among these traits were fruit component weight, endosperm moisture content, and fruit production (Baudouin et al. 2006).

12.5.3 Physical Maps and Synteny Studies

Physical maps provide the probable order of distinct genomic DNA segments down to the chromosomes. Physical maps can be applied in genome-wide gene discovery, expressed sequence tag (EST) mapping (functional genomics), or comparative genomics (synteny studies). Physical maps for coconut and oil palm were attempted by mapping the genomic DNA data obtained by cosmid clones onto the molecular linkage maps. These are to be mapped in association to earlier mapped polymorphic AFLP markers expecting high-density molecular linkage reference maps for both coconut and oil palm.

12.5.4 Association Studies

An alternative approach for developing segregating populations for linkage mapping is association mapping which is done by correlating quantitative traits and markers in natural populations. Association studies require a high density of molecular markers and thus require a high-throughput marker system such as diversity array technology (DArT) (Jaccoud et al. 2001). Association mapping is based on linkage disequilibrium and the search for special connection between tightly linked markers. The French Agricultural Research Centre for International Development (CIRAD) has scientists in Vanuatu that are examining the intensity of the linkage disequilibrium as a preliminary step to association mapping.

12.5.5 Applications of DNA Markers in Coconut Breeding

Developing varieties for tolerance/resistance to biotic and abiotic stresses is a priority in coconut breeding. Molecular marker technology has been used in the recent past to assist conventional breeding to develop such cultivars.

Coconut mite is an important pest of coconut palm cultivation, and the main problems caused by the pest are flower abortion and a reduction in coconut water, copra, weight, size, and commercial value of the fruit (Moore et al. 1989). Shalini et al. (2007) reported molecular markers associated with coconut mite (*Aceria guerreronis* Keifer) resistance, and a combination of five markers (three SSR and two RAPD) explained 100% of the association with mite resistance in coconut.

Phytoplasma, which is an intracellular pathogen, is one of the most prominent pathogens affecting coconut. Among the diseases caused by phytoplasma are lethal diseases such as lethal yellowing and wilt diseases which cause devastating losses for the industry. Additional to standard screening and breeding techniques, molecular techniques were used to define the resistant genotypes to lethal yellowing, and some encouraging results were recorded (Cardena et al. 2003; Konan-Noël et al. 2007). Given the importance of phytoplasma, the transcriptome of infected coconut leaves was analyzed. In reaction to phytoplasma assault, a key set of genes associated with defense were discovered in coconut, along with new candidate genes for the defense response with unknown function. These data can be used to support breeding strategies (Nejat et al. 2015).

Among the most destructive pests that attack the palm leaves is the coconut hispine beetle (CHB) (*Brontispa longissimi* Gestro). Yan et al. (2015) have sequenced and analyzed the transcriptome of CHB. In order to develop an effective pest control approach, the data produced in the study is likely to be useful for gene identification and to elucidate the molecular process of the biological and ecological aspects of the pest.

Added to the biotic factors, drought is a multidimensional stress factor of great importance for the sustainability of production as it causes a significant reduction in the yield of coconut nuts and even the death of palms. However, studies focused on the identification of genes and breeding programs that target drought-tolerant coconut genotypes are still scarce.

12.5.6 Mutant Selections and Cloning

Some useful rare mutants have been identified in coconut germplasm collections, and these would be of tremendous use for the development of new products. Coconuts with edible husk and sweet or flavored kernel and some aromatic coconuts are such mutations which may contribute to the sustainability of the coconut industry. The Makapuno coconut which was originally reported in the Philippines and subsequently reported in several other countries under different names (Kopyor, Indonesia; ThairuThengai or NeiThengai or Ghee Thengai, India; Coco Gra, Seychelles; Dikiri Pol, Sri Lanka; Sap, Vietnam; Mapharao Khati, Thailand; Niu Garuk, Papua New Guinea) is due to a single recessive mutation resulting in different endosperm characteristics. Makapuno-type mutants, with soft jelly-like endosperms, can be utilized to develop new varieties and for product diversification (Nguyen et al. 2016). Certain Dwarf varieties have shown extreme dwarfism (Lakshadweep coconut in India) compared to the existing Dwarfs. Such plants could be useful for coconut breeding programs to produce short-statured palms for ease of picking (Nair et al. 2016).

In addition to the cloning and the characterization of genes linked to certain unique features mentioned above, a mutation in the galactosidase gene has been identified as the possible reason for the "curd" coconut phenotype. Complementary DNA (cDNA) and genomic sequences encoding galactosidase gene alleles have been cloned and characterized from normal and mutant coconut (Phoeurk et al. 2018). A pleasant pandan-like aroma is reported to be a unique characteristic due to the presence of 2-acetyl-1-pyrroline (2AP) in the coconut liquid endosperm. De novo assembly of transcriptome data derived from *coconut* endosperm enabled the identification of a gene *amino aldehyde dehydrogenase 2* (*CnAMADH2*), an ortholog of the rice aromatic gene. Detailed nucleic acid sequence identified a G-to-C substitution in the 14th exon, resulting in a change in the amino acid composition (alanine-to-proline), which is associated with 2AP synthesis in the aromatic green Dwarf (Saensuk et al. 2016).

Successful cloning of resistance gene analogues (RGAs) using a comparative genomic approach has identified putative RGAs in the Chowghat Green Dwarf (CGD) cultivar which is resistant to coconut root (wilt) disease. Quantitative real-time polymerase chain reaction (PCR) analysis showed a higher expression rate of RGAs in disease-resistant plants compared to a susceptible counterpart (Rachana et al. 2016).

The storage of lipids in coconut endosperm is associated with oleosins which stabilize the structure of the oil bodies. Complimentary DNA sequences of three L-class oleosin isoforms (*CnOLE500a*, *CnOLE500c*, and *CnOLE300a*) have been cloned and characterized. Solid endosperms of 5- to 6-month-old coconuts showed very little oleosin proteins, while there is a 20-fold increase of protein accumulation at the 6–7-month-old maturity stage. A continuous increase in the oleosin proteins has been detected with fruit maturity, and a further12-fold increase was recorded at the 8–9-month maturity stage (Vargas et al. 2018). The information generated through these studies and many more studies in coming years will provide insights for future breeding programs for the development of cultivars with elite traits, not just pest and disease resistance but also for value-added products.

12.5.7 Genome-Associated Breeding

With the advent of new-generation sequencing (NGS) technologies, several complex genomes have been sequenced with the speed and efficiency of next-generation sequencers. With this type of facility in 2017, a group of Chinese researchers published a draft of the coconut genome using the Illumina HiSeq 2000 platform (Xiao et al. 2017). The findings of the genome annotation disclosed that 72.75% of the coconut genome is made up of transposable elements. The genetic information generated will make it possible to develop a functional genomics and marker-assisted breeding approach for this species soon.

The different studies involving molecular, genomic, and transcriptomic markers of the coconut have generated large amount of data. Most of this is deposited in a public database called Tropgene: http://tropgenedb.cirad.fr/tropgene/JSP/index.jspor GenBank Databases http://www.ncbi.nlm.nih.gov/nucleotide and https://www.ncbi.nlm.nih.gov/sra (Perera et al. 2016). The accessibility of this data has allowed new functional markers to be developed quickly and economically. Some of these studies are EST-SSR (expressed sequence tag-simple sequence repeat) markers (Preethi et al. 2016); SNAP (single nucleotide amplified polymorphism) markers (Maskromo 2015); and SCoT (start codon targeted polymorphism) markers (Rajesh et al. 2015).With the large variety of existing markers, breeders should be cautious when using only the markers results. The combined information (molecular markers and phenotype) should always be preferred for success (Perera et al. 2016).

12.6 A Role for Rapid Clonal Multiplication

12.6.1 Experiences to Date

Over the past few decades, cloning of coconut via somatic embryogenesis has been addressed by several researchers worldwide. In vitro propagation of coconut stagnated without much improvement for a certain period, mainly in the 1990s. Many published reports have emphasized the importance and feasibility of coconut tissue culture; however, at the same time, these reports have expressed the constraints associated with coconut tissue culture protocols (Bandupriya et al. 2016; Nguyen et al. 2015). The main constraints were highlighted by the researches of eight different countries who presented their research findings at the Workshop on Somatic

Embryogenesis for Rapid Coconut Multiplication at the International Coconut Biodiversity for Prosperity Conference held in India in 2010.These constraints are the slow response of cultures under in vitro conditions, variation in tissue response under in vitro conditions, low embryogenic potential of callus, formation of incomplete somatic embryos lacking shoot meristems, low plant regeneration frequency, and intense tissue browning.

In the first decade of twenty-first century, a tremendous improvement to the protocol has occurred by addressing the abovementioned constrains, which have enabled scientists to fine-tune propagation protocols. The introduction of repeated cycles of callogenesis and secondary somatic embryogenesis has increased the effectiveness of the protocol for plumule micropropagation (Perez-Nunez et al. 2006), which has allowed commercial propagation of certain genotypes (Lédo et al. 2019). This 50,000-fold increase is a remarkable success, compared to somatic embryos obtained only through primary somatic embryogenesis (Perez-Nunez et al. 2006).

Plumule explants have been preferred by lab researchers due to their better response compared to other explants, but they are not at all the most convenient for breeders. The production of clones of palms with unknown performance due to the cross-pollination in coconut is the major disadvantage of using plumule explants. However, this technique is useful for propagating the best Dwarf coconut genotypes, such as green Malayan Dwarf Coconut, which show high self-fertilization rates. Moreover, plumule culture can be applied to propagate superior populations such as genotypes that are resistant to diseases. Plumule culture may also help to propagate progenies of elite palms under controlled pollination. It would be particularly useful for the multiplication of Makapuno and Dikiri coconut, which are currently propagated by embryo rescue at a high cost. According to the reports from Indonesia, Kopyor coconut palms are raised through embryo culture at a cost of US$ 70 plant^{-1} (Sisunandar et al. 2018). Application of plumule culture would substantially reduce that cost as the number of plants produced per plumule would vastly increase.

Scientists are continuously working on selecting a better explant for clonal propagation, knowing that Tall varieties and their hybrids are best micropropagated through somatic tissues. Perera et al. (2007) reported micropropagation of coconut from unfertilized ovary tissue (Fig. 12.3), which is a somatic tissue, and the results were encouraging and consistent. There are also now protocols which cause minimal damage to the mother palm from which explants are collected (Bandupriya et al. 2016). As for plumule explants, similar callus multiplication cycles can be introduced to increase the callus yield obtained from unfertilized ovary explants. Callus multiplication and plant regeneration were observed in palms in both Tall and Dwarf × Tall cultivars, although results revealed significant palm to palm variation in in vitro response (Vidhanaarachchi et al. 2013). From one inflorescence, the number of explants that can be cultured is low because the average number of female flowers that exist is around 30. This is one of the drawbacks of using the unfertilized ovary as the explant as it limits large-scale culturing to produce a higher number of clonal plants.

Fig. 12.3 Images of different steps in micropropagation of coconut through ovary culture. (**a**) An inflorescence at −4 stage (inflorescence is fully covered by the outer spathes). (**b**) A female flower flanked by two male flowers. (**c**) Embryogenic callus [EC] formed from unfertilized ovary explants. (**d**) EC with developing somatic embryos. (**e**) EC with developing shoots. (**f**) Individual plantlets with complete shoot and root structures. (**g**) Plantlets under individual propagators. (**h**) Plants at acclimatization in ex vitro conditions

12.6.2 *Clonal Multiplication and Embryo Culture in Plant Breeding*

Not only varietal improvement matters for the increased productivity of coconut, but also it is important to make available plenty of improved planting material for replanting and for new coconut lands. Most coconut plantations worldwide require replanting either because of aging or due to loss of diseased palms. The availability of seedlings at present is below the reported demand for seedlings. For example, at present the annual coconut seedling demand in Sri Lanka is four million seedlings. Nevertheless, only 1.2 million seed nuts are produced annually through conventional breeding, of which 0.2 million are hybrid seed nuts. Despite the higher demand for hybrid seedlings, the acceleration of seedling production cannot be achieved due to certain limitations. Therefore, alternative approaches for the propagation of improved planting material should be considered. The supply of high-quality planting material with high production potential, especially to smallholders, will lower the burden of domestic consumption which increases the availability of nuts to industries. A successful clonal propagation method will pave the way for the enhancement of the coconut industry by making available sufficient planting material to meet demand, provided that the genetic quality and stability of the clones can be assured.

The multiplication of disease-tolerant planting material is one of the possible applications of in vitro clonal propagation. Coconut breeders are fighting hard to combat the losses due to severe disease conditions such as lethal yellowing disease, other phytoplasma diseases prevailing in India and Sri Lanka, cadang-cadang disease, etc. The planting of seed nuts from diseased plants spreads the disease. Identifying disease-resistant or disease-tolerant palms in a heavily infested plantation may enable the multiplication of resistant or tolerant palms. Similarly, palms adopted for adverse weather conditions could be multiplied using in vitro micropropagation.

Another possible role of coconut in vitro culture is the creation of bi-clonal seed gardens (Perera 2018). Seed gardens can be developed using two selected superior coconut palms instead of the current practice of using two populations. Clonal propagation has the advantage of selecting individual palms having quality secondary characteristics, i.e., better oil quality, reduced palm height, etc., and expressed uniformly in a progeny. Thus the likelihood of success of bi-clonal seed gardens is high (Oropeza et al. 2018). This would also be valuable as plants regenerated from in vitro culture techniques, especially through somatic embryogenesis, may exhibit somaclonal variations which are heritable (Breiman et al. 1987), so it is very important to check resultant plants for their uniformity throughout the regeneration process and in field conditions.

Besides clonal propagation, embryo culture technology is also important for the development of current and future coconut breeding programs. Coconut germplasm (the genetic resources collected and conserved for the purpose of breeding) is a potential treasure trove of coconut genetic variability. Conservation of coconut germplasm is extremely important for mining the specific desirable traits for breeding and future activities. The embryo culture technique facilitates collection, exchange, and conservation of coconut germplasm. The use of DNA-based germplasm characterization studies has revealed that the genetic base of the available coconut gene pool is narrow, and therefore further improvement of coconut by utilizing only locally available materials is limited (Perera et al. 1999). Therefore, having identified the importance of the incorporation of exotic germplasm, several exotic varieties were exchanged through the germplasm exchange programs facilitated by COGENT. For example, the Coconut Research Institute of Sri Lanka has acquired 19 exotic varieties from four countries (four from India, six from Papua New Guinea, eight from Ivory Coast, and one from Indonesia). These varieties were brought into the country as embryos; plants were raised in vitro, acclimatized, and field planted. Now the palms are at bearing stage and can be used in breeding programs. Moreover, coconut embryo culture offers a potential alternative strategy for the conservation of coconut germplasm by preserving thousands of plantlets in a small space under specific conditions that reduce the growth of cultures to maintain a minimum growth (slow growth method) or by completely halting the growth of the plant embryo (cryopreservation method). Instead of having different coconut accessions only under field conditions, which are prone to pest and disease attacks and adverse weather conditions, embryos have the potential to be stored safely under laboratory conditions with minimum maintenance. A proven protocol has been developed for cryopreserving coconut embryos (Sisunandar 2010), and more recently a collaborative project between the Rural Development Administration (RDA) of Korea and Bioversity International has been developing a cryopreservation protocol for coconut using meristem tips.

Moreover, the production of coconut clones may have a direct impact on the conservation of germplasm. As an example, the International Centre for Tropical Agriculture (CIAT) has a collection of 300 embryos from palms that survived red ring disease (caused by a nematode). Researchers are trying to multiply these embryos using plumule culture technology. As previously discussed, the conserved material will not have the same genotype as the resistant/resilient parent if the fruit was collected after open pollination, and multiplied plants from collected embryos may not be resistant either. This same problem occurs when trying to select palms tolerant to phytoplasma or any other disease or any other favorable character preferred by coconut breeders. Therefore, callus lines obtained from somatic tissues from the best palms that hold superior characters need to be preserved carefully for future use allowing the multiplication of the best clones. This could be achieved with cryopreservation. Initial steps on embryogenic calli cryopreservation have been attempted using unfertilized ovary calli through an encapsulation-dehydration technique (Iroshini 2017; Iroshini et al. 2017).

12.6.3 Sustainability of Clonal Multiplication

It is worth comparing the micropropagation system of oil palm (*Elaeis guineensis* Jacq.) with that of coconut in order to determine the sustainability of coconut micropropagation. Even though large-scale in vitro multiplication of oil palm was initiated in the 1970s, the occurrence of abnormal inflorescence, called mantled inflorescence, on palms that resulted from micropropagation delayed the success story. Nearly 30 years passed before this problem was solved. That is why it is very important to do field evaluations of coconut palms derived through micropropagation. Although preliminary studies show the capability of regenerating genetically stable coconut plants, which show normal growth (Fig. 12.4) after acclimatization and field transfer (Armendariz et al. 2006; Bandupriya et al. 2017; Fernando et al. 2004), however, thorough field evaluations should be carried out evaluating the plants generated before being distributed among farmers. Even though oil palm

Fig. 12.4 In vitro micropropagated coconut plant under field conditions. Left: A plumule-derived plant just after field planting. Right: A plumule-derived plant at fruit bearing stage

micropropagation is claimed to be successful, so far out of the 250 million hybrid seedling requirement, only three to five million clonal ramets are sold annually in Malaysia (Soh et al. 2011). The price of a clonal seedling is also a consideration. The unit price of a clonal oil palm varies between three to five dollars, while the price of a palm seed is generally lower than a dollar. At this stage, there is a price difference between a seed-germinated oil palm plant and a tissue-cultured clonal plant, and this is likely to be the same for coconut. Researchers need to work hard at making clonal production more efficient and therefore a reality. At present, only Mexico sells coconut clonal seedlings. However, due to institutional policy, the prices are not disclosed.

12.7 Homozygous Lines Through Haploid Plant Production

Doubled haploids are effective in plant breeding because of the 100% homozygosity they reach within one generation after inducing the haploids in comparison to the tedious and time-consuming process of achieving homozygous lines through a conventional selfing method. Haploids can be developed in two ways in higher plants: anther or microspore culture which is termed as androgenesis and ovary or megaspore culture termed as gynogenesis. The latter has not been used widely in crop plants due to the presence of only a few numbers of megaspores and the difficulties associated with the fine dissection of the gametes.

The culturing of anthers aiming for haploid plants was first initiated in coconut in 1981. However, not much progress was made until the first decade of the millennium 2000. The first successful production of somatic embryos with both shoot and root apices, followed by plantlet production, was reported in Sri Lanka (Perera et al. 2008a, 2009).

Critical factors for androgenesis such as the microspore developmental stage, stress pretreatment and the duration of the treatment, the composition of the basal culture medium, type and concentration, and the combination of growth regulators have been experimentally determined (Perera et al. 2008a, 2009). Even though plant regeneration from anther-derived embryos is reproducible, the success rate is still low, and the maximum plantlet regeneration observed so far is 7% (Perera et al. 2009). Not only the genotype but also individual palms determine the efficiency of in vitro performance of haploid plant production in coconut (unpublished data). It has been observed that only certain individual palms show a high response to androgenesis, while others are recalcitrant. The identification of mother palms responsive to androgenesis is a way to achieve efficient production of haploids in coconut, and such study is underway for screening many individuals (unpublished data). Flow cytometric analysis showed equal proportions of haploid and dihaploid plants (50% each) among the regenerated plantlets (Perera et al. 2008a, b). Moreover, testing with a diagnostic SSR marker (CNZ43) proved that all the diploid plants were homozygous and derived from microspores (Perera et al. 2008b). Histological studies on coconut calli and embryos developed from anthers revealed that calli and embryos have originated only from microspores in the pollen sacs (Perera et al. 2008a). The occurrence of albino plants among regenerated plants from androgenic calli or embryos is a common phenomenon in many crops including rice (Guiderdoni et al. 1992), wheat (Shimada 1981), and barley (Day and Ellis 1985), although, no albino plants were observed in the protocol described for coconut androgenesis (Fig. 12.5).

The development of an anther culture protocol for haploid plant production was a major feat in coconut biotechnology despite of the low frequency of anther-derived plant regeneration. Certain constraints such as a high genotypic effect of palms for androgenesis, shoot necrosis, and poor development of plantlets need to be addressed in order to fine-tune the protocol. Once the protocol is optimized, this will enable the production of continuous doubled haploid (DH) coconut plants. Coconut breeders need to develop speed and efficient methodologies to develop new varieties. Application of DHs in a cross-pollinating plant like coconut improves the selection efficiency by overcoming the inbreeding depression. A lack of homozygous inbred lines in Tall varieties prevents the stable propagation of elite genotypes of coconut by the seed. The use of DH technology provides an opportunity to establish completely homozygous plants in only one generation. This will enable the use of DH plants, either directly as homozygous lines or as parents for gearing up the production of improved hybrid varieties of coconut. These DHs have the potential to increase the selection efficiency since the process fixes the recessive alleles of the improved characters. Moreover, deleterious recessive alleles could be eliminated at the very early stages of the breeding program, since DHs are fully homozygous for

Fig. 12.5 Stages of coconut androgenesis. (**a**) Cultured anthers. (**b**) An anther bearing an embryo. (**c**) Somatic embryos developed from anther-derived callus. (**d**) Developing multiple shoots from somatic embryos. (**e**) A complete plantlet derived from anther. (**f**) A complete plant at acclimatization

each locus. Since DHs are regenerated by individual microspores, the meiotic recombination that occurs during the gametophyte division is fixed as new recombinant products in a homozygous state. Therefore, DHs increase the genetic gain in the resulting progeny, one of the main advantages obtained through DHs. At the same time, already existing characteristics of the hybrid plants, which have been developed through conventional breeding programs, can be fixed using both anther and microspore culture. This is an important way to fix and express desirable recessive traits generated through mutation induction or hybridization techniques, even though the approaches are difficult to perform in a long life cycle like coconut.

Populations generated through DHs have facilitated the generation of molecular marker maps in several species including rice (Maluszynska 2003). Homozygosity, uniformity and the ability to replicate indefinitely over several locations make DHs ideal for genetic mapping research. Linkage maps of coconut have the potential to be improved through similar approaches with DHs. The identification of genes for resistance to bacterial blight, rice blast, and sheath blight has been accomplished in rice with this technology (Mishra et al. 2016). Similarly, linkage maps of coconut, when developed through DHs, may lead to the identification of molecular markers linked with important genes such as those for disease resistance.

12.8 Genetic Transformation of Coconut

During the last two decades, novel tools for direct gene transfer, termed genetic engineering (synonyms: gene transformation), paved a path for new directions in plant breeding. Genetic engineering is the insertion of genetic information, from one plant species to another plant species which are genetically distinct, through nonconventional methods. This helps introduce new characteristics into plants that are otherwise not possible to introduce by conventional breeding and opens up new avenues for breeding efforts. In genetic transformation methods, a foreign gene with a known gene function is introduced into plant cells, which is then produced into a complete plant in vitro.

The alteration of the genetic makeup of a plant by genetic transformation displays unique advantages over conventional breeding in transferring the desired genes while preventing the transfer of undesirable genes into the progeny. Genetic engineering permits the transfer of one or more desired genes directly. Improvements in traits can also be achieved by removing or silencing responsible genes or gene promoters.

The approaches that have been recorded for the genomic improvement of palm species using this technique, including coconut, are far behind that of most other crop species such as vegetables, cereals, and temperate fruit crops. This delay may be due to the inherent difficulties associated with palms and because most of the economically important palm species are grown in developing countries where there are limited resources for research.

Research on the genetic engineering of coconut was first reported by Samosir et al. in 1999 with the attempts to transform embryogenic calli and young leaf tissues with the GUS gene, and the two constitutive promoters, namely, Act1 and Ubi, were identified as the most expressed promoters. After a long silence, in 2011 Andrade-Torres and colleagues reported the *Agrobacterium*-mediated gene transfer into a range of coconut explants (immature male flowers collected from −1 or −2 inflorescences, zygotic embryos excised from mature nuts, embryogenic calli derived from plumule explants, roots and leaves of plantlets obtained through somatic embryogenesis) (Andrade-Torres et al. 2011). Interestingly, endogenous GUS-like activity was observed in the calli which were not subjected to transformation showing the unsuitability of gusA gene in transformant selection in coconut. Alternatively, two other genes, red and green fluorescent protein genes, have been tested for their ability as reporter genes, yielding positive results. Furthermore, it was revealed that combining bioballistics to generate micro-wounds in explants with vacuum infiltration and co-culture with *Agrobacterium tumefaciens* facilitated efficient gene transfer better than when these treatments are used independently.

The application of genetic engineering in coconut provides the possibility to develop new cultivars with added characteristics. There would be a reduction in the time and cost of selecting desired traits and mutants, improvement of the accuracy, and finally the increase of genetic base of coconut. As discussed only a few studies have been carried out on coconut genetic transformation. The three main components of any efficient transformation system are in vitro regeneration, an efficient DNA delivery system, and functional DNA inside the transformant. Research on coconut genetic transformation was initially hampered by a lack of an efficient in vitro plant regeneration protocol. The situation has now changed, with an in vitro method available, so it is time to focus on developing a competent gene transfer system for coconut. The plant regeneration protocol itself may even be improved using genetic transformation to introduce SE-related genes, even from other species. Genetic transformation holds a great opportunity to combat the challenges coconut faces, for example, developing disease- or biotic stress-resistant plants by incorporating resistant genes from other species. Currently, there is a tremendous development in the genetic studies of crop species. With the advent of new-generation sequencing (NGS) technologies, several complex genomes have been sequenced because of the speed and efficiency of next-generation sequencers. Since most of this data is available for global scientific communities through freely available databases, this information would be important for future crop improvement programs in coconut, and an acceleration of research on coconut transformation could be expected.

Genome editing, a new technology developed in the last decade, allows the creation of specific desirable mutations by accurately manipulating genomic sequences with the help of sequence-specific nucleases. A new stage of crop improvement and plant breeding has been started by combining three techniques: plant tissue culture, genome editing, and genetic transformation. Genome editing relies on plant tissue culture technology to produce complete plants upon gene editing. Clustered regularly interspaced short palindromic repeats (CRISPRs/Cas9) is one of the widely using genome-editing systems. Unlike genetic transformation, no foreign DNA is used to make the genetic change in the plant which receives the modification. The modification is done in the plants' own genome without inserting new genes. Genome-editing approaches may now be possible in coconut with the recent developments in coconut micropropagation, and this is a very promising technology, especially when considering complex crops such as coconut.

12.9 Conclusion

Genetic improvement for desirable traits is identified as vital for improving the productivity of coconut worldwide. Biotechnological tools are immensely important to overcome the shortcomings of traditional approaches.

Coconut breeding programs urgently need to be revived. Many countries have been using the same coconut hybrids for more than 20 years. With links to farmers and the industry, large field experiments need to be put in place, to produce improved varieties, to select the palms for breeding/cloning within the best progenies, and, subsequently, to characterize these plants and confirm their value. Molecular techniques can help to select more efficiently the best progenies at all stages, including the nursery before field planting. Molecular markers have been widely used in genetic diversity studies, and a few genome maps are shedding light on the path toward marker-assisted selection. A draft coconut genome sequence has been recently published, while the studies on the transformation of coconut are also in progress. The in vitro techniques, embryo culture and somatic embryogenesis, have proved their value, while doubled haploids also provide an attractive option.

It is essential to avoid falling for the mistake of applying sophisticated laboratory methods to questionable biological materials, whose performance and characteristics are not appropriately evaluated in the field. For instance, it is not enough to identify, in farmers' fields (or in poorly managed heterogeneous experiments), a few coconut palms that have apparent characteristics for cloning. This will not revolutionize coconut cultivation and will not satisfy farmers.

The best clones will be found in field experiments specifically designed for clonal selection: after a first choice of the best progenies, the best trees within these progenies will have to be used for true-to-type clonal propagation.

In the strategy document recently released by the COGENT network, the idea of a "globally coordinated breeding program" was abandoned: "Breeders prefer to make their own choices, closely related to the specificities of their countries" (Duong et al. 2018). Instead COGENT proposed a focus on three main breeding targets (Bourdeix et al. 2018b):

1. In each country, to reach a situation where farmers will have the choice between a set of at least six varieties, of which Talls, Dwarfs, and hybrids, and there are eventually other varietal types.
2. Help farmers to strengthen their knowledge on coconut biology and breeding.
3. Better assess farmer and consumer preferences. Coconut is not only agricultural but a cultural entity within a human community.

A network of specific field experiments should be put in place in order for the new biotechnologies to be efficiently used in coconut palm breeding. For boosting coconut breeding and making it more efficient, there is a need to commit a team of international experts to conduct a full review of the major programs. Special attention should be paid to costing and funding issues for the various breeding operations and to the balance between field trials and use of new technologies. This will help to revive field experiments and to advise decision-makers on the best way to integrate biotechnologies into those programs. With these precautions, the application of these new biotechnological tools will accelerate coconut breeding programs and help to increase coconut productivity and sustainability worldwide.

References

Andrade-Torres A, Oropeza C, Sáenz L et al (2011) Transient genetic transformation of embryogenic callus of Cocos nucifera. Biologia 66:790–800

Armendariz BHC, Orapeza C, Chan JL et al (2006) Pollen fertility and female flower anatomy of micropropagated coconut palms. Rev Fitotec Mex 29:373–378

Ashburner GR, Thompson WK, Halloran GM (1997) RADP analysis of South Pacific coconut palm populations. Crop Sci 37:992–997

Breiman A, Rotem-Abarbanell A, Karp A et al (1987) Heritable somaclonal variation in wild barley (*Hordeum spontaneum*). Theor Appl Genet 74(1): 104–112
Bandaranayake CK, Kearsey MJ (2005) Genome mapping, QTL analysis and MAS: importance, principle, constraints and application in coconut. Int Plant Genet Resour Newsl 142:47–54
Bandupriya HDD, Fernando SC, Vidhanaarachchi VRM (2016) Micropropagation and androgenesis in coconut: an assessment of Sri Lankan implication. Cocos 22:31–47
Bandupriya HDD, Iroshini WWMA, Vidhanaarachchi VRM et al (2017) Genetic fidelity testing using SSR marker assay confirms trueness to type of micropropagated coconut (*Cocos nucifera* L.) plantlets derived from unfertilized ovaries. Open Plant Sci J 10:46–54
Baudouin L, Lebrun P (2002) The development of a microsatellite kit and dedicated software use with coconuts. Burotrop Bull 17:16–20
Baudouin L, Lebrun P, Konan JL et al (2006) QTL analysis of fruit components in the progeny of a Rennell Island Tall coconut (*Cocos nucifera* L.) individual. Theor Appl Genet 112:258–268
Bourdeix R (1988) Effectiveness of mass selection on the yield component of coconut. Oleagineux 43:283–295
Bourdeix R, Santos G, Labouisse JP et al (2005) Useful definition of terms and nomenclature. In: Batugal P, Ramanatha Rao V, Oliver J (eds) Coconut genetic resources. IPGRI, Rome, pp 9–10
Brown WL (1983) Genetic diversity and genetic vulnerability – an appraisal. Econ Bot 37:4–12
Batugal P, Bourdeix R, Baudouin L (2009) Coconut breeding. In: Jain SM, Priyadarshan PM (eds) Breeding plantation tree crops: Tropical species. Springer, New York, pp. 327–375
Batugal P, Konan JL, Sanaoussi A et al (2005) Multilocation coconut hybrid trials in three African and three LAC countries. Coconut Genetic Resources, 326
Bourdeix R, Kumar V (2018) Compact and "super" dwarf varieties. Coconut planting material for the Pacific region. Retreived from : http://replantcoconut.blogspot.com/2017/12/compact-dwarf-varieties.html
Bourdeix R, Baudouin L, Santos GA (2018a) 2.1.3 International Coconut nomenclature - Chapter 2. Where we are today. In: Bourdeix R, Prades A (eds) A global strategy for the conservation and use of coconut genetic resources 2018–2028. Bioversity International, Montpellier, France, pp 39–40
Bourdeix R, Yace G, Sileye T (2018b) 3.7.1 Global objectives in terms of planting material - Chapter 3. Where we need to be to secure diversity and promote use. In: Bourdeix R, Prades A (eds.) A global strategy for the conservation and use of coconut genetic resources 2018-2028. Bioversity International, Montpellier, France, pp 155–156
Bourdeix R, Meunier J, N'Cho YP (1991a) Coconut (Cocos nucifera L.) selection strategy II - Improvement of Tall x Tall hybrids. Oléagineux 46(7):267–282
Bourdeix R, Meunier J, N'Cho YP (1991b) Coconut (Cocos nucifera L.) selection strategy III-Improvement of Dwarf x Tall hybrids. Oléagineux 46(10):361–374
Bourdeix R, N'Cho YP, Le Saint JPOF (1990) A coconut (Cocos nucifera L.) selection strategy. I. Rundown of achievements. Oléagineux 45:359–371
Bourdeix R, Sangare A, Le Saint JP et al (1989) Efficacité des tests hybrides d'aptitude individuelle à la combinaison chez le cocotier: premiers résultats. Oléagineux 44(5):209–214
Bourdeix R (1999) Selection and breeding. In: Olher JG (ed) Modern coconut management: palm cultivation and products. Intermediate Technology Publications, London, United Kingdom, pp 117–195
Cardena R, Ashburner GR, Oropeza C (2003) Identification of RAPDs associated with resistance to lethal yellowing of the coconut (Cocos nucifera L.) palm. Scientia Horticulturae 98:257–263
Comstock RE, Robinson HF, Harvey PH (1949) A breeding procedure to make maximum use of both general and specific combining ability. Agron J 41:360–367
Dasanayake PN, Everard JMDT, Karunanayake EH et al (2003) Characterization of coconut germplasm by microsatellite markers. Trop Agric Res 15:51–61
Day A, Ellis TH (1985) Deleted forms of plastid DNA in albino plants from cereal anther culture. Curr Genet 9:671–678

Duran Y, Rohde W, Kullaya A et al (1997) Molecular analysis of East African tall coconut genotypes by DNA marker technology. J Genet Breed 51:185–193
Duong NTK, Yace G, Pereira MG (2018) 3.7.4 International breeding experiments - Chapter 3. Where we need to be to secure diversity and promote use In: Bourdeix R, Prades A (eds) A global strategy for the conservation and use of coconut genetic resources 2018–2028. Bioversity International, Montpellier, France, pp 159–162
Everard JMDT (1996) Use of molecular markers for breeding of the coconut palm (Cocos nucifera L.). University of New England, Armidale
Fernando SC, Weerakonn LK, Perera PIP et al (2004) Genetic fidelity and *ex vitro* performance of tissue cultured coconut plants. In: Peiris TSG, Ranasinghe CS (eds) Proceedings of the international conference of the coconut research Institute of Sri Lanka- Part II. CRI, Lunuwila
Guiderdoni E, Galinato E, Luistro J et al (1992) Anther culture of tropical japonica × indica hybrids of rice (Oryza sativa L.). Euphytica 62:219–224
Herran A, Estioko L, Becker D et al (2000) Linkage mapping and QTL analysis in coconut (*Cocos nucifera* L.). Theor Appl Genet 101:292–300
Iroshini WWMA (2017) Studies on cryopreservation of embryogenic callus from unfertilized ovaries using the encapsulation- dehydration technique and post thaw plant regeneration in coconut (Cocos nucifera L.). Department of Plant Sciences. University of Colombo, Colombo
Iroshini WWMA, Jayasekera GAU, Perera SACN et al (2017) Genetic stability of coconut embryogenic calli after cryopreservation by encapsulation-dehydration technique. In: Proceedings of the 6th YSF symposium Peradeniya, Sri Lanka
Jaccoud D, Peng K, Feinstein D et al (2001) Diversity arrays: a solid state technology for sequence information independent genotyping. Nucleic Acids Res 29:25e
Konan KJN, Koffi KE, Konan JL et al (2007) Microsatellite gene diversity in coconut (Cocos nucifera L.) accessions resistants to lethal yellowing disease. Afr J Biotechnol 6:341–347
Lebrun P, N'Cho YP, Seguin M et al (1998) Genetic diversity in coconut (Cocos nucifera L) revealed by restriction fragment length polymorphism (RFLP) markers. Euphytica 101:103–108
Lebrun P, Baudouin L, Bourdeix R et al (2001) Construction of a linkage map of the Rennell Island tall coconut type (Cocos nucifera L.) and QTL analysis for yield characters. Genome 44:962–970
Ledo AS, Passos EEM, Fontes HR et al (2019) Advances in Coconut palm propagation. Rev Bras Frutic 41
Liyanage DV (1954) Controlled pollination of coconut palms. Ceylon Coconut Q 5:135–138
Liyanage DV (1958) Varieties and forms of coconut palms grown in Ceylon. Ceylon Coconut Q 9:1–10
Liyanage DV, Wickramaratne MRT, Jayasekara C (1988) Coconut breeding in Sri Lanka: a review. Cocos 6 (1): 1 – 26
Marechal H (1926) Observations and preliminary experiments on the coconut with a view to developing improved seednuts for Fiji. Agri J Fiji 1:16–45
Maskromo I (2015) Karakterisasi dan pemanfaatan plasma nutfah melalui pendekatan pemuliaan dan molekuler untuk peningkatan hasil buah kopyor dan kualitas benih kopyor. Bogor Agricultural University, Bogor
Meerow AW, Wisser RJ, Brown JS et al (2003) Analysis of genetic diversity and population structure within Florida coconut (Cocos nucifera L.) using microsatellite DNA, with special emphasis on the Fiji Dwarf cultivar. Theor Appl Genet 106:715–726
Moore D, Alexander L, Hall RA (1989) The coconut mite, Eriophyes guerreronis Keifer in St. Lucia: yield losses and attempts to control it with acaricide, polybutene and Hirsutella fungus. Trop Pest Manag 35:83–89
Maluszynska J (2003) Cytogenetic tests for ploidy level analyses: Chromosome counting. In: Maluszynski M, Kasha KJ, Forster BP, Szarejko I (eds) Doubled haploid production in crop plants. Springer, Netherlands, pp 391–395

Nair RV, Thomas RJ, Jacob PM (2010) Breeding for resistance to coconut root (wilt) disease. In: Thomas GV, Chandramohanan R, Jacob PM, Krishnakumar V (eds) Coconut root (wilt) management. CPCRI, Kasaragod, pp 58–71

Nair RV, Jerard BA, Thomas RJ (2016) Coconut breeding in India. In: Al-Khayri JM, Jain SM (eds) Advances in plant breeding strategies: agronomic, abiotic and biotic stress traits. Springer, Dordrecht, pp 257–279

Nejat N, Cahill DM, Vadamalai G et al (2015) Transcriptomics-based analysis using RNA-Seq of the coconut (Cocos nucifera) leaf in response to yellow decline phytoplasma infection. Mol Genet Genomics 290:1899–1910

Nguyen QT, Bandupriya HDD, Villalobo AL et al (2015) Tissue culture and associated biotechnological interventions for the improvement of coconut (Cocos nucifera L.): a review. Planta 242:1059–1076

Nguyen QT, Bandupriya HDD, Foale M et al (2016) Biology, propagation and utilization of elite coconut varieties (makapuno and aromatics). Plant Physiol Biochem 109:579–589

Nuce de Lamothe MD, Rognon F (1986) Cocotiers hybrides ou cocotiers grands, un choix basé sur des résultats. Oléagineux (France) 41:549–555

Nunez TC, de Paz VM (2004) Growth and development of F2 pure macapuno palms. Philipp J Crop Sci 19:69

Oropeza C, Engelman F, Cueto CA et al (2018) 2.2.4 In vitro culture and cryopreservation – Chapter 2. Where we are today. In: Bourdeix R, Prades A (eds) A global strategy for the conservation and use of coconut genetic resources 2018–2028. Bioversity International, Montpellier, pp 50–53

Patel JS (1938) The coconut: a monograph. Government Press, Madras

Perera L (2018) Where we need to be to secure diversity and promote use. In: Bourdeix R, Prades A (eds) A global strategy for the conservation and use of coconut genetic resources 2018–2028. Bioversity International, Montpellier

Perera L, Russell JR, Provan J et al (1998) Evaluating genetic relationships between indigenous coconut (*Cocos nucifera* L.) accessions from Sri Lanka by means of AFLP profiling. Theor Appl Genet 96:545–550

Perera L, Russell JR, Provan J et al (1999) Identification and characterization of microsatellite loci in coconut (Cocos nucifera L.) and the analysis of coconut populations in Sri Lanka. Mol Ecol 8:344–346

Perera L, Russell JR, Provan J et al (2000) Use of microsatellite DNA markers to investigate the level of genetic diversity and population genetic structure of coconut (Cocos nucifera L). Genome 43:15–21

Perera L, Russell JR, Provan J et al (2003) Studying the genetic relationships among coconut varieties/populations using microsatellite markers. Euphytica 132:121–128

Perera PIP, Hocher V, Verdeil JL et al (2008a) Androgenic potential of coconut (*Cocos nucifera* L.). Plant Cell Tissue Org Cult 92:293–302

Perera PIP, Perera L, Hocher V et al (2008b) Use of SSR markers to determine the anther-derived homozygous lines in coconut. Plant Cell Rep 27:1697–1703

Perera PIP, Yakandawala DMY, Hocher V et al (2009) Effect of growth regulators on microspore embryogenesis in coconut anthers. Plant Cell Tissue Org Cult 96:171–180

Perera L, Manimekalai R, Sudarsono S et al (2016) Coconut. Biotechnology of Plantation Crops. Central Plantation Crop Research Institute, Kasaragod, pp 219–239

Perez-Nunez MT, Chan JL, Saenz L et al (2006) Improved somatic embryogenesis from coconut (Cocos nucifera L.) plumule explants cultured in vitro. In vitro cellular & developmental biology. Plant 42:37–43

Petchpiroon C, Thirakul A (1994) Coconut breeding programme in Thailand. Workshop on standardization of coconut breeding research techniques. IPGRI/COGENT, GTZ, Marc Delorme Coconut Station, Port Bouet, Cote d'Ivoire

Phoeurk C, Somana J, Sornwatana T et al (2018)Three novel mutations in α-galactosidase gene involving in galactomannan degradation in endosperm of curd coconut. Phytochemistry 156:33–42

Preethi P, Rajesh MK, Rahul CU et al (2016) Identification and utilization of informative EST-SSR markers for genetic purity testing of coconut hybrids. J Plant Crop 44:77–84

Perera PIP, Hocher V, Verdeil JL et al (2007) Unfertilized ovary: a novel explant for coconut (Cocos nucifera L.) somatic embryogenesis. Plant Cell Rep 26(1): 21–28

Rachana KR, Naganeeswaran SA et al (2016) Cloning, characterization and expression analysis of NBS-LRR-type resistance gene analogues (RGAs) in coconut. Acta Bot Croat 75:1–10

Rajesh MK, Sabana AA, Rachana KE et al (2015) Genetic relationship and diversity among coconut (*Cocos nucifera* L.) accessions revealed through scot analysis. Biotechnology 5:999–1006

Rivera R, Edwards KJ, Barker JHA et al (1999) Isolation and characterization of polymorphic microsatellites in Cocos nucifera L. Genome 42:668–675

Rohde W, Becker D, Kullaya A et al (1999) Analysis of coconut germplasm biodiversity by DNA marker technologies and construction of a first genetic linkage map. In: Oropeza C, Verdeil JL, Ashburner GR, Cardena R, Santamaria JM (eds) Current advances in coconut biotechnology. Kluwer Academic Publishers, Dordrecht, pp 99–120

Ruas M (2018) Genetic resources information management – chapter 2. Where we are today. In: Prades RBA (ed) a global strategy for the conservation and use of coconut genetic resources 2018–2028. Bioversity International, Montpellier, France, pp 68–74

Mishra R, Rao GJN (2016) In-vitro Androgenesis in Rice: Advantages, Constraints and Future Prospects. Rice Science 23 (2):57–68

Sisunandar A, Sopade PA, Yohannes MS et al (2010) Dehydration improves cryopreservation of coconut (Cocos nucifera L.). Cryobiology 61(3):289–296

Saensuk C, Wanchana S, Choowongkomon K et al (2016) De novo transcriptome assembly and identification of the gene conferring a "pandan-like" aroma in coconut (Cocos nucifera L.). Plant Sci 252:324–334

Santos GA (1990) Activities in coconut genetic resources collection, conservation and genetic improvement in the Philippines. Philipp J Coconut Stud 15:16–20

Santos PHAD (2016) Melhoramento Genético Do Coqueiro (Cocos nucifera L.): Capacidade Combinatória E Diversidade Genética Via Rad-Sequencing. Campos dos Goytacazes – RJ. Universidade Estadual do Norte Fluminense Darcy Ribeiro

Santos GA, Rivera RL (1994) Coconut breeding programme of the Philippines. Workshop on standardization of coconut breeding research techniques. IPGRI/COGENT, GTZ, Marc Delorme Coconut Station, Port Bouet, Cote d'Ivoire

Shalini KV, Manjunantha S, Lebrun P et al (2007) Identification of molecular markers associated with mite resistance in coconut (Cocos nucifera L.). Genome 50:35–42

Shimada T (1981) Haploid plants regenerated from the pollen callus of wheat (Triticum aestivum L.). Jpn J Genet 56:581–588

Sisunandar A, Alkhikmah A, Husin A et al (2018) Ex vitro rooting using a mini growth chamber increases root induction and accelerates acclimatization of Kopyor coconut (Cocos nucifera L.) embryo culture-derived seedlings. In Vitro Cell Dev Biol Plant 54:508–517

Soh AC, Wong G, Tan CC et al (2011) Commercial-scale propagation and planting of elite oil palm clones: research and development towards realization. J Oil Palm Res 23:935–952

Santos GA, Batugal P, Othman A et al (1996) Manual on standardized research techniques in coconut breeding (STANTECH). COGENT/IPGRI–APO, Serdang, Selangor, Malaysia pp 46

Teulat B, Aldam C, Trehin R et al (2000) An analysis of genetic diversity in coconut (Cocos nucifera L.) populations from across the geographic range using sequence-tagged microsatellites (SSRs) and AFLPs. Theor Appl Genet 100:764–771

Vargas AG, Cabanos CS, Garcia RN et al (2018) Cloning, molecular analysis, and developmental expression of 3 oleosin cDNA isoforms in coconut (Cocos nucifera L.). J Hortic Sci Biotechnol 93:255–263

Vidhanaarachchi VRM, Fernando SC, Perera PIP et al (2013) Application of un-fertilized ovary culture to identify elite mother palms of *Cocos nucifera* L. with regenerative potential. J Natl Sci Found Sri 41(1)
Xiao Y, Xu P, Fan H et al (2017) The genome draft of coconut (*Cocos nucifera*). Giga Sci 6:1–11
Yan W, Liu L, Li CX et al (2015) Transcriptome sequencing and analysis of the coconut leaf beetle, *Brontispa longissima*. Genet Mol Res 14:8359–8365

Correction to: In Situ and Ex Situ Conservation of Coconut Genetic Resources

Roland Bourdeix, Steve Adkins, Vincent Johnson, Lalith Perera, and Sisunandar

Correction to:
Chapter 4 in: S. Adkins et al. (eds.), *Coconut Biotechnology: Towards the Sustainability of the 'Tree of Life'*, https://doi.org/10.1007/978-3-030-44988-9_4

This book was inadvertently published with the below mentioned incorrect citation and its corresponding reference, this has now been corrected.

On Page 65, Section *4.5.1 Frozen Zygotic Embryos and Plumules*

The reference citation in the last line has been corrected from (Wilms et al. 2018) to (Wilms et al. 2019).

On Page 75, References section

The reference "Wilms H, Rhee JH, Rivera RL et al (2018) Developing coconut cryopreservation protocols and establishing cryo-genebank at RDA; a collaborative project between RDA and Bioversity International. III international symposium on plant cryopreservation 1234, pp 343–348" has now been corrected to "Wilms H, Rhee JH, Rivera RL et al (2019) Developing coconut cryopreservation protocols and establishing cryo-genebank at RDA; a collaborative project between RDA and Bioversity International. III international symposium on plant cryopreservation 1234, pp 343–348".

The updated online version of this chapter can be found at
https://doi.org/10.1007/978-3-030-44988-9_4

S. Adkins et al. (eds.), *Coconut Biotechnology: Towards the Sustainability of the 'Tree of Life'*, https://doi.org/10.1007/978-3-030-44988-9_13

Index

S. Adkins et al. (eds.), *Coconut Biotechnology: Towards the Sustainability of the 'Tree of Life'*, https://doi.org/10.1007/978-3-030-44988-9

The manufacturer's authorised representative in the EU is Springer Nature Customer Service Centre GmbH, Europaplatz 3, 69115 Heidelberg, Germany. If you have any concerns regarding our products, please contact ProductSafety@springernature.com

Printed and bound by CPI Group (UK) Ltd, Croydon, CR0 4YY
30/09/2026
02225557-0006